Advanced Chemical and Biological Strategies for Detection and Remediation of Water Contaminants

Water contamination is a pressing global concern, impacting public health, ecosystems, and economic development. This edited volume presents cutting-edge advancements in chemical and biological strategies for identifying, analyzing, and eliminating contaminants from water systems. It covers topics ranging from micro- and nanoplastics, engineered bacteria, fluorescent sensors, and supramolecular chemistry to AI-driven monitoring tools and nano-photocatalysis.

This book is structured to provide a holistic understanding of emerging pollutants—such as pharmaceuticals, endocrine disruptors, microplastics, and heavy metals—and advanced solutions tailored for both urban and rural water treatment systems. With inputs from interdisciplinary experts, this book bridges the gap between academic research and practical implementation, offering guidance to researchers, engineers, policymakers, and students.

Smart Technologies for Water Engineering & Sustainability

With the latest advancements in AI and associated smart technologies, there have been conscious efforts towards a sustainable environment. Water Engineering is one of these areas influenced by recent technologies including smart materials, efficient water management, advanced water systems like dams, irrigation, and so on. This series is a compilation of such books which work at the intersection of latest technology, sustainability, and water engineering, making them extremely relevant in today's world.

Integrated Smart Technologies for Sustainable Water Systems using Computer Applications
Edited by Srikanta Kumar Mohapatra, Monika Sethi, Shivani Sood, Abhineet Anand, and Jayashree Mohanty

Smart Technologies for Water Sustainability and Policy
Edited by Neha Sharma, Abhineet Anand, Abhishek Kumar, Rajeev Tiwari, Anupam Tiwari, and Prateek Thapar

Advanced Chemical and Biological Strategies for Detection and Remediation of Water Contaminants
Edited by Sneha Gautam, Biswanath Mahanty, and Raju Nandhakumar

Advanced Chemical and Biological Strategies for Detection and Remediation of Water Contaminants

Edited by
Sneha Gautam, Biswanath Mahanty,
and Raju Nandhakumar

CRC Press
Taylor & Francis Group
Boca Raton London New York

CRC Press is an imprint of the
Taylor & Francis Group, an **informa** business

Designed cover image: © Sneha Gautam

First edition published 2027
by CRC Press
4 Park Square, Milton Park, Abingdon, Oxon, OX14 4RN

and by CRC Press
2385 NW Executive Center Drive, Suite 320, Boca Raton FL 33431

CRC Press is an imprint of informa UK Limited

British Library Cataloguing-in-Publication Data
A catalogue record for this book is available from the British Library

ISBN: 9781041243731 (hbk)
ISBN: 9781041243748 (pbk)
ISBN: 9781003741626 (ebk)

DOI: 10.1201/9781003741626

Typeset in Times
by codeMantra

Contents

About the Editors

Sneha Gautam is a Professor (Research) and Head of the Water Institute and Division of Civil Engineering at the Karunya Institute of Technology and Sciences, Coimbatore, India. He is also a Visiting Professor in the Department of Climate and Energy Systems Engineering, Ewha Womans University, Seoul, South Korea. He earned his Ph.D. (Engineering) from the Indian Institute of Technology, Kharagpur, and his M.Tech. (Environmental Science and Engineering) from IIT (ISM) Dhanbad. After completing his Ph.D., he worked as a researcher at INCLEN and the University of California, Berkeley, followed by a postdoctoral fellowship funded by the Ministry of Science and Technology, Taiwan, and further postdoctoral research at National Chiao Tung University (NCTU), Taiwan. His name has been consecutively listed (2021–2025) among the world's top 2% scientists by Stanford University. He has received several prestigious recognitions, including the Young Scientist Award (2023) from the Academy of Sciences, Chennai, the Young Scientist Award (2022), the IEI Special Appreciation Award (2021), the IEI Young Achiever Award (2020), the IEI Young Scientist Award (2019), the Outstanding Reviewer Award, and an International Travel Grant from the Government of India. With over 210 research articles in reputed international journals (h-index: 48; i10-index: 164; citations: 8561+), he has established a strong global research network, and his work has been featured in *Al Jazeera*, *The Times of India*, and *Trinity*. His research interests encompass the atmospheric environment, focusing on climate–weather interactions, air and water pollution measurement and modeling, bioaerosols, personal exposure assessment, urban and rural air quality, microplastics, emerging contaminants in air and water, and sensor-based environmental monitoring technologies.

Biswanath Mahanty is Professor in the Department of Biotechnology at the Karunya Institute of Technology and Sciences, Coimbatore, India. He earned his Ph.D. in Biotechnology from the Indian Institute of Technology Guwahati and has academic and research experience in India and South Korea, including appointments at Inha University and the University of Ulsan. His research interests include bioprocess engineering, environmental biotechnology, mathematical modeling, artificial intelligence and machine learning applications in biotechnology, bioremediation, wastewater treatment, sustainable biopolymers, and biomass valorization. He has made significant contributions by integrating data-driven modeling and optimization techniques with microbial and biochemical systems for sustainable environmental and industrial applications.

He has authored more than 100 research publications, including articles in leading journals such as *Bioresource Technology*, *Chemical Engineering Science*, *Science of the Total Environment*, *Journal of Environmental Management*, *Food Chemistry*, and *Biotechnology and Bioengineering*. His recent work focuses on explainable artificial intelligence, hybrid dynamic modeling, bioprocess optimization, lignocellulosic biorefineries, polyhydroxyalkanoates, pullulan biosynthesis, and environmental remediation technologies. In addition to his research contributions, he serves as a reviewer for numerous high-impact international journals. His interdisciplinary expertise and extensive publication record make him a valuable contributor to emerging areas of biotechnology, environmental engineering, and sustainable process development.

Raju Nandhakumar specializes in bio-organic and supramolecular chemistry. He received his UG, PG, and Ph.D. degrees in Chemistry from Bharathiar University, Coimbatore, Tamil Nadu, India. Presently, he is the Professor of Chemistry in the Division of Physical Sciences, Karunya Institute of Technology and Sciences (Deemed to be University). His research interests include carbon-based materials, fluorescent chemosensors, and forensic sciences. He has successfully completed several projects from various funding agencies and guided ten doctoral students. To his credit, he has written seven books, ten book chapters, two international patents (granted), 13 national patents (granted), and published more than 200 research papers (including five review articles) in peer-reviewed international and national journals.

Contributors

N. Anand
Division of Civil Engineering
Karunya Institute of Technology and Sciences
Coimbatore, Tamil Nadu, India

Lazarus Godson Asirvatham
Centre for Research in Material Science and Thermal Management
Karunya Institute of Technology and Sciences
Coimbatore, Tamil Nadu, India

Kandavalli Babu
Division of Mechanical Engineering
Karunya Institute of Technology and Sciences
Coimbatore, Tamil Nadu, India

Shishir Kumar Behera
Industrial Ecology Research Group
School of Chemical Engineering
Vellore Institute of Technology, Vellore, Tamil Nadu, India

Vidhya Bhojan
Karunya Institute of Technology and Sciences
Coimbatore, Tamil Nadu, India

A.V. Roger Krupa Bhushan
Division of Food Processing Technology
Karunya Institute of Technology and Sciences
Coimbatore, Tamil Nadu, India

Abubakar M. Bilyaminu
Department of Engineering Systems and Services
Faculty of Technology, Policy and Management
Delft University of Technology
Delft, The Netherlands

Shaswata Biswal
Department of Chemistry
Faculty of Engineering and Technology (ITER)
Siksha 'O' Anusandhan (Deemed to be University)
Bhubaneswar, Odisha, India

Kumarasamy Chandra Prakash
Department of Chemistry
Kongunadu Arts and Science College
Coimbatore, Tamil Nadu, India

Achlesh Daverey
Doon University
Dehradun, Uttarakhand, India

Lokashree Dharmalingam
Karunya Institute of Technology and Sciences
Coimbatore, Tamil Nadu, India

A. Diana Andrushia
Division of Electronics and Communication
Karunya Institute of Technology and Sciences
Coimbatore, Tamil Nadu, India

Ajitha E.
Karunya Institute of Technology and Sciences
Coimbatore, Tamil Nadu, India

Anina James
Dilshad Garden
Delhi, India

S. Vincent Sam Jebadurai
Department of Civil Engineering
Karunya Institute of Technology and Sciences
Coimbatore, Tamil Nadu, India

Adeline Juanita
Karunya Institute of Technology and Sciences
Coimbatore, Tamil Nadu, India

Cyril Samuel J.S.
Division of Civil Engineering
Karunya Institute of Technology and Sciences
Coimbatore, Tamil Nadu, India

Kuppusamy Kanagaraj
Department of Chemistry and Chemical Sciences
Central University of Jammu
Jammu, India

Damian Kasperczyk
Ekoinwentyka Ltd.
Ruda Śląska, Poland

Nimish Khandelwal
Department of Natural and Applied Sciences
TERI School of Advanced Studies
New Delhi, India

Roshini Praveen Kumar
Division of Civil Engineering
Karunya Institute of Technology and Sciences
Coimbatore, Tamil Nadu, India

Soham Kundu
Adamas University
Kolkata, West Bengal, India

G. Lawrance
Division of Mechanical Engineering
Karunya Institute of Technology and Sciences
Coimbatore, Tamil Nadu, India

Biswanath Mahanty
Karunya Institute of Technology and Sciences
Coimbatore, Tamil Nadu, India

Jyotiraditya Mallik
Sister Nivedita University
Kolkata, West Bengal, India

Satya Sundar Mohanty
Karunya Institute of Technology and Sciences
Coimbatore, Tamil Nadu, India

Sushma M.
Division of Physical Sciences
Karunya Institute of Technology and Sciences (Deemed to be University)
Coimbatore, Tamil Nadu, India

Raju Nandhakumar
Karunya Institute of Technology and Sciences
Coimbatore, Tamil Nadu, India

Shanmugam Paramasivam
Department of chemistry, Faculty of Science and Technology
Thammasat University
Pathum Thani, Thailand

Umapriya R.
Karunya Institute of Technology and Sciences
Coimbatore, Tamil Nadu, India

Pandian Bothi Raja
School of Chemical Sciences
Universiti Sains Malaysia
Penang, Malaysia

Bairi Levi Rakshith
Centre for Research in Material Science and Thermal Management
Karunya Institute of Technology and Sciences
Coimbatore, Tamil Nadu, India

Eldon R. Rene
Department of Water Supply, Sanitation and Environmental Engineering
IHE Delft Institute for Water Education
Delft, The Netherlands

Naresh Kumar Sahoo
Siksha 'O' Anusandhan (Deemed to be University)
Bhubaneswar, Odisha, India

Satyanjib Sahoo
Siksha 'O' Anusandhan (Deemed to be University)
Bhubaneswar, Odisha, India

Tapas Ranjan Sahoo
Kalinga Institute of Technology
Bhubaneswar, Odisha, India

Gayatree Sanjibita Sahoo
Department of Chemistry
Faculty of Engineering and Technology (ITER)
Siksha 'O' Anusandhan (Deemed to be University)
Bhubaneswar, Odisha, India

Nageswar Sahu
Karunya Institute of Technology and Sciences
Coimbatore, Tamil Nadu, India

Shampa Sen
Edumatter, Kolkata, West Bengal, India
West Bengal State University
Kolkata, West Bengal, India

Sanjana Thanjan Shaju
Karunya Institute of Technology and Sciences
Coimbatore, Tamil Nadu, India

Dhanush S.
Karunya Institute of Technology and Sciences
Coimbatore, Tamil Nadu, India

Kavanya S.
Karunya Institute of Technology and Sciences
Coimbatore, Tamil Nadu, India

Arumugam Thangamani
Department of Chemistry
Karpagam Academy of Higher Education
Coimbatore, Tamil Nadu, India

Jesu Dakshin Thippiah K.
Division of Mechanical Engineering
Karunya Institute of Technology and Sciences
Coimbatore, Tamil Nadu, India

Tharmaraj Vairaperumal
Institute of Clinical Medicine, College of Medicine, National Cheng Kung University
Taiwan

Shelly Verma
Army Cadet College Wing, Indian Military Academy
Dehradun, Uttarakhand, India

Mohd. Zafar
University of Technology and Applied Sciences—Sur
Oman

Seyed Morteza Zamir
Department of Biochemical Engineering
Faculty of Chemical Engineering
Tarbiat Modares University
Tehran, Iran

Micro- and Nanoplastics in Drinking Water

1

Sources, Risks, and Remediation Approaches

Sanjana Thanjan Shaju, Lokashree Dharmalingam, Ajitha E., Kavanya Srinivasan, Vidhya Bhojan, and Raju Nandhakumar

1.1 INTRODUCTION

The "Great Acceleration" of plastic production commenced around 1950, with universal output of nearly 2 million tonnes every year.[1] Production increased to 322 million tonnes by 2015 and achieved about 460 million tonnes in 2019 (Organisation for Economic Co-operation and Development, 2022), with trade-linked flows nearing 400 million tonnes in 2022. Consequently, this exponential development driven by cost-effectiveness, versatility, and prevalent applications, global recycling rates remain below 10%. Projections indicate that plastic consumption may double within two decades and approach quadrupling by 2050.[2] With less than 10% recycled, almost 79% of plastic waste builds up in landfills or the environment, and 12% is incinerated,[1] discharging pollutants like carbon monoxide, nitrogen oxides, and dioxin-related molecules. Unproductive waste management has intensified ecological contamination and public health threats. Environmental weathering activities, including photodegradation, chemical oxidation, mechanical abrasion, and pathogen activity, destroy macroplastics into microplastics (MPs) and nanoplastics (NPs). Exposure to UV radiation, heat, and wetness accelerates this breakdown and produces hydrophilic organic fragments.[3] Common polymers like polypropylene (PP), polystyrene (PS), and low-density polyethylene (LDPE) degrade under simulated sunlight, producing MPs and NPs, with chemical, physical, and biological eroding driving their establishment in terrestrial and aquatic systems.[4]

Microplastics are termed as plastic particles varying from 1 μm to 5 mm, whereas NPs are less than 1 μm (1–1,000 nm).[5,6] These particles may possibly form homoaggregates/heteroaggregates with minerals,

DOI: 10.1201/9781003741626-1

stones, and organic matter.[6] They occur in several morphologies, foams, pellets, fragments, films, and fibres, and consist of polymers such as PS, polyethylene (PE), polyvinyl chloride (PVC), and PP originating from industrial activities, domestic packaging, and household sources. Primary plastics are deliberately produced at a microscopic scale, although secondary plastics evolve from breakup of larger materials.[5] Due to their high surface-area-to-volume ratio, MPs and NPs adsorb pesticides, heavy metals, and organic contaminants, playing as vectors for pollutant transport. They are transported through rivers, lakes, and reservoirs into drinking water systems, linking global plastic pollution to human exposure pathways. Municipal drinking water treatment procedures like ozonation additionally fragment particles, increasing counts below 20 μm in treated water. Their extensive occurrence, contaminant-binding capacity, and dubious toxicological implications render MPs and NPs in drinking water an alarming concern. Information of marine plastic pollution reported in the 1970s, including documentation of plastic pellet collection in the Sargasso Sea. The creation of the term "microplastics" by Thompson et al. (2004) marked the launch of focused investigation. Nevertheless, between 2010 and 2015, research stretched beyond marine systems to freshwater, air, and soil.[7,8] In 2018, MPs were noticed in tap and bottled water,[9,10] intensifying public health concerns. International bodies such as the United Nations Environment Programme (UNEP) and the European Commission afterwards incorporated MPs into monitoring frameworks. Figure 1.1 compares the size ranges of MPs and NPs with common environmental and biological structures, including their morphologies.

Latest researchers emphasize NP detection employing pyrolysis–GC/MS and nanoparticle tracking analysis alongside toxicity examination.[5] Evidence of MPs and NPs in human placenta and feces[11] has a paradigm shift in research towards human exposure and health hazard evaluation. Literature reports mainly based on publications increased sharply between 2020 and 2024, centring on methodological standardization, microbiome disruption, oxidative stress pathways, and drinking water threat evaluation.[12] Regulatory frameworks are developing but remain restricted by toxicological and analytical uncertainties. The World Health Organization decided that the current data do not validate adverse health impacts from MPs in drinking water but stressed the requirement for further research. The European Union adopted Regulation EU 2023/2055 limiting intentionally appended synthetic polymer MPs, combined with circular economy strategies to moderate plastic leakage (European Commission,). In the United States, the Food and Drug Administration (FDA) accepts the presence of MPs and NPs in food practices but considers existing exposure levels unlikely to create some health risks while persisting to monitor emerging evidence. Although increasing regulatory attention, regulated methodologies for size classification, sampling, extraction, and polymer identification are still lacking, restricting comparability across

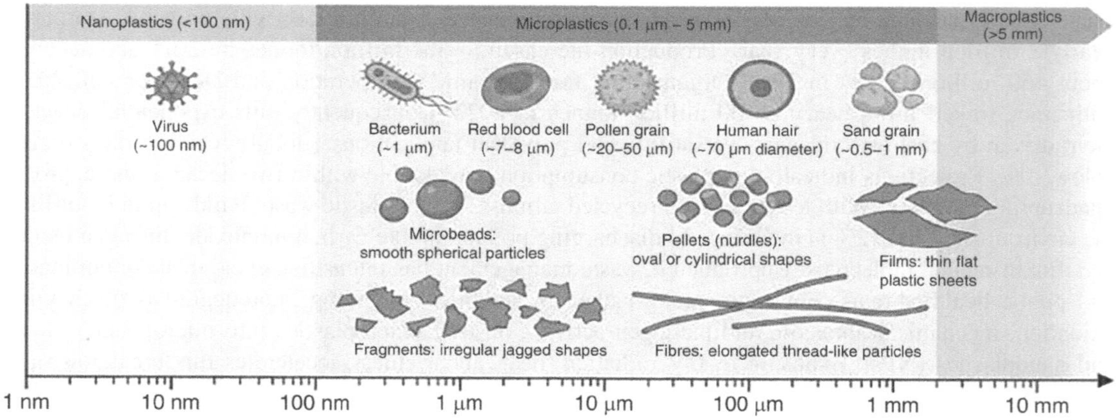

FIGURE 1.1 Size range comparison of nano-, micro-, and microplastics in relation to common biological and environmental structures and representation of particle morphologies, including microbeads, pellets, films, fibres, and fragments.

studies.[8,13] Unpredictable size thresholds, particularly for nanoparticles, and inadequate analytical capacity to detect nanofractions undermine global monitoring endeavours.[14,15] Toxicological models also ineffectively address polymer-specific toxicity, size-dependent effects, and co-contaminant interactions.[12,15] Consequently, regulatory thresholds remain temporary. Collective and coordinated international efforts to standardize monitoring protocols, standardize detection models, and standardize risk assessment frameworks are essential to enable robust, evidence-based regulation.[12,14]

1.2 SOURCES OF MPS AND NPS IN DRINKING WATER

MPs pass into drinking water systems via several sources and pathways. Initial inputs incorporate fragmentation of microfibre release, industrial pellet loss (nurdles), and larger plastic debris (e.g. bottles and packaging) during the synthetic textile washing. These particles are transported through the surface waters that amass MPs from litter, neighbouring catchments, and atmospheric deposition; wastewater treatment plant effluents that discharge depositing particles; and stormwater runoff that assembles the urban and agricultural plastic debris. Groundwater also becomes ruined through leachate infiltration from dump sites and landfills. Finally, MPs reach drinking water treatment resources, where removal efficiencies differ and smaller particles sneak in through to the tap. The possible sources of MPs and NPs in drinking water are shown in Figure 1.2.

Within the vicinity of the distribution networks, further contamination may occur. Under stagnant conditions and disinfectant exposure, PVC and comparable polymeric pipes can leach organic compounds and MPs, with total organic carbon levels enhancing proportionally with stagnation time ($R^2 \approx 0.96$), confirming pipe-derived release.[17] Plastic pipes are also subject to organic migration, surface wear, oxidation,

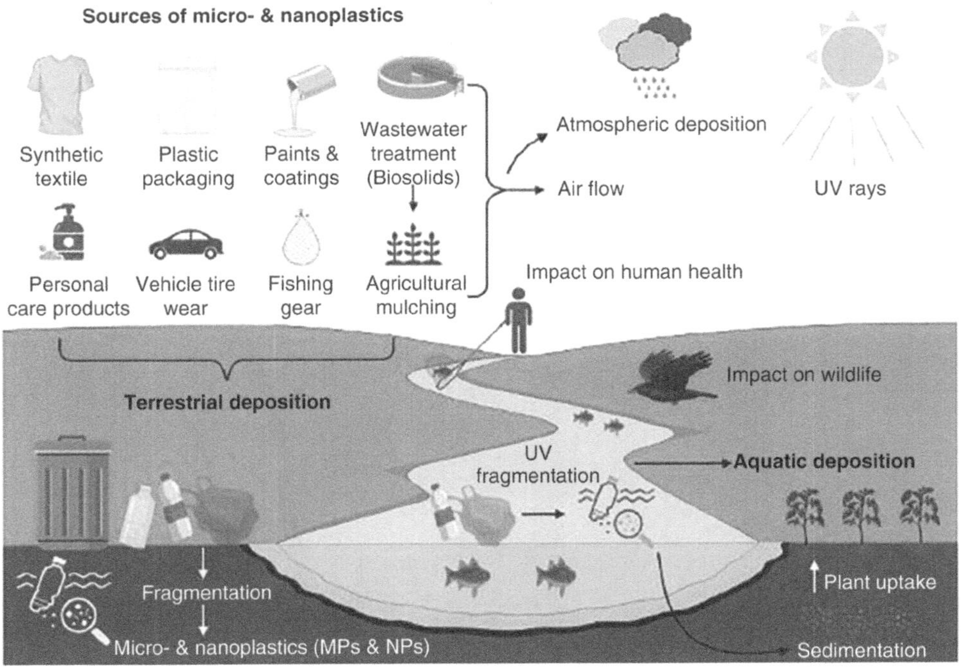

FIGURE 1.2 Sources of MPs and NPs [reproduced under the terms of the Creative Commons CC BY 4.0 licence].[16]

and additive depletion, whereas hydraulic disturbances like water hammer events can dislodge accumulated substances.[18] Furthermore, biofilms within pipes act both as secondary sources and sinks. The MPs retained in extracellular polymeric matrices or corrosion layers may afterward be released during chemical or hydraulic disturbances. Experimental evidence signals that microbial corrosion can encourage MP discharge by approximately 2.1-fold, alongside improved leaching of microbial by-products and plastic-associated chemicals.[19,20]

Point-of-use (POU) exposure to MPs and NPs occurs at the tap, i.e. the final interface between consumers and drinking water systems. Outside source water treatment, exposure is induced by the distribution network, where leaks, pipe corrosion, abrasion, and hydraulic disturbances can create further particles. Plumbing biofilms additionally act as reservoirs, retaining and intermittently releasing MPs/NPs. Therefore, even treated water may influence directly human ingestion during drinking and food preparation. Commercial bottled water is a well-documented POU exposure path. Concentrations have been estimated at $2.4 \pm 1.3 \times 10^5$ particles/L, with NPs comprising ~90% of the total.[21] In a survey of 259 bottles across 11 brands, 93% contained MPs, averaging 10.4 particles/L (≥100 μm), rising to 325 particles/L when smaller particles (6.5–100 μm) were included.[9] Contamination predominantly originates from packaging and bottling practices, abrasion in reused bottles, including PP cap fragments, bottle-wall shedding, and production-line inputs.[9,22,23] Excessive particle loads compared to municipal tap water emphasize packaging and handling as serious contamination points.[9,24] Mitigation strategies include non-plastic packaging, avoiding even minimal heat exposure and the use of active and effective home filtration systems.[25,26] Household filtration components can both introduce and remove MPs/NPs. A detailed comparative study reported 9,262 particles largely, with higher concentrations in untreated tap water (1.13 MPs/L) and purifier-treated water (1.41 MPs/L) compared to ceramic filters (0.49 MPs/L). Degradation of seals, filter housings, and backwashing processes may liberate fragments in non-membrane systems. Conversely, pitcher filters demonstrated 97–99% MP removal with marginal media release. Effective mitigation requires sub-micron-rated membrane systems and strict adherence to maintenance and replacement programmes. Figure 1.3 illustrates the POU exposure to MPs and NPs with the mitigation cycle and the source–pathway–receptor framework.

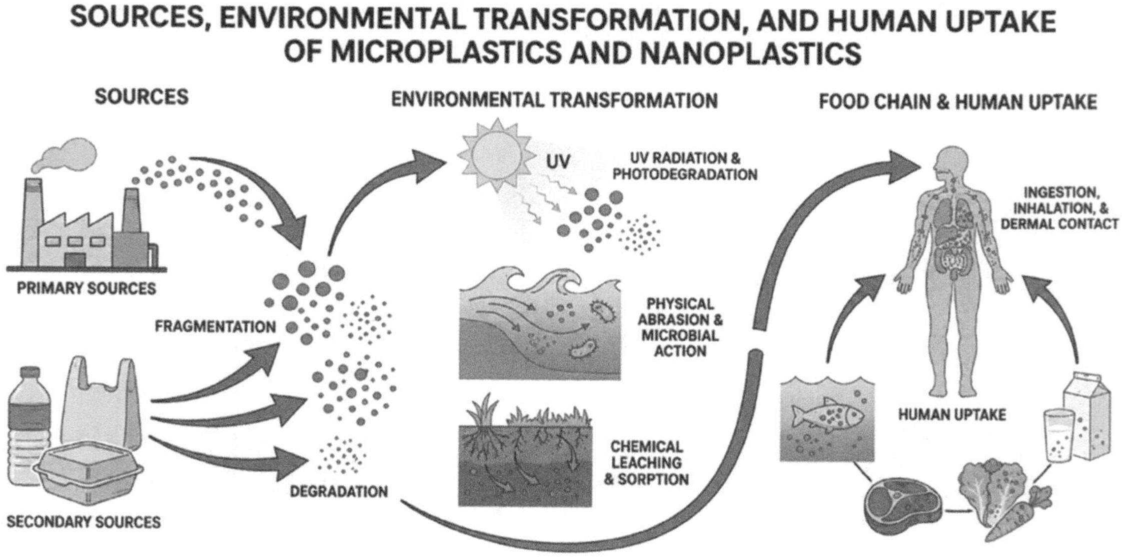

FIGURE 1.3 Sources, environmental transformation, and human exposure routes of MPs and NPs, showing primary and secondary sources, fragmentation into smaller particles, and transfer through the food chain.

1.3 OCCURRENCE AND DETECTION

1.3.1 Global Occurrence and Contamination Patterns

The condition of MPs in various water sources across the globe, between municipal (tap) drinking water, raw water (surface and groundwater), and bottled water, including the provincial differences and concentration limits. Surface and groundwater bodies such as rivers, lakes, and aquifers play as major sinks for MPs spawned from tyre and road-mark abrasion,[27] personal care products, synthetic fibre shedding, marine coatings, and urban activities. These particles enter across treated wastewater effluent, atmospheric deposition, and overland runoff.[28] Rain-induced amalgamated sewer overflows additionally discharge partially treated sewage encompassing plastics into surface waters, serving as episodic secondary ataxia.[29,30] Their transport and fate rest on particle size, seasonality, morphology, hydrodynamics, wind, and even any small external forces. Distribution patterns are also greatly influenced by socio-economic factors, such as physicochemical conditions, water body type, and flow regimes.[31] A meta-analysis of 53 studies reported excessive impurity in Asian waters (2.6×10^3 items·m^{-3}) compared with Europe (1.4×10^2 items·m^{-3}) and North America (2.7×10^2 items·m^{-3}).[31] Throughout the drinking water treatment and distribution, MP concentrations normally decline; however, unknown and new particles may be initiated through filtration processes, pipe corrosion, abrasion, and treatment by-products. MPs are extensively reported in municipal tap water, typically ranging from less than one to several hundred particles/litre depending on treatment technologies and relevant analytical methods.[10,32,33] Concentrations across freshwater and drinking water sources change by up to ten orders of magnitude (10^{-2}–10^8 particles/m^3). An international survey reported fibres in 83% of tap water samples, which estimates the daily ingestion of 14 and 10 particles for men and women, respectively. Similarly, the concentrations in China reached up to 440 MPs/L. The World Health Organization (2019) highlighted many concerns regarding both treatment efficiency and human exposure. Bottled water creates additional contamination from source water, caps, treatment processes, storage conditions, and packaging materials. MPs have been found in all tested brands worldwide, with levels changing by methodology and packaging type. In India (Kollam), concentrations ranged from 3 ± 1.7 to 9 ± 1.0 items/L,[32] while Australian samples reported up to 13 particles/L, largely PP with a mean size of 77 µm.[34] Estimated 24-hour exposure in India is ~153 particles/person through drinking water. In Malaysia, concentrations ranged from 8 to 22 particles/L (mean 11.7 ± 4.6), with 31% of particles between 100 and 300 µm; packaging materials and caps were recognized as the primary contributors.

1.3.2 Polymer Distribution and Particle Characteristics

The MP characteristics, such as shape, polymer type, size, and density, govern their transport, treatment removal efficiency, and human exposure potential. Among raw, municipal, and bottled water, PS, PE, polyethylene terephthalate (PET), PP, and PVC are most detected. Koelmans et al. reported the prevalence sequence PE ≈ PP > PS > PVC > PET in drinking water. Correspondingly, Sefiloglu et al. identified PE (91%), PVC (96%), and PET (57%) as predominant polymers using pyrolysis–GC/MS, with PS and PP following. PE and PP are frequently detected due to prevalent use, fragmentation ability, and low density (buoyancy), during which denser PVC and PET tend to settle regardless of frequent occurrence.[33] Morphologically, MPs appear as fibres, films, fragments, pellets, and foams, with fibres and fragments dominating. In bottled water, fragments accounted for 66% of particles that are greater than 100 µm, followed by fibres. Hence, the shape influences fate. The fibres remain suspended and disperse extensively; fragments and films deposit more quickly, and pellets/beads, which are more often primary MPs, accumulate fastly. Particle sizes in bottled water are commonly 6.5–100 µm.

Detection of MPs/NPs in drinking water is critically challenging due to low concentrations in the range of 0.1–10 particles/L and matrix interferences (Ivleva, 2023). Sampling usually requires filtration of

10–100 L through 1–10 μm membranes utilizing non-plastic equipment, with severe contamination control and procedural blanks to prevent airborne false positives. Insignificant enzymatic or H_2O_2 digestion removes organic matter, though critical treatments usually damage the NPs, which are less than 100 nm. Fourier-transform infrared (FTIR) and Raman spectroscopy are the commonly used standard techniques for polymer identification for those which are less than 1 μm. μ-FTIR is optimal for particles greater than 10 μm but less sensitive for smaller or highly absorbing materials. μ-Raman offers superior spatial resolution (~0.4–1 μm) but is constrained by fluorescence and laser-induced heating. Emerging approaches include hyperspectral FT-Raman and machine-learning-based mapping, though matrix interference and lack of regulated and standardized spectral libraries, which, however, remain challenges. For NPs which are less than 1 μm, Scanning Electron Microscope (SEM) and Transmission Electron Microscope (TEM) afford high-resolution morphology, with TEM detecting particles down to approximately 10 nm, though polymer recognition is indirect unless coupled with spectroscopy.[35] Nanoparticle tracking analysis (NTA) estimates hydrodynamic size and concentration (10^6–10^9 particles/mL) based on Brownian motion and has earned attention for drinking water applications. Nonetheless, separation from colloids naturally requires ultracentrifugation and spectral confirmation. Recent advances integrate dielectrophoresis with Raman spectroscopy for material compositional tracking. Thermal techniques such as pyrolysis–GC/MS (Py–GC/MS) quantify polymer mass rather than particle counts by detecting distinguishing monomer fragments. Detection limits typically vary from low microgram/L to nanogram/L, depending on the preparation methodologies. However, overlapping pyrolysis products and dissolved organic matter can make quantification difficult, particularly in biological matrices. Thermal extraction desorption (TED)–GC/MS supports the analysis of semi-volatile polymer fragments and additives, reporting MP concentrations of 0.04–1 μg/L in wastewater effluent and correlating well with FTIR data.[36]

1.4 RISKS AND HEALTH IMPLICATIONS

1.4.1 Physical Impacts

Micro- and nanoplastics (MNPs) enter the body through inhalation, ingestion, and dermal exposure, translocate amidst epithelial barriers, and accumulate in organs such as the kidney, liver, brain, and lung.[11,37,38] PP, PET, and PS have been identified in human tissues, with higher burdens reported in diseased livers. Animal studies confirm renal damage and hepatic oxidative stress. Toxicity is size-dependent; miniature particles instantly enter cells and may cross the blood–brain barrier, advocating neuroinflammation.

1.4.2 Chemical Impacts

MNPs leach additives like bisphenols and phthalates, which are not chemically bound to polymers and transfer under physiological or environmental conditions.[39] Leaching is driven by diffusion processes and influenced by particle size and environmental factors. Phthalates (DIBP, DBP, DEP) play a role as endocrine disruptors associated with cardiovascular, reproductive, carcinogenic, and metabolic risks.[40,41] Bisphenol A (BPA) exposure causes endocrine disruption, oxidative stress, and hepatic injury.

1.4.3 Biological Impacts

MNPs create reactive oxygen species (ROS), causing oxidative damage and activating inflammatory pathways.[42] NPs trigger mitochondrial dysfunction and apoptosis, and their presence in atherosclerotic

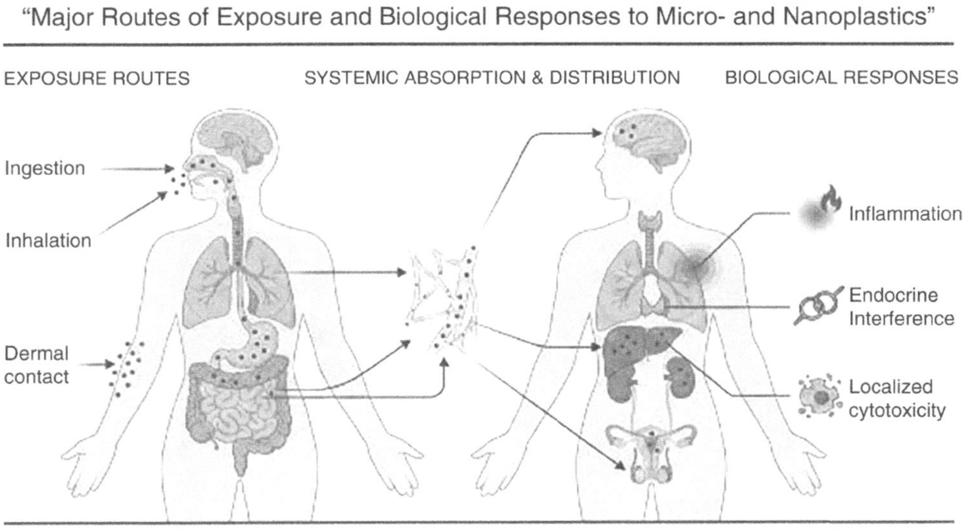

FIGURE 1.4 Major routes of exposure: ingestion, inhalation, and dermal contact.

plaques is linked with a 4–5-fold increased cardiovascular risk.[5] They also interrupt the gut microbiota by reducing diversity and encouraging pathogenic strains, contributing to inflammatory and metabolic disorders. Figure 1.4 illustrates the major routes of exposure.

1.4.4 Interaction with Other Contaminants and Epidemiological Evidence

Hydrophobic and irregular MNP surfaces promote microbial adhesion and biofilm formation with remarkable colonization than surrounding aquatic matrices, allowing MNPs to play a vital role as vectors for pathogens and chemical contaminants.[43] Their sorption of persistent organic pollutants (POPs), metals, and other contaminants changes the bioavailability and migration behaviour in water systems. Human exposure occurs through inhalation, ingestion, and drinking water, with particles identified in blood, stool, and placental tissues.[44] Although MNPs are frequently detected in drinking water, epidemiological evidence remains insufficient to determine definitive health effects. NPs are often excluded from exposure assessments, and dose–response relationships at environmental concentrations are, however, not yet defined (World Health Organization,). Addressing these toxicological and analytical gaps is a necessary requirement for robust risk assessment. Figure 1.5 demonstrates the MPs as carriers of multiple contaminants.

1.5 REMEDIATION AND TREATMENT APPROACHES

MPs and NPs from drinking water systems are becoming progressively recognized as pivotal because they persist, have the capability to bioaccumulate, and pose health risks that have been documented, such as chronic inflammation and endocrine disruption. Standard technologies for treating water, such as coagulation, sedimentation, flocculation, and membrane filtration, achieve incomplete removal efficiencies

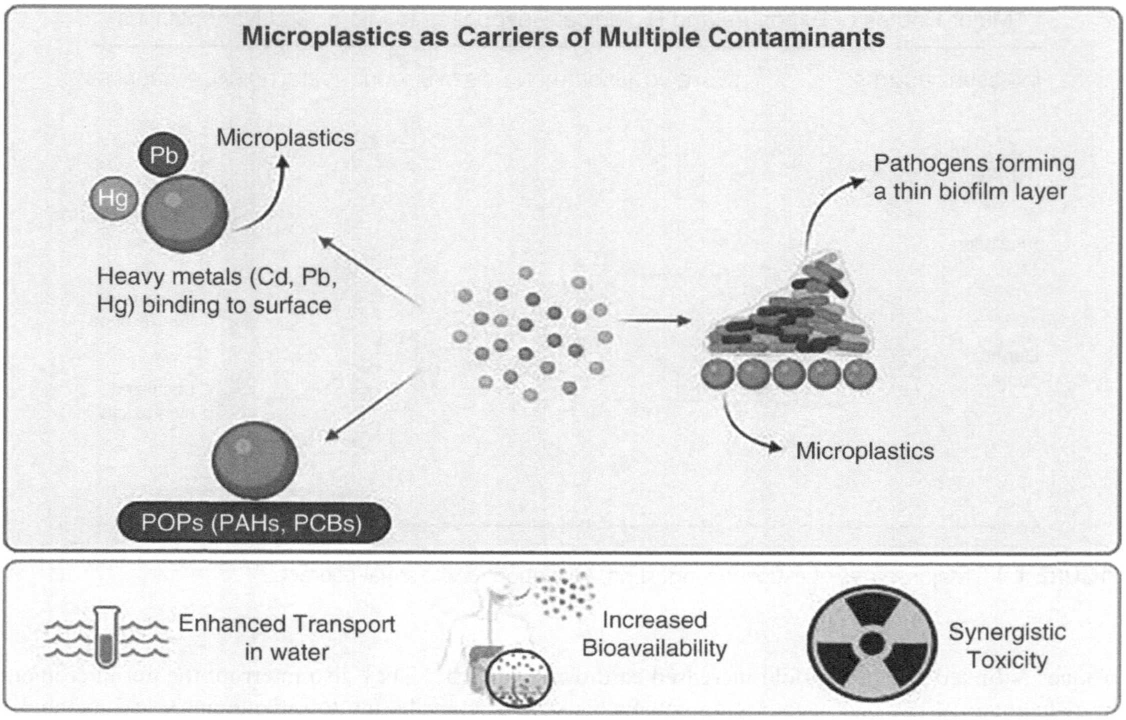

FIGURE 1.5 Microplastics as vectors of pollutants and biofilms in aquatic systems.

ranging from 50% to 90%, mainly for larger MPs. Nevertheless, NPs which are less than 1 µg usually do not pass through these traditional barriers due to their high mobility, small size, and surface reactivity. Recent reports (2023–2025) emphasize hybrid and POU treatment systems integrating cutting-edge adsorption with photocatalysis plus nanocomposite filtration which eventually achieve more than 95% removal efficiency for plastic particles larger than 1 µg. However, ongoing innovations search to enhance selectivity and capture at the nanoscale.

1.5.1 Conventional Water Treatment

Conventional processes in drinking water and wastewater treatment plants such as coagulation, filtration, sedimentation, and flocculation primarily remove turbidity and suspended solids, with irregular performance against MNPs. Particles above 10 µm are typically removed at 50–90% efficiency, whereas smaller MPs and most NPs endure due to low density, surface charge, and small size limiting the aggregation and settling.[45] Coagulation–flocculation–sedimentation removes 10–99% of MPs which are greater than 10 µm, achieving up to 89.2% polyvinyl chloride removal at optimal alum doses (20–50 mg/L), though efficiency declines at neutral pH or suboptimal dosing. Ballasted flocculation has improved PVC removal to 95% by enhancing settling velocity. Rapid sand filtration removes 70–97% of MPs under 10 µm through straining and electrostatic interactions, with iron-loaded media enhancing retention up to 99% for larger particles; however, efficiency drops to 30–50% for fibres and NPs. Disinfection steps modify rather than fully remove plastics: chlorination induces surface cracking with limited mass loss (2–10%), potentially creating smaller fragments, while ozonation achieves 50–80% degradation of PS NPs but may release additives such as BPA and even increase small-particle counts due to fragmentation.[45,46] Figure 1.6 illustrates the MP presence and movement from raw water through water treatment processes and into distribution systems, showing how particles persist beyond drinking water treatment plants and how additional

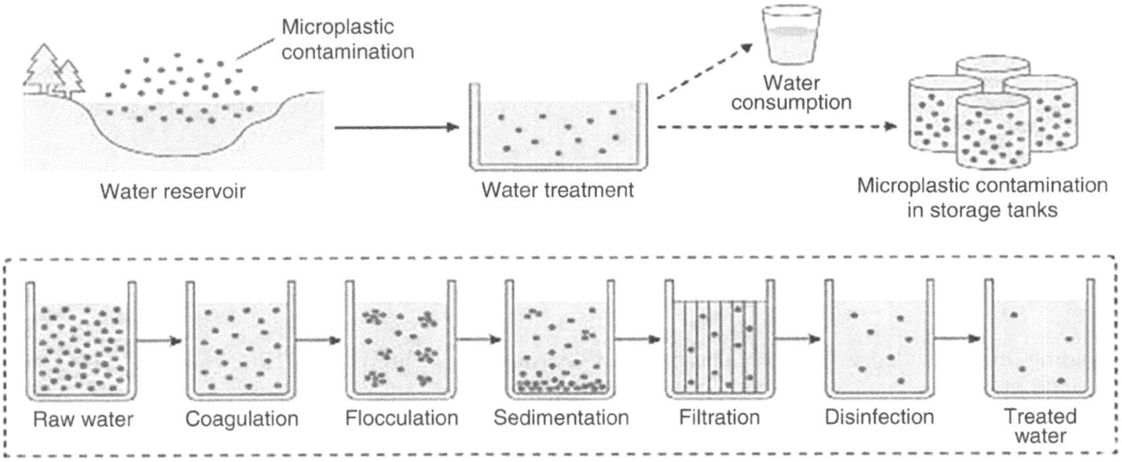

FIGURE 1.6 Microplastic transport from raw water through treatment and distribution systems.

TABLE 1.1 Reported removal efficiencies of micro- and nanoplastics in drinking water treatment processes

TREATMENT	*MP/NP SIZE RANGE*	*REMOVAL EFFICIENCY*	*KEY MECHANISM*	*REFERENCE*
CFS	>10 μm	10–99%	Sweep flocculation	45
RSF	<10 μm	70–97%	Electrostatic adhesion	47
Chlorination	1–100 μm	2–10% degradation	Surface cracking	45
Ozonation	<1 μm	50–80% degradation	Chain scission	48

MPs may enter from ageing pipes and storage tanks before reaching both tap and bottled water. These limitations highlight the need for advanced or hybrid strategies targeting nanoscale plastics (Table 1.1).

1.5.2 Advanced Treatment Technologies

Advanced technologies significantly improve MP/NP removal, achieving 85–99.9% efficiency for particles less than 1 μm. Membrane systems show high rejection rates: UF (37–100%), NF (~99%), and reverse osmosis (RO) (54–99.8%) with optimized and hybrid configurations reaching up to 99.9% removal whereas energy demand and reducing fouling.[42] Granular activated carbon removes 70–95% MPs through hydrophobic and π–π interactions, improved by surface modification. Advanced oxidation processes reduce 50–90% of MPs/NPs but may need post-treatment.[46,49] Magnetic nanomaterials and hybrid systems (e.g. MBR-RO, CFS-AOP) further improve removal to 95–99.9% through integrated physical, chemical, and biological mechanisms (Table 1.2).

1.5.3 Point-of-Use Systems

POU systems act as an ultimate barrier against residual MPs after drinking water treatment plant (DWTP) treatment (70–90% removal).[26] Membrane-based units and RO achieve greater than 99–99.8% removal, ceramic filters remove 72–95%, granular activated carbon (GAC/IX) filters remove 70–90% (less effective <1 μm), and distillation achieves ≥99% removal.[26,53] No international limits exist for MPs/NPs in bottled water. The World Health Organization (WHO) (2019; 2023) considers particles <10 μm low risk

TABLE 1.2 Advanced and hybrid treatment technologies for micro- and nanoplastic removal

TECHNOLOGY	*MP/NP SIZE*	*REMOVAL EFFICIENCY*	*OPTIMIZATION*	*REFERENCE*
UF/NF/RO	<1–100 μm	85–99.9%	AI pressure control	50
GAC	<1 μm	70–95%	$ZnCl_2$ modification	51
AOP (UV/H_2O_2)	1–50 μm	30–95%	Catalyst doping	49
Magnetic	10–100 μm	92–99%	Fe_3O_4 coating	52
MBR-RO hybrid	<5 μm	98–99.9%	Bio-adsorption	52

pending further evidence; California proposes <12 particles/L (<1 mm); the EU 2023 and ISO 24187:2023 provide regulatory frameworks; and IBWA (2025) recommends advanced treatments such as GAC and RO to minimize contamination.

1.6 REGULATORY AND POLICY FRAMEWORK

Regulatory approaches to MPs and NPs in drinking water vary considerably across territories. The WHO considers current concentrations (0.1–10 particles/L) to be a low health risk due to inadequate toxicity data and poor gastrointestinal absorption. Nevertheless, it recommends incorporating MP monitoring into water safety plans and strengthening detection methods for 1–5,000 μm particles. The United States Environmental Protection Agency (USEPA) is assessing risks under the Safe Drinking Water Act. Still, no federal standards exist as of 2025, though California mandates Raman-based annual monitoring (1–5,000 μm; 0.1 particles/L detection limit). In India, the Bureau of Indian Standards (BIS) does not report MNPs under IS 10500:2012 although reported levels are 0.5–8 particles/L in drinking water. On the contrary, the European Union (EU) Drinking Water Directive (2020/2184) mandates harmonized monitoring through automated Raman microspectroscopy (20–5,000 μm; 0.1 particles/L detection limit) and aligns with the Zero Pollution Action Plan targeting a 30% reduction in MPs by 2030.

Monitoring frameworks give emphasis to spectroscopic techniques but remain technically limited. WHO and USEPA recommend FTIR and Raman spectroscopy in a contamination-controlled environment, despite Raman detection limits (~0.5 μm) restricting NP analysis.[25,54] The EU methodology involves semi-annual reporting and <10% interlaboratory variability, while India currently needs standardized protocols despite pilot FTIR initiatives. At the global level, the International Organization for Standardization (ISO 2418:2023) prescribes 1 μm filtration with FTIR/Raman analysis (<15% variability) (International Organization for Standardization), and the International Atomic Energy Agency (IAEA) NUTEC Plastics programme supports harmonized methodologies and laboratory capacity building. The UNEP Global Plastics Treaty extends calls for enforceable monitoring, global databases, technology transfer (e.g. nanofiltration removing >99% of MNPs), and development of cost-effective detection tools such as laser-induced breakdown spectroscopy (~50 nm range) to promote equitable global standardization.[33,55]

1.7 CASE STUDIES

1.7.1 High MP Levels in Bottled Water (Regional Comparisons)

Bottled water is a noteworthy exposure pathway for MPs, with contamination arising from storage, packaging, transport, and bottling. Comparative surveys frequently show higher MP counts than municipal supplies. A 2024 systematic review reported MPs in 56–100% of sampled bottles, with detection frequency effectively dependent on analytical method and size limit. Spectroscopic/visual methods usually report mean concentrations in the tens of particles per litre (e.g. 10–100 particles/L for MPs >100 μm), while techniques detecting smaller particles or NPs generate markedly higher counts. Polyethylene terephthalate generally dominates polymer profiles, with PP often linked to caps; reported values must therefore indicate size thresholds and methods.[10] Chinese surveys reported mean values around 72.3 ± 44.6 particles/L for the analysed size scale, whereas NP-sensitive methods identified 110,000–400,000 particles/L in limited experiments, underscoring method dependence. European studies show comparable heterogeneity, with some German investigations reporting multi-thousand particles/L affected by size limits and cap abrasion. Studies from Latin America and Nigeria largely report low tens to low hundreds of particles/L, regularly identifying PET and PP, though results are highly method-dependent and not generalizable at a continental scale. Membrane processes (MF/UF/NF/RO) in bottling plants remove >80–95% of MPs above pore size, but NP removal and long-term integrity involve site-specific validation.

1.7.2 Municipal Tap Water and Aquifer Exposure

Municipal tap water contamination arises from source water inputs, incomplete treatment removal, and distribution system wear. Chinese surveys report low-to-moderate levels (tens of particles/L for reported size windows), commonly PP & PE, linked to riverine inputs and infrastructure.[56] Bangkok surface waters show high MP loads, but treated water levels depend on treatment configuration and detection range. In Czech plants, treated water ranged 338–628 particles/L contained by study size thresholds, reflecting urban inputs and treatment efficiency,[32] whereas German groundwater-fed systems reported very low levels (mean 0.7 $MPs \cdot m^{-3}$; 0–7 $MPs \cdot m^{-3}$).[57] North American studies largely report low tens of particles/L, with PP & PE widespread and possible distribution contributions. Membrane systems (UF/NF/RO) effectively remove MPs above cutoff sizes but require validation for NP and long-term performance. In community aquifer-dependent regions, groundwater MPs begin from soil infiltration, agricultural runoff, and septic leakage.[21] Concentrations are typically lower than in surface waters but differ with hydrogeology, depth, and method. Noticeable MPs have been reported in deep Korean agricultural aquifers and shallow coastal wells in Sinaloa, often reflecting agricultural signatures. Textile effluents and plastic mulch are reported local sources, though robust attribution involves polymer/additive fingerprinting. Pilot studies suggest biochar and engineered media can retain 86–93% of MPs in controlled columns. However, field-scale permeable reactive barrier performance remains site-specific.

1.8 RECENT TRENDS AND GOVERNMENT SCHEMES/PROJECTS

Research on MNPs in drinking water has hurried since 2023 owing to improved analytical methods, notably, Raman/SRS microscopy, greater awareness of possible human health hazards and evolving policy attention.[21] Global annual plastic production is currently 400 million tonnes and, under business-as-usual scenarios, is predicted to rise substantially by 2050 (scenario ranges commonly reported at 800–1,100 $Mt.yr^{-1}$), thereby increasing the potential of secondary MNPs in the environment (Our World in Data; OECD Global Plastics Outlook). Surveys and systematic reviews constantly document prevalent detection of MPs in tap and bottled water, but reported frequencies and concentrations vary strongly with analytical method and particle size window.[24] For example, SRS-based imaging detected 1.1×10^5–3.7×10^5 particles/L in some bottled waters, the popular being NPs. Note that such counts refer to particles ≥100 nm and are not directly comparable to studies using μ-FTIR/visual microscopy. These observations stimulate intensified method standardization, targeted remediation, policy interventions, and exposure assessment. Nonetheless, human health implications continue to be an area of active research rather than settled fact.[33]

1.8.1 Emerging Nanoplastics Risks, Detection, and Remediation Advances

Growing evidence highlights the requirement to distinguish NPs from MPs (1–5 mm) due to their higher specific surface area, altered surface chemistry, and enhanced cellular uptake. NPs can absorb larger amounts of hydrophobic contaminants (e.g. triclosan) under typical experimental conditions, though outcomes depend collectively on polymer type, size, and surface functionalization. Advanced analytical tools now facilitate improved nanoscale detection. Hyperspectral stimulated Raman scattering microscopy identified ~240,000 particles/L in bottled water (≈90% nanoscale),[21] while surface-enhanced Raman spectroscopy (SERS) using AgNP@TiO_2 arrays detects particulates in the tens-of-nanometre range under laboratory conditions. ISO 24187:2023 creates principles for harmonized MP analysis to enhance inter-study comparability, although nanoscale-specific standardization remains under development (ISO 24187:2023, 2023). Concurrently, biomonitoring and experimental data recommend emerging health and ecological concerns. MNPs (PE, PVC) were spotted in 58.4% of carotid atheroma samples, with plaque-associated MNP occurrence linked to an altered hazard ratio of 4.53 (95% CI 2.00–10.27) for major cardiovascular events. Experimental studies show inflammatory and endocrine-disrupting ability, yet strong human causal hazard quantification remains inadequate. Ecologically, NPs bioaccumulate in seafood, with mussel-based estimates suggesting approximately 2 mg NP ingestion per annum among European consumers. Reviews report MNP occurrence across at least 15 human biological matrices, highlighting the need for toxicological models integrating size, morphology, and polymer composition. Traditional drinking water and wastewater treatment, as already discussed, generally remove 70% to >90% of larger MPs, but NP removal remains unpredictable. Upcoming technologies aim to address this research gap. Magnetic metal–organic framework composites (e.g. nano-Fe@ZIF-8) exhibit rapid co-removal of PS microspheres and accompanying endocrine disruptors under much controlled conditions. Electrocoagulation achieves high MP elimination efficiencies in bench-scale trials, and machine-learning models are being exploited to normalize dosing and operational control. Biotechnological approaches (biofilms, enzymatic degradation) show promise but currently exhibit modest degradation rates even under normal environmental conditions. Engineered systems remain fundamentally laboratory-scale. Biochar-based sorbents and permeable reactive barriers can retain significant fractions of MNPs in soil and groundwater systems, though performance is feedstock- and hydraulics-dependent. Complementary green infrastructure, including permeable pavements, has confirmed significant but site-specific reductions in particulate and tyre-wear runoff contributions.

1.8.2 Government Schemes and Projects

Since 2023, national and regional authorities have extended monitoring, regulatory, and pilot remediation efforts highly targeting MNPs, with emphasis on wastewater management, source control, and drinking water protection (European Commission). These proposals associate with Sustainable Development Goal (SDG) 6 and the 2022 mandate of the United Nations Environment Assembly to negotiate a legally binding global plastics agreement (United Nations Environment Programme), though treaty dialogues remain ongoing. Standardization under ISO 24187:2023 is improving analytical comparability, while recent EU regulatory measures and national food and water monitoring programmes are testing implementation pathways. Overall, policy developments favour lifecycle management, preventive governance, and circular economy approaches, but enforcement and global harmonization prevail. In the United States, the Environmental Protection Agency (EPA)'s National Strategy to Prevent Plastic Pollution establishes a lifecycle framework highlighting source reduction (e.g. textiles, tyre wear) and infrastructure upgrades for wastewater and stormwater capture, without setting quantitative NP removal targets (U.S. EPA, Office of Circular Economy). The suggested Microplastics Safety Act (U.S. Congress, 2025) directs Health and Human Service and Food and Drug Administration (HHS) and Food and Drug Administration (FDA) to assess human exposure and potential endocrine/developmental risks, informing forthcoming regulatory thresholds. Locally, the Chesapeake Bay Program's Microplastics Monitoring Framework advances standardized estuarine monitoring. In Europe, Horizon Europe supports MNP research and remediation innovation under broader Zero Pollution and Circular Economy Action Plans, while OECD guidance emphasizes source control and nature-based runoff management strategies (Organisation for Economic Co-operation and Development). At the state level, California's Ocean Protection Council funds stormwater Best Management Practices (BMP) research and statewide plastics checking to monitor scalable mitigation and drinking water protection strategies.[58,59]

1.9 FUTURE PERSPECTIVES

By 2030, synchronised implementation of an integrated policy framework and progressive remediation technologies could possibly reduce human ingestion of MNPs through drinking water, provided universal commitments are met and technological uptake augumented.[60,61] NP toxicokinetics, enduring health impacts, and cost-effective remediation options in low-resource settings are some of the critical and significant knowledge gaps that remain unfilled. Global experts have recommended attaining international harmonization of ISO analytical standards for data comparability and regulatory coherence across borders. In addition, public–private partnerships should be encouraged to fasten the advancement and use of scalable detection and treatment technologies like SERS sensors and AI-integrated filtration systems. These actions will effectively close the gap between scientific knowledge and practical application, thus pushing forward global initiatives for safe as well as sustainable water systems in a world increasingly threatened by plastic pollution.

REFERENCES

1. Geyer, R.; Jambeck, J. R.; Law, K. L. Production, Use, and Fate of All Plastics Ever Made. *Sci. Adv.* **2017**, *3* (7), e1700782. https://doi.org/10.1126/sciadv.1700782.
2. Dokl, M.; Copot, A.; Krajnc, D.; Fan, Y. V.; Vujanović, A.; Aviso, K. B.; Tan, R. R.; Kravanja, Z.; Čuček, L. Global Projections of Plastic Use, End-of-Life Fate and Potential Changes in Consumption, Reduction, Recycling and Replacement with Bioplastics to 2050. *Sustain. Prod. Consum.* **2024**, *51*, 498–518. https://doi.org/10.1016/j.spc.2024.09.025.
3. Pfohl, P.; Santizo, K.; Sipe, J.; Wiesner, M.; Harrison, S.; Svendsen, C.; Wohlleben, W. Environmental Degradation and Fragmentation of Microplastics: Dependence on Polymer Type, Humidity, UV Dose and Temperature. *Micropl. Nanopl.* **2025**, *5*, *7M.* https://link.springer.com/article/10.1186/s43591-025-00118-9 (accessed 2026-03-01).
4. Yousafzai, S.; Farid, M.; Zubair, M.; Naeem, N.; Zafar, W.; Asam, Z. ul Z.; Farid, S.; Ali, S. Detection and Degradation of Microplastics in the Environment: A Review. *Environ. Sci. Adv.* **2025**, *4* (8), 1142–1165. https://doi.org/10.1039/D5VA00064E.
5. Arif, Y.; Mir, A. R.; Zieliński, P.; Hayat, S.; Bajguz, A. Microplastics and Nanoplastics: Source, Behavior, Remediation, and Multi-Level Environmental Impact. *J. Environ. Manage.* **2024**, *356*, 120618. https://doi.org/10.1016/j.jenvman.2024.120618.
6. Gigault, J.; El Hadri, H.; Nguyen, B.; Grassl, B.; Rowenczyk, L.; Tufenkji, N.; Feng, S.; Wiesner, M. Nanoplastics Are Neither Microplastics Nor Engineered Nanoparticles. *Nat. Nanotechnol.* **2021**, *16*, 501–507. https://www.nature.com/articles/s41565-021-00886-4 (accessed 2026-03-01).
7. Sharma, S.; Chatterjee, S. Microplastic Pollution, a Threat to Marine Ecosystem and Human Health: A Short Review. *Environ. Sci. Pollut. Res. Int.* **2017**, *24* (27), 21530–21547. https://doi.org/10.1007/s11356-017-9910-8.
8. Andrady, A. L. Microplastics in the Marine Environment. *Mar. Pollut. Bull.* **2011**, *62* (8), 1596–1605. https://doi.org/10.1016/j.marpolbul.2011.05.030.

9. Mason, D.; McConachie, H.; Garland, D.; Petrou, A.; Rodgers, J.; Parr, J.R. Predictors of Quality of Life for Autistic Adults. *Autism Res.* **2018**, *11*, 1138–1147. https://onlinelibrary.wiley.com/doi/full/10.1002/aur.1965 (accessed 2026-03-01).
10. Kosuth, M.; Mason, S. A.; Wattenberg, E. V. Anthropogenic Contamination of Tap Water, Beer, and Sea Salt. *PLoS One* **2018**, *13* (4), e0194970. https://doi.org/10.1371/journal.pone.0194970.
11. Ragusa, A.; Svelato, A.; Santacroce, C.; Catalano, P.; Notarstefano, V.; Carnevali, O.; Papa, F.; Rongioletti, M. C. A.; Baiocco, F.; Draghi, S.; D'Amore, E.; Rinaldo, D.; Matta, M.; Giorgini, E. Plasticenta: First Evidence of Microplastics in Human Placenta. *Environ. Int.* **2021**, *146*, 106274. https://doi.org/10.1016/j.envint.2020.106274.
12. Thompson, R. C.; Courtene-Jones, W.; Boucher, J.; Pahl, S.; Raubenheimer, K.; Koelmans, A. A. Twenty Years of Microplastic Pollution Research—What Have We Learned? *Science.* **2024**, *386*, eadl2746. https://www.science.org/doi/10.1126/science.adl2746 (accessed 2026-03-01).
13. Marcharla, E.; Vinayagam, S.; Gnanasekaran, L.; Soto-Moscoso, M.; Chen, W.-H.; Thanigaivel, S.; Ganesan, S. Microplastics in Marine Ecosystems: A Comprehensive Review of Biological and Ecological Implications and Its Mitigation Approach Using Nanotechnology for the Sustainable Environment. *Environ. Res.* **2024**, *256*, 119181. https://doi.org/10.1016/j.envres.2024.119181.
14. Huang, X.; Zhang, M.; Su, X.; Feng, Y. CO_2-Responsive Polymer Promoted by Polyether to Efficient Viscosity Increase for CO2 Plugging. *Polymer* **2024**, *306*, 127227. https://doi.org/10.1016/j.polymer.2024.127227.
15. Sangkham, S.; Faikhaw, O.; Munkong, N.; Sakunkoo, P.; Arunlertaree, C.; Chavali, M.; Mousazadeh, M.; Tiwari, A. A Review on Microplastics and Nanoplastics in the Environment: Their Occurrence, Exposure Routes, Toxic Studies, and Potential Effects on Human Health. *Mar. Pollut. Bull.* **2022**, *181*, 113832. https://doi.org/10.1016/j.marpolbul.2022.113832.
16. Bhardwaj, G.; Abdulkadhim, M.; Joshi, K.; Wankhede, L.; Kumar Das, R.; Kaur Brar, S. Exposure Pathways, Systemic Distribution, and Health Implications of Micro- and Nanoplastics in Humans. *Appl. Sci.* **2025**, *15*, 8813.
17. Temam, R.; Mortula, M. M. Leaching of Microplastics from PVC Pipes under Stagnant Conditions. *Desalination Water Treat.* **2023**, *311*, 243–252. https://doi.org/10.5004/dwt.2023.29979.
18. Świetlik, J.; Magnucka, M. Aging of Drinking Water Transmission Pipes during Long-Term Operation as a Potential Source of Nano- and Microplastics. Int. *J. Hyg. Environ. Health* **2025**, *263*, 114467. https://doi.org/10.1016/j.ijheh.2024.114467.
19. Moyal, J.; Dave, P. H.; Wu, M.; Karimpour, S.; Brar, S. K.; Zhong, H.; Kwong, R. W. M. Impacts of Biofilm Formation on the Physicochemical Properties and Toxicity of Microplastics: A Concise Review. *Rev. Environ. Contam. Toxicol.* **2023**, *261* (1), 8. https://doi.org/10.1007/s44169-023-00035-z.
20. Xiu, F.-R.; Chen, C.; Song, Z.; Wang, S.; Qi, Y. A Novel Resource Conversion Strategy of Waste Plastic Express Packaging Bags Using Supercritical Water-Ammonia Process: Optimization by Response Surface Methodology. *Process Saf. Environ. Prot.* **2022**, *168*, 142–149. https://doi.org/10.1016/j.psep.2022.09.072.
21. Qian, N.; Gao, X.; Lang, X.; Deng, H.; Bratu, T. M.; Chen, Q.; Stapleton, P.; Yan, B.; Min, W. Rapid Single-Particle Chemical Imaging of Nanoplastics by SRS Microscopy. *Proc. Natl. Acad. Sci. U. S. A.* **2024**, *121* (3), e2300582121. https://doi.org/10.1073/pnas.2300582121.
22. Winkler, A.; Santo, N.; Ortenzi, M. A.; Bolzoni, E.; Bacchetta, R.; Tremolada, P. Does Mechanical Stress Cause Microplastic Release from Plastic Water Bottles? *Water Res.* **2019**, *166*, 115082. https://doi.org/10.1016/j.watres.2019.115082.
23. Polychronopoulos, N. D.; Benos, L.; Vlachopoulos, J. Mathematical Modelling of Coalescence of Viscous Particles: An Overview. *Can. J. Chem. Eng.* **2023**, 101 (9), 5231. https://onlinelibrary.wiley.com/doi/abs/10.1002/cjce.24910 (accessed 2026-03-01).
24. Gambino, I.; Bagordo, F.; Grassi, T.; Panico, A.; De Donno, A. Occurrence of Microplastics in Tap and Bottled Water: Current Knowledge. *Int. J. Environ. Res. Public. Health* **2022**, *19* (9), 5283. https://doi.org/10.3390/ijerph19095283.
25. *World health statistics 2019: monitoring health for the SDGs, sustainable development goals.* https://www.who.int/publications/i/item/9789241565707 (accessed 2026-03-01).
26. Cherian, T.; Ragavendran, C.; Vijayan, S.; Kurien, S.; Peijnenburg, W. J. G. M. A Review on the Fate, Human Health and Environmental Impacts, as Well as Regulation of Antibiotics Used in Aquaculture. *Environ. Adv.* **2023**, *13*, 100411. https://doi.org/10.1016/j.envadv.2023.100411.
27. Tamis, J. E.; Jongbloed, R. H.; Rozemeijer, M. J. C.; Grundlehner, A.; de Vries, P.; Van Gerven, A.; Jak, R. G.; Piet, G. J. Assessing the Potential of Multi-Use to Reduce Cumulative Impacts in the Marine Environment. *Front. Mar. Sci.* **2024**, *11*, 1420095. https://doi.org/10.3389/fmars.2024.1420095.
28. Wang, S.; Abbas, J.; Sial, M. S.; Álvarez-Otero, S.; Cioca, L.-I. Achieving Green Innovation and Sustainable Development Goals through Green Knowledge Management: Moderating Role of Organizational Green Culture. *J. Innov. Knowl.* **2022**, *7* (4), 100272. https://doi.org/10.1016/j.jik.2022.100272.

29. Di Nunno, F.; Granata, F.; Parrino, F.; Gargano, R.; de Marinis, G. Microplastics in Combined Sewer Overflows: An Experimental Study. *J. Mar. Sci. Eng.* **2021**, *9*, 1415. https://www.mdpi.com/2077-1312/9/12/1415 (accessed 2026-03-02).
30. Forrest, K.; Barricella, A. C.; Pohar, S. A., Hinman, A. M.; Amack, J. D. Understanding Laterality Disorders and the Left-Right Organizer: Insights from Zebrafish. *Front. Cell Dev. Biol.* **2022**, *10*, 1035513. https://www.frontiersin.org/journals/cell-and-developmental-biology/articles/10.3389/fcell.2022.1035513/full (accessed 2026-03-02).
31. Cai, M.; Xia, B.; Huang, W.; Mai, L.; Liu, M. Micro- and Nano-Plastics: An Emerging Contaminant in Marine and Freshwater Ecosystems. *J. Oceanol. Limnol.* **2025**, *43*, 343–345. https://doi.org/10.1007/s00343-025-5343-2.
32. Pivokonsky, M.; Cermakova, L.; Novotna, K.; Peer, P.; Cajthaml, T.; Janda, V. Occurrence of Microplastics in Raw and Treated Drinking Water. *Sci. Total Environ.* **2018**, *643*, 1644–1651. https://doi.org/10.1016/j.scitotenv.2018.08.102.
33. Koelmans, A. A.; Mohamed Nor, N. H.; Hermsen, E.; Kooi, M.; Mintenig, S. M.; De France, J. Microplastics in Freshwaters and Drinking Water: Critical Review and Assessment of Data Quality. *Water Res.* **2019**, *155*, 410–422. https://doi.org/10.1016/j.watres.2019.02.054.
34. Samandra, S.; Johnston, J. M.; Jaeger, J. E.; Symons, B.; Xie, S.; Currell, M.; Ellis, A. V.; Clarke, B. O. Microplastic Contamination of an Unconfined Groundwater Aquifer in Victoria, Australia. *Sci. Total Environ.* **2022**, *802*, 149727. https://doi.org/10.1016/j.scitotenv.2021.149727.
35. Ośko, J.; Kadac-Czapska, K.; Jażdżewska, K.; Nowak, N.; Kowalczyk, P.; Grembecka, M. Nanoplastics: From Separations to Analysis—Challenges and Limitations. *Separations* **2025**, *12* (7), 185. https://doi.org/10.3390/separations12070185.
36. Dümichen, E.; Eisentraut, P.; Bannick, C. G.; Barthel, A.-K.; Senz, R.; Braun, U. Fast Identification of Microplastics in Complex Environmental Samples by a Thermal Degradation Method. *Chemosphere* **2017**, 174, 572–584. https://doi.org/10.1016/j.chemosphere.2017.02.010.
37. Nihart, A. J.; Garcia, M. A.; El Hayek, E.; Liu, R.; Olewine, M.; Kingston, J. D.; Castillo, E. F.; Gullapalli, R. R.; Howard, T.; Bleske, B.; Scott, J.; Gonzalez-Estrella, J.; Gross, J. M.; Spilde, M.; Adolphi, N. L.; Gallego, D. F.; Jarrell, H. S.; Dvorscak, G.; Zuluaga-Ruiz, M. E.; West, A. B.; Campen, M. J. Bioaccumulation of Microplastics in Decedent Human Brains. Nat. *Med.* **2025**, *31*, 1114–1119. https://www.nature.com/articles/s41591-024-03453-1 (accessed 2026-03-02).
38. Jenner, L. C.; Rotchell, J. M.; Bennett, R. T.; Cowen, M.; Tentzeris, V.; Sadofsky, L. R. Detection of Microplastics in Human Lung Tissue Using μFTIR Spectroscopy. *Sci. Total Environ.* **2022**, *831*, 154907. https://doi.org/10.1016/j.scitotenv.2022.154907.
39. Van Do, M.; Le, T. X. T.; Vu, N. D.; Dang, T. T. Distribution and Occurrence of Microplastics in Wastewater Treatment Plants. *Environ. Technol. Innov.* **2022**, *26*, 102286. https://doi.org/10.1016/j.eti.2022.102286.
40. Washington, H.; Piccolo, J. J.; Kopnina, H.; O'Leary Simpson, F. Ecological and Social Justice Should Proceed Hand-in-Hand in Conservation. *Biol. Conserv.* **2024**, *290*, 110456. https://doi.org/10.1016/j.biocon.2024.110456.
41. Campanale, C.; Massarelli, C.; Savino, I.; Locaputo, V.; Uricchio, V. F. A Detailed Review Study on Potential Effects of Microplastics and Additives of Concern on Human Health. *Int. J. Environ. Res. Public. Health* **2020**, *17* (4), 1212. https://doi.org/10.3390/ijerph17041212.
42. Prata, J. C.; da Costa, J. P.; Lopes, I.; Duarte, A. C.; Rocha-Santos, T. Environmental Exposure to Microplastics: An Overview on Possible Human Health Effects. *Sci. Total Environ.* **2020**, *702,* 134455. https://doi.org/10.1016/j.scitotenv.2019.134455.
43. Kirstein, I. V.; Wichels, A.; Krohne, G.; Gerdts, G. Mature Biofilm Communities on Synthetic Polymers in Seawater - Specific or General? *Mar. Environ. Res.* **2018**, *142*, 147–154. https://doi.org/10.1016/j.marenvres.2018.09.028.
44. Roslan, N. S.; Lee, Y. Y.; Ibrahim, Y. S.; Tuan Anuar, S.; Yusof, K. M. K. K.; Lai, L. A.; Brentnall, T. Detection of Microplastics in Human Tissues and Organs: A Scoping Review. *J. Glob. Health* **2024**, *14*, 04179. https://doi.org/10.7189/jogh.14.04179.
45. Mao, Y.; Chen, Z.; Xue, S.; Lu, Y.; Cao, K.-F.; Wu, Y.-H.; Hu, H.-Y. Synergistic Ferrate(VI) and Chlorine for Reclaimed Water Disinfection: Microbial Control and Chlorine Decay Mitigation. *Sci. Total Environ.* **2024**, *947*, 174521. https://doi.org/10.1016/j.scitotenv.2024.174521.
46. Shen, M.; Song, B.; Zeng, G.; Zhang, Y.; Huang, W.; Wen, X.; Tang, W. Are Biodegradable Plastics a Promising Solution to Solve the Global Plastic Pollution? *Environ. Pollut.* **2020**, *263*, 114469. https://doi.org/10.1016/j.envpol.2020.114469.
47. Chabi, K.; Li, J.; Ye, C.; Kiki, C.; Xiao, X.; Li, X.; Guo, L.; Gad, M.; Feng, M.; Yu, X. Rapid Sand Filtration for <10 Mm-Sized Microplastic Removal in Tap Water Treatment: Efficiency and Adsorption Mechanisms. *Sci. Total Environ.* **2024**, *912*, 169074. https://doi.org/10.1016/j.scitotenv.2023.169074.

48. Ziembowicz, S.; Kida, M. The Effect of Water Ozonation in the Presence of Microplastics on Water Quality and Microplastics Degradation. *Sci. Total Environ.* **2024**, *929*, 172595. https://doi.org/10.1016/j.scitotenv.2024.172595.
49. Zakib Fazil, A.; Dhawala Wijeratna, D. D.S.; Gomes, P. I. A. Predicting Microplastic Transport in Open Channels with Different Bed Types and River Regulation with Machine Learning Techniques. *Environ. Pollut.* **2025**, *384*, 126912. https://www.sciencedirect.com/science/article/abs/pii/S0269749125012850 (accessed 2026-03-02).
50. Acarer, S. Abundance and Characteristics of Microplastics in Drinking Water Treatment Plants, Distribution Systems, Water from Refill Kiosks, Tap Waters and Bottled Waters. *Sci. Total Environ.* **2023**, *884*, 163866. https://doi.org/10.1016/j.scitotenv.2023.163866.
51. Ramirez Arenas, L.; Ramseier Gentile, S.; Zimmermann, S.; Stoll, S. Nanoplastics Adsorption and Removal Efficiency by Granular Activated Carbon Used in Drinking Water Treatment Process. *Sci. Total Environ.* **2021**, *791*, 148175. https://doi.org/10.1016/j.scitotenv.2021.148175.
52. Pasanen, A.; Karjalainen, M. K.; FinnGen; Zhang, G.; Tiensuu, H.; Haapalainen, A. M.; Ojaniemi, M.; Feenstra, B.; Jacobsson, B.; Palotie, A.; Laivuori, H.; Muglia, L. J.; Rämet, M.; Hallman, M. Meta-Analysis of Genome-Wide Association Studies of Gestational Duration and Spontaneous Preterm Birth Identifies New Maternal Risk Loci. *PLoS Genet.* **2023**, *19* (10), e1010982. https://doi.org/10.1371/journal.pgen.1010982.
53. Amirah Mohd Napi, N. N.; Ibrahim, N.; Adli Hanif, M.; Hasan, M.; Dahalan, F. A.; Syafiuddin, A.; Boopathy, R. Column-Based Removal of High Concentration Microplastics in Synthetic Wastewater Using Granular Activated Carbon. *Bioengineered* **2023**, *14* (1), 2276391. https://www.researchgate.net/publication/375523849_Column-based_removal_of_high_concentration_microplastics_in_synthetic_wastewater_using_granular_activated_carbon (accessed 2026-03-02).
54. Cowger, W.; Booth, A. M.; Hamilton, B. M.; Thaysen, C.; Primpke, S.; Munno, K.; Lusher, A. L.; Dehaut, A.; Vaz, V. P.; Liboiron, M.; Devriese, L. I.; Hermabessiere, L.; Rochman, C.; Athey, S. N.; Lynch, J. M.; De Frond, H.; Gray, A.; Jones, O. A. H.; Brander, S.; Steele, C.; Moore, S.; Sanchez, A.; Nel, H. Reporting Guidelines to Increase the Reproducibility and Comparability of Research on Microplastics. *Appl Spectrosc.* **2020**, *74* (9), 1066-1077 https://journals.sagepub.com/doi/10.1177/0003702820930292 (accessed 2026-03-02).
55. Mitrano, D. M.; Wohlleben, W. Microplastic Regulation Should Be More Precise to Incentivize Both Innovation and Environmental Safety. *Nat. Commun.* **2020**, *11*, 5324. https://www.nature.com/articles/s41467-020-19069-1 (accessed 2026-03-02).
56. Xue, Q.; Wang, N.; Yang, H.; Yang, J.; Bai, H. Detection of Microplastics Based on Spatial Heterodyne Raman Spectroscopy. *Spectrochim. Acta. A. Mol. Biomol. Spectrosc.* **2022**, *283*, 121712. https://doi.org/10.1016/j.saa.2022.121712.
57. Mintenig, S. M.; Löder, M. G. J.; Primpke, S.; Gerdts, G. Low Numbers of Microplastics Detected in Drinking Water from Ground Water Sources. *Sci. Total Environ.* **2019**, *648*, 631–635. https://doi.org/10.1016/j.scitotenv.2018.08.178.
58. State of California Coast and Ocean Annual Report.
59. *Science for people and nature | San Francisco Estuary Institute.* https://www.sfei.org/ (accessed 2026-03-02).
60. *The Organisation for Economic Co-operation and Development | OECD.* https://www.oecd.org/en.html (accessed 2026-03-02).
61. *World Development Report 2025: Standards for Development.* World Bank. https://www.worldbank.org/en/publication/wdr2025 (accessed 2026-03-02).

Engineered Bacteria for Bioremediation of Organic and Inorganic Water Pollutants

2

Adeline Juanita, Dhanush S., Umapriya R., and Satya Sundar Mohanty

2.1 INTRODUCTION

The fast pace of urban growth, industrial processes, and modern agricultural practices has significantly enhanced the discharge of both organic and inorganic contaminants with considerable chemical diversity over the past two decades. Surface waters, groundwater, and municipal wastewater streams today carry complex mixtures of compounds previously unidentified or considered of minor environmental significance. Most of these contaminants are persistent, tend to bioaccumulate, and even show biological activity at trace concentrations, with ecosystem and human health implications (Morin-Crini et al., 2022). More specifically, pharmaceutical compounds, personal care products, pesticides, industrial additives, microplastics, metalloids, and nanomaterials represent emerging contaminants that have gained detection in environmental water compartments around the world. Several such compounds are known to be endocrine disruptors, carcinogens, or mutagens, although most of their toxicity profiles are not well characterized. The continued input into aquatic systems along with incomplete removal through current wastewater treatment (WWT) infrastructure has increased the interest in advanced remediation strategies.

2.1.1 Emerging Organic and Inorganic Water Pollutants: Sources and Environmental Significance

Emerging water pollutants can be broadly defined as chemical or biological substances that are not usually monitored or regulated yet are increasingly recognized as having the potential to threaten human health and the environment. According to the United States Environmental Protection Agency, an emerging contaminant is a chemical or material resulting from a new source or pathway of exposure for which there is no developed health-based standard and yet has the potential to cause real or perceived environmental or

DOI: 10.1201/9781003741626-2

public health risks (US EPA, 2012). Recent global assessments have indicated that effluents from municipal WWT plants are, in fact, a main route through which pollutants reach the surface waters. Domestic discharges, hospital wastes, pharmaceutical manufacturing discharges, agricultural runoff, and landfill leachates are seen as the main contributors to the presence of such substances in aquatic ecosystems (Morin-Crini et al., 2022).

Contrary to conventional pollutants, emerging contaminants have structural diversity and functional specificity, such as hormonal activity, antimicrobial effect, and neurotoxicity, that enable biological effects at nanogram-per-litre concentrations.

Organic emerging pollutants include pharmaceuticals, personal care products, surfactants, industrial chemicals, pesticides, polycyclic aromatic hydrocarbons (PAHs), and microplastics. However, toxic metals, metalloids, radionuclides, rare earth elements, and engineered nanoparticles are inorganic emerging contaminants. Most of these compounds are resistant to natural attenuation and conventional treatment procedures, which enables them to persist in water bodies, accumulate in sediments and biota, and biomagnify in food webs. Their universal detection across continents indicates that it is a global problem and also evidences the inadequacy of the currently available technologies to face the contaminant profile of the present time.

2.1.2 Limitations of Conventional Wastewater Treatment and the Need for Biological Solutions

Traditional WWT systems were typically designed to remove bulk organic matter, suspended solids, nutrients, and pathogenic microorganisms through combinations of physical, chemical, and biological processes. Primary treatment of wastewater uses separation techniques such as screening, sedimentation, and skimming, while secondary treatment relies on microbial metabolism that metabolizes biodegradable organic load. Tertiary or advanced treatment processes include adsorption, membrane filtration, oxidation, or disinfection that further refines the effluents (Shah & Gandhi, 2021). However, several limitations arise with the remediation of emerging contaminants. As the structures of emerging pollutants are complex, poorly biodegradable, and present in lower concentrations, they evade removal by conventional biological processes. Furthermore, physicochemical treatments require high energy inputs and may cause secondary pollution, such as sludge, and also yield toxic transformation by-products. These limitations not only raise the cost of operations but also make the operations of WWT infrastructures non-sustainable (Shah & Gandhi, 2021).

Within this framework, bioremediation has recently received great attention as an environmentally compatible and energy-efficient alternative. Bioremediation can degrade, transform, or immobilize contaminants under mild conditions by exploiting the metabolic versatility of microorganisms. Unlike conventional treatment technologies, biological approaches may be designed for complex pollutant mixtures, functioning with minimal chemical inputs. However, the efficiency of native microbial communities is often limited by a lack of degrading potential, toxic effects of pollutants, and environmental stress. These challenges have catalysed the development of engineered bacterial systems that integrate advances in genetic engineering, synthetic biology, and metabolic optimization for enhanced efficiency, specificity, and robustness in pollutant removal.

2.1.3 Use of Engineered Bacteria as a Biological Strategy for Remediation

Microorganisms generally possess exceptional capabilities in degrading, detoxifying, and accumulating toxic organic and inorganic contaminants. The use of genetically modified organisms (GMOs), particularly engineered, plays a pivotal role in the field of bioremediation. These engineered bacteria are designed in a way that efficiently removes toxins that indigenous bacteria find difficult to break down, thus offering a targeted and effective approach to environmental cleanup. For instance, the use of genetically modified *E. coli* has several advantages such as precision, efficiency, and versatility. The selection of bacteria for remediation is based on parameters such as molecular nature, gene and enzyme induction, metabolite production, growth efficiency, and survival rate. These modified microbes can adapt to adverse conditions. For example, when the moisture rate increases, anaerobic conditions persist, which slows down the degradation rate. When temperature drops, microbial degradation of heavy metal is slow but not completely destroyed due to their inhibited metabolic activities, as the microbial transport routes are frozen by the sub-zero water. Similarly, at higher temperatures, the solubility of heavy metals increases, which in turn increases their availability and the rate of microbial biodegradation. Also, the rate of microbial biodegradation is determined by the pollutant's chemical structure, bioavailability, concentration, toxicity, and stability. The degradation of the n-alkanes is easier when compared to the branched alkanes, aromatics with low molecular weight, hydrocarbons with high molecular weight, and the asphaltenes. In certain cases, toxins will be converted to nontoxic substances by the bioremediation process, and this is capable by a variety of bacteria species, such as *Achromobacter*, *Pseudomonas*, *Aspergillus niger*, *Rhodococcus*, and *Bacillus subtilis*. Similarly, petroleum-contaminated soil was tested using nitrate-reducing bacterial strains such as *Brevibacillus* sp. and *Pseudomonas* sp. Other species such as *Bacillus*, *Corynebacterium*, *Staphylococcus*, *Streptococcus*, *Shigella*, *Escherichia*, and *Enterobacter* were also found to be the best hydrocarbon-degrading bacteria (Sharma et al., 2024).

2.2 CLASSIFICATION AND CHARACTERISTICS OF WATER POLLUTANTS

The aquatic ecosystem and, in turn, human health are continuously threatened by a wide range of pollutants, which are most commonly classified based on their chemical nature, sources, biological effects, and persistence (Shekhar et al., 2024). As illustrated in Figure 2.1, the major classification includes organic pollutants that are carbon-based compounds from domestic, industrial, and agricultural sources and inorganic pollutants, which are metals and non-carbon compounds (Ahamad et al., 2020). Other minor categorizations consist of physical pollutants, pathogens, nutrient pollutants, and emerging contaminants like pharmaceuticals and microplastics that escape the earlier treatment strategies (Zhou et al., 2019). This widespread classification tends to assist in targeted remediation strategies, as organic pollutants may be biodegradable, whereas inorganic pollutants are likely to persist and resist biodegradation.

2.2.1 Organic Pollutants: Types and Environmental Impacts

Organic pollutants in aquatic environments include pesticides, PAHs, pharmaceuticals and personal care products (PPCPs), surfactants, phenols, and volatile organic compounds (VOCs) (Wołowicz & Munir, 2025). Among these, PPCPs are of concern as they get introduced continuously into the rivers through industrial discharges, effluents, and surface runoffs. They sometimes reach concentrations of approximately 10 μg/L, exceeding the environmental safety limits (Brauns et al., 2024). Once in an aquatic system, the microbial degradation takes place, which consumes the available oxygen and creates a hypoxic condition. This depletion of oxygen reduces the benthic population by 40–60% (Ahmad et al., 2024). Several pollutants act as endocrine disruptors that interfere with hormone regulation in aquatic organisms, causing a decline of 25% in fish populations (Grzegorzek et al., 2024). PAHs also pose threat by inducing oxidative stress in corals, resulting in bleaching and physical damage to tissues (Xiang et al., 2019). Many organic pollutants are lipophilic in nature, which leads to their persistence in sediments and living organisms. This leads to bioaccumulation and biomagnification across trophic levels. Compounds

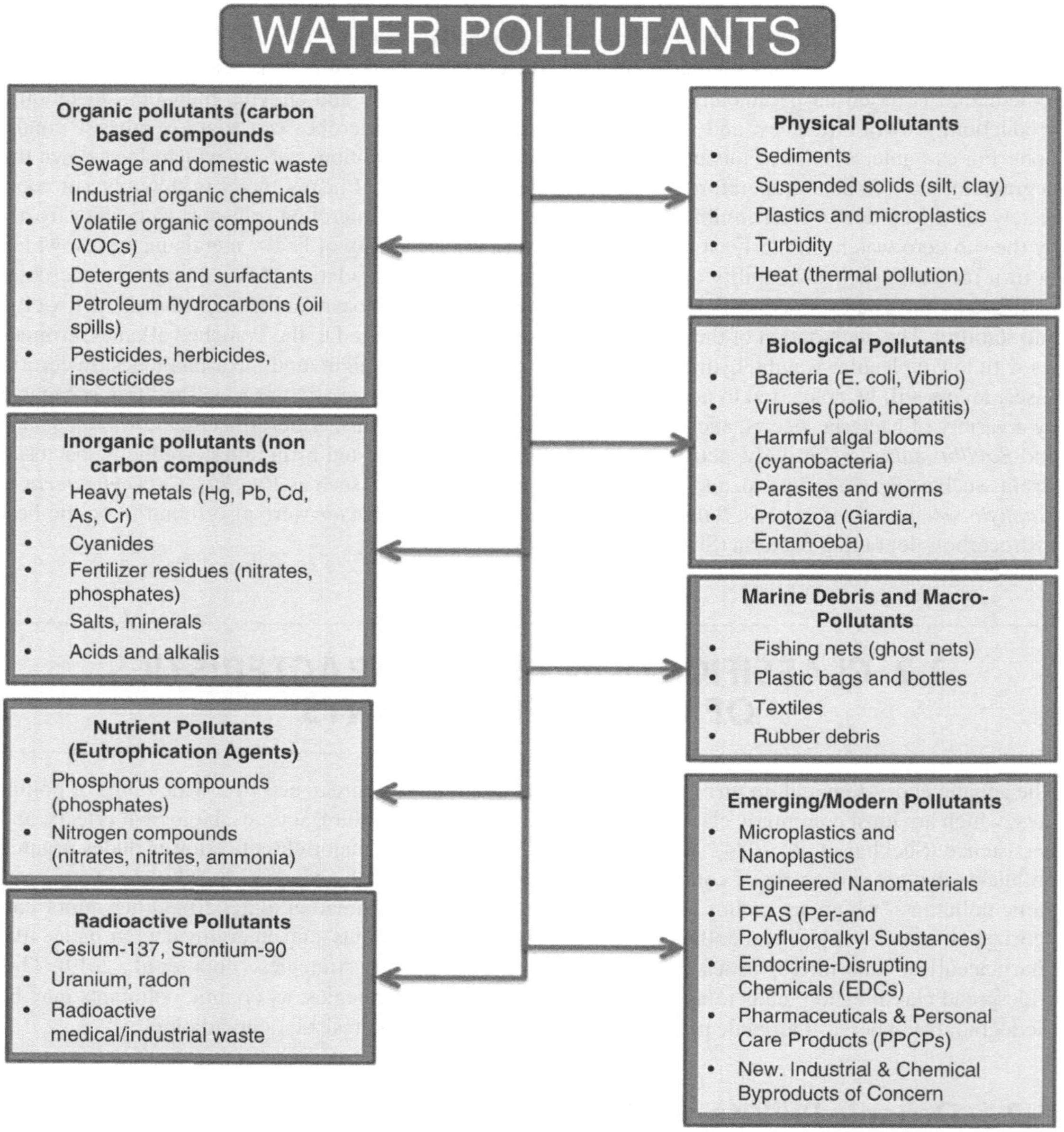

FIGURE 2.1 Overview of major water pollutant categories including organic, inorganic, nutrient, radioactive, physical, biological, marine debris, and emerging contaminants.

like polychlorinated biphenyls (PCBs), polybrominated diphenyl ethers (PBDEs), and per- and polyfluoroalkyl substances (PFAS) are now widely detected in aquatic organisms and cause sublethal effects, including altered behaviour, impaired reproduction, and physiological stress in fish (Nayak et al., 2023; Miller et al., 2024; Porras-Rivera et al., 2024).

2.2.2 Inorganic Pollutants

The rise in total dissolved solids and disruption of water chemistry are due to contamination by inorganic substances like heavy metals (arsenic, cadmium, chromium VI, lead, and mercury), excess nutrients, and

TABLE 2.1 Influence of environmental conditions in organic and inorganic effluents

PARAMETER	EFFECT ON ORGANIC POLLUTANTS	EFFECT ON INORGANIC POLLUTANTS
pH	Alters solubility and degradation rates	Controls metal solubility and precipitation
Temperature	Faster degradation at higher temperatures	Increases metal mobility and chemical reactions
Redox/DO	Affects oxidation/reduction of organics	Controls metal speciation
Organic matter	Can bind organics, slowing or promoting degradation	Can bind metals, reducing or increasing mobility
Salinity/ions	Can change solubility and partitioning	Competing ions can release or bind metals
Turbidity/sediments	High turbidity reduces photodegradation	Metals can adsorb to sediments; disturbance can release them

dissolved salts (Narayanan et al., 2025). These originate from human activities like mining (accounting for 40% of inputs), metal-plating, and agriculture. In some industrial areas the arsenic levels have been reported to be 10–50 times the safety limits (Hama Aziz et al., 2023). These contaminants cause serious health issues like cadmium harming the kidneys, lead and mercury affecting the nervous system, and arsenic and chromium VI are found to be carcinogenic (Sanusi et al., 2025). Metals can also accumulate in aquatic organisms and further biomagnify in food chains, posing health issues to humans consuming those seafood (Tripathi et al., 2025) (Table 2.1).

2.2.3 Physicochemical Behaviour of Pollutants in Water

The physicochemical interactions between contaminants and water constituents determine whether pollutants remain dissolved, become particle-bound, volatilize into the atmosphere, or undergo chemical transformation. Factors controlling these interactions include pH, temperature, ionic strength, dissolved organic matter (DOM), and suspended particulate matter (SPM). These impact the pollutant mobility, persistence, and potential risks (Zhang et al., 2023).

2.2.3.1 Partitioning and Sorption Dynamics

The pollutants get distributed between water and solid phases such as SPM and sediments through partitioning and sorption processes (Niu et al., 2025). PAHs are organic contaminants which strongly get bound to particles, especially when turbidity is high, thus reducing their dissolved availability (He et al., 2021). DOM has the ability to attach pollutants as well as improve their transport through colloids (Zhou et al., 2019). The behaviour of metal is dependent on pH; the acidic conditions increase metal solubility, whereas alkaline conditions favour precipitation and reduce mobility (Hama Aziz et al., 2023; Jee et al., 2024). Higher salinity and ionic strength also contribute to sorption of neutral organic compounds (Mahanty et al., 2023).

2.2.3.2 Influence of Environmental Conditions

High temperatures increase the rate of reactions, volatilization, and change in pollutants (Olasupo et al., 2024). The mobility of redox-sensitive elements is regulated by redox conditions; the oxygen-rich waters promote metal immobilization through precipitation, while low-oxygen conditions release metals back into the water (Jones et al., 2015). Turbidity increases the transport of pollutants bound to the particles, while high electrical conductivity is linked to greater mobility of emerging contaminants like pharmaceuticals (Lednicka et al., 2022; de Sousa et al., 2014). The nanoscale interactions between pollutants, DOM, and minerals also increase the persistence of the contaminant (Xu & Tsang, 2022).

2.2.4 Biodegradation Challenges and Persistence

Organic pollutants, such as PAHs, pesticides, pharmaceuticals, and emerging polluters, are all chemically stable and slow to degrade, and their half-lives exceed one year (Wang et al., 2024). Even though microbes have the capacity to degrade these ingredients, in most cases, the concentration of these compounds in the environment is too low to trigger efficient microbial processes that lead to poor degradation rates (Choudhary et al., 2025). Particle sorption reduces bioavailability, while low temperatures, nutrient imbalance, and metal toxicity reduce microbial activity (Maria et al., 2023; Pande et al., 2022).

The inorganic pollutants like mercury, lead, cadmium, and arsenic do not degrade but transform their form by chemical reactions that influence mobility. They co-exist with organic pollutants and hence are hard to remove because metals prevent biodegradation and organic ligands increase metal mobility (Domingos et al., 2015). Sediments serve as pollutant reservoirs, which emit contaminants under varying conditions. Even though promising results have been indicated in controlled studies using advanced bioremediation strategies, large-scale applications have not been done (Mishra et al., 2025). On the whole, a combination of chemical, biological, and environmental limitations explains the persistence of pollutants and their constant risks to aquatic ecosystems and human health.

2.3 PRINCIPLES OF MICROBIAL BIOREMEDIATION

Microbial bioremediation is a biologically based process that exploits the intrinsic metabolic versatility of microorganisms to degrade, transform, immobilize, or detoxify a wide array of organic and inorganic contaminants in aquatic environments. Apart from physicochemical remediation approaches, which rely on external energy inputs and chemical reagents, microbial bioremediation operates via naturally occurring enzymatic reactions integrated into cellular metabolism and thus provides a number of advantages in terms of energy efficiency, sustainability, and environmental compatibility. The increasing incidence of recalcitrant and emerging pollutants in aquatic systems has accordingly made microbial bioremediation one of the core methods for long-term management of water quality.

Fundamentally, microbial bioremediation is based on the ability of microorganisms to use contaminants either as primary sources of carbon, energy, or electron acceptors or to fortuitously transform them through co-metabolic processes. Due to their metabolic versatility, fast growth rates, and high genetic adaptability, bacteria have emerged as the most studied and deployed group of microorganisms in bioremediation technologies (Bala et al., 2022). In some specific cases, however, fungi, algae, and archaea play a significant role. The efficiency of microbial remediation is therefore determined by the interplay of microbial metabolic pathways with physicochemical pollutant characteristics and environmental conditions, making a mechanistic understanding pre-eminently necessary for its successful application.

2.3.1 Aerobic and Anaerobic Metabolic Pathways

Microbial degradation of organic pollutants proceeds, depending on oxygen availability and the overall redox conditions of the contaminant environment, either via aerobic or anaerobic metabolic routes. In aerobic conditions, molecular oxygen serves as the terminal electron acceptor and is integrated into pollutant molecules through oxygenase-catalysed reactions. Aerobic degradation has been particularly effective for hydrocarbon, PAH, pesticide, and aromatic pollutants, in which monooxygenases and dioxygenases initiate ring hydroxylation, followed by sequential cleavage reactions that shunt intermediates into the tricarboxylic acid cycle of central metabolism (Elyamine et al., 2021). A number of publications in the literature show that different aerobic bacteria, representing the genera *Pseudomonas, Mycobacterium, Rhodococcus, Sphingomonas,* and *Bacillus*, are able to harbour catabolic gene clusters that are responsible for the ring activation, cleavage, and mineralization of complex organic pollutants (Bala et al., 2022). Importantly, in the process of aerobic degradation of PAHs of high molecular weight, such as pyrene, sequential dioxygenation, cis-dihydrodiols, and ortho- or meta-ring cleavage reactions proceed and eventually result in complete mineralization to carbon dioxide and water (Elyamine et al., 2021).

By contrast, anaerobic microbial degradation takes place under oxygen-limited or anoxic conditions, such as in sediments, groundwater aquifers, and sludge systems, with the involvement of alternative electron acceptors, for example, nitrate, sulphate, ferric iron, or carbon dioxide. Anaerobic pathways play a vital role in the degradation of chlorinated solvents, azo dyes, and some aromatic hydrocarbons resistant to aerobic oxidation. Usually, slower compared with the aerobic processes, anaerobic degradation is vital in subsurface remediation scenarios and often complements aerobic treatment under sequential or coupled remediation strategies (Bala et al., 2022). Aerobic degradation pathways of high-molecular-weight PAHs such as pyrene have been extensively characterized, with strain-specific enzymatic routes identified in bacteria including *Mycobacterium vanbaalenii* PYR-1 and *Mycobacterium* sp. KMS and anaerobic degradation pathways of PAHs have also been reported in diverse microorganisms, where reductive activation and alternative electron acceptors drive the transformation of pyrene under oxygen-limited conditions, as demonstrated in *Pseudomonas* sp. JP1, *Klebsiella* sp. LZ6, and the fungus *Coriolopsis byrsina* APC5. The following figure illustrates aerobic and anaerobic microbial degradation pathways of pyrene, highlighting strain-specific enzymatic routes and the metabolic diversity underlying PAH bioremediation (Figure 2.2i and ii).

2.3.2 Mechanisms of Biosorption, Bioaccumulation, and Biotransformation

Microbial remediation involves different mechanisms that are interlinked with each other, including biosorption, bioaccumulation, and biotransformation. Biosorption refers to the passive association of pollutants – especially heavy metals and dyes – with microbial cell surfaces through physicochemical interactions between functional groups, such as carboxyl, hydroxyl, phosphate, and amino moieties located in cell walls and extracellular polymeric substances. It is a metabolism-independent process, occurring even in both living and inactivated biomass, making the process highly effective for inorganic contaminants (Srinivasan et al., 2025).

However, bioaccumulation involves active, energy-dependent transport of contaminants into microbial cells, where they may be sequestered, complexed, or compartmentalized to reduce toxicity. This is well documented in microorganisms under trace metal and metalloid exposure, where intracellular chelators and binding proteins play a role in detoxification and tolerance (Srinivasan et al., 2025).

Biotransformation is an important mechanism in microbial bioremediation, wherein pollutants are enzymatically transformed by microorganisms into less toxic or mobile forms or, however, more bioavailable ones. Redox reactions, such as the reduction of hexavalent chromium to trivalent chromium or methylation and demethylation of mercury species, are most common in inorganic pollutants. These changes greatly reduce bioavailability and ecological risk when the complete removal is not possible (Srinivasan et al., 2025). The following figure illustrates the various bacterial mechanisms involved in heavy metal handling, including bioaccumulation, biosorption, and biotransformation along with certain other mechanisms such as detoxification, bioleaching, and bioprecipitation. Each mechanism

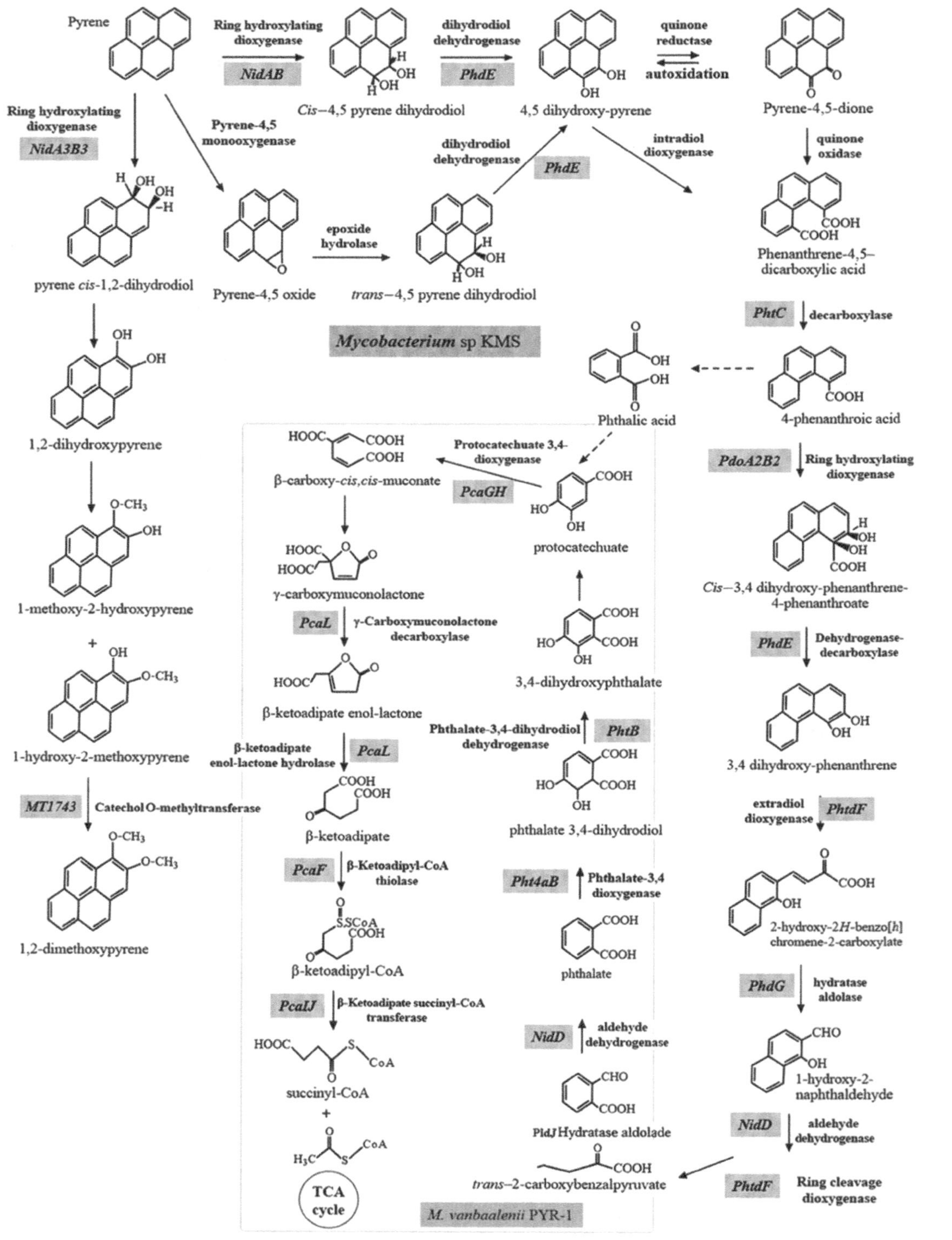

FIGURE 2.2: (i) Aerobic pyrene degradation by *M. vanbaalenii* PYR-1 (uncoloured) and *Mycobacterium* sp. KMS (coloured); half-coloured steps are shared, and solid/dashed arrows indicate single/multiple steps. (ii) Anaerobic pyrene degradation pathways by *Pseudomonas* sp. JP1 (A), *Klebsiella* sp. LZ6 (B), and *Coriolopsis byrsina* APC5 (C); solid and dashed arrows indicate single and multiple steps (Elyamine et al., 2021).

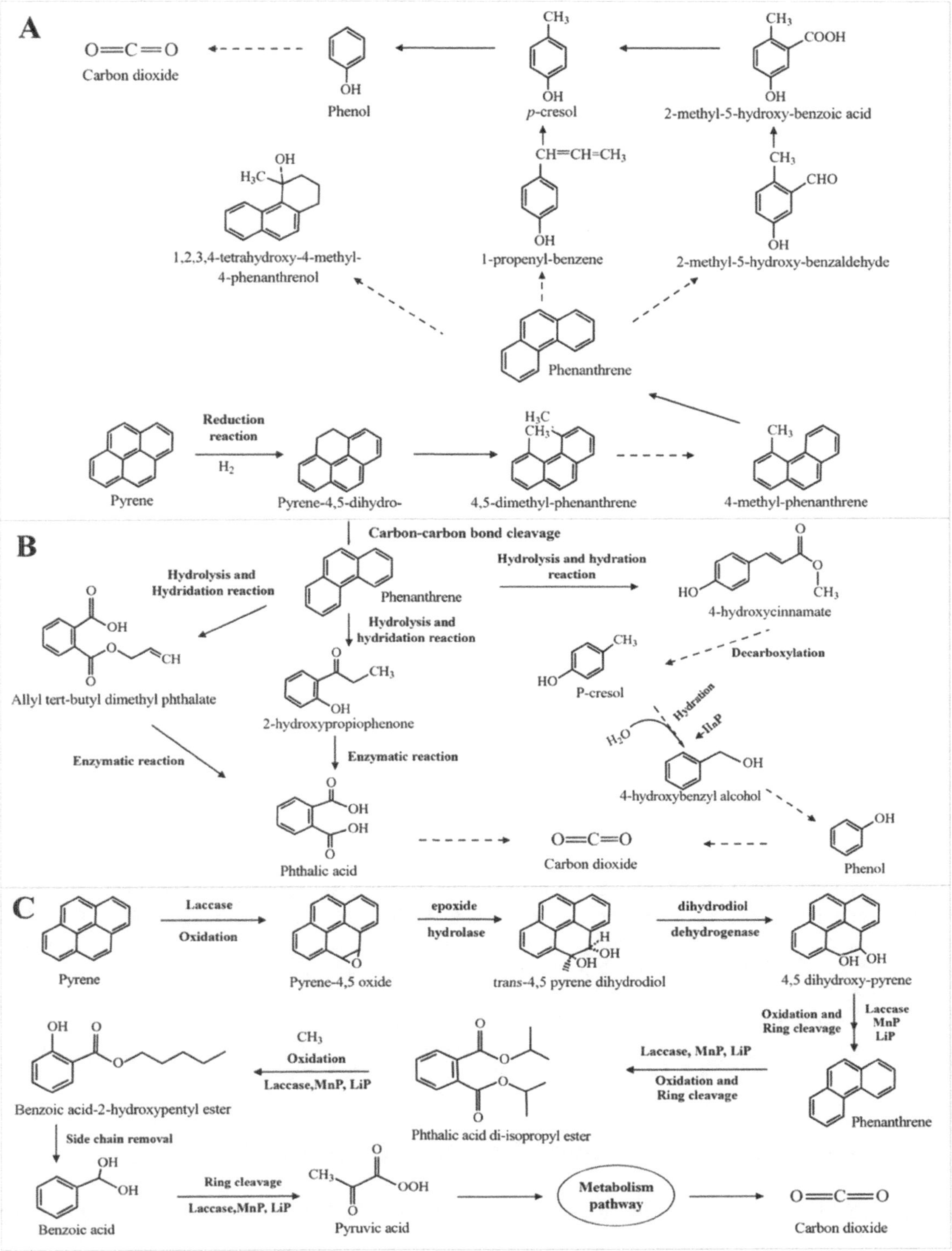

FIGURE 2.2: Continued

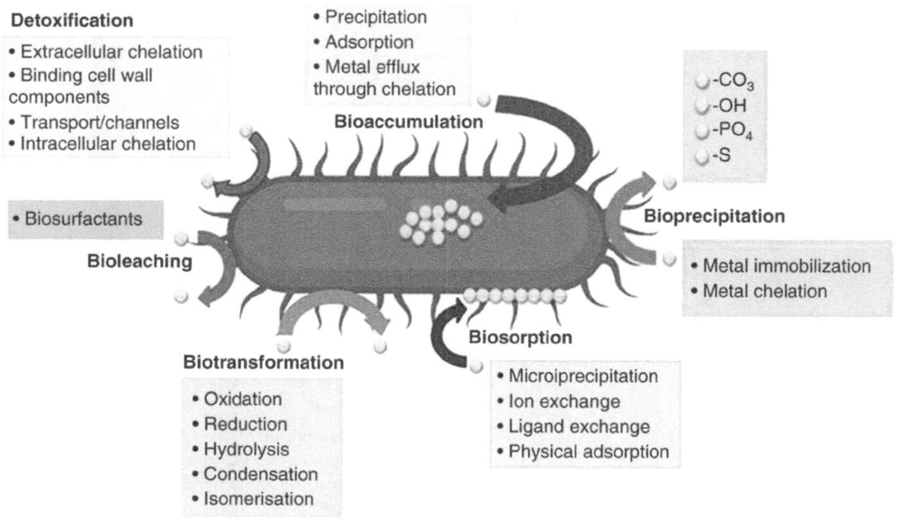

FIGURE 2.3 Mechanisms of microbial heavy metal interaction, including detoxification, bioaccumulation, bioprecipitation, biosorption, biotransformation, and bioleaching.

employs specific biochemical or physical strategies to interact with and neutralize metals to remediate heavy metals (Pande et al., 2022) (Figure 2.3).

2.3.3 Factors Influencing Microbial Degradation Efficiency

The efficacy of microbial bioremediation depends on an interaction among environmental, chemical, and biological factors. In fact, physicochemical parameters such as pH, temperature, salinity, and redox potential govern microbial growth and enzymatic activity directly. Their deviation from optimum range may drastically lower the degradation rate because it will affect enzyme stability and membrane transport processes (Bala et al., 2022).

Nutrient availability, especially the balance of carbon, nitrogen, and phosphorus, is crucial for maintaining microbial metabolism. Nutrient limitation rather than the absence of microbes controls degradation in many aquatic systems, which calls for nutrient additions to improve degradation rates. Other key factors influencing microbial interactions with contaminants include bioavailability, in terms of solubility, hydrophobicity, and sorption onto sediments or organic matter (Bala et al., 2022).

In fact, biological factors – microbial population density, genetic diversity, and interspecific interactions – play a determinative role. Microbial consortia often outperform single strains because of synergistic metabolism, cross-feeding, and division of labour (DOL), thus allowing the more complete degradation of complex mixtures of pollutants (Bala et al., 2022).

2.3.4 Limitations of Native Microbial Communities

Despite their theoretical remediation potential, native microbial communities usually have certain limitations that restrain their performance in contaminated aquatic environments. Most emerging pollutants are of a xenobiotic nature and lacking naturally evolved degradation pathways; thus, their transformation is incomplete or very slow (Bala et al., 2022). High contaminant concentrations will commonly have cytotoxic effects on microbial growth and enzymatic activity, while fluctuating environmental conditions further destabilize microbial performance.

2.4 GENETIC AND SYNTHETIC ENGINEERING STRATEGIES FOR BACTERIAL BIOREMEDIATION

The intrinsic bioremediation potential of native bacterial communities is often limited by narrow metabolic capacity, low tolerance to toxic pollutants, and poor adaptability to fluctuating environmental conditions. Although naturally occurring microorganisms can survive in contaminated environments, their efficiency in accomplishing detoxification is often not enough to tackle emerging organic and inorganic water pollution that is getting increasingly complex and concentrated. Overcoming these challenges, genetic and synthetic biology-based engineering approaches have emerged as potent tools to improve bacterial degradation efficiency, stress tolerance, and functional stability in contaminated aquatic systems (Thai et al., 2023; Garg et al., 2025).

Recent advances in molecular biology, genome editing, and systems biology have provided detailed and precise manipulation of bacterial genomes, including rational design for the improvement of pollutant uptake, transformation, and detoxification by microorganisms. These range from classical recombinant DNA methods to advanced CRISPR-based genome editing and fully synthetic regulatory circuit design. Collectively, such strategies aim at reprogramming bacterial metabolism, enhancing resistance mechanisms, and enabling controlled and predictable performance of bioremediation under real environmental conditions.

2.4.1 Conventional Genetic Engineering Approaches

Classical genetic engineering is the first approach developed for improving bacterial bioremediation efficiency. It typically involves the transfer of pollutant-degrading genes or operons into bacterial hosts by means of usually plasmid-based expression systems or chromosomal integration. Many genes encoding oxygenases, reductases, hydrolases, and transport proteins have been transferred to competent bacterial hosts to increase substrate range and degradation efficiency.

For instance, recombinant bacteria expressing hydrocarbon monooxygenases, dehalogenases, or azo-reductases showed advanced removal of petroleum hydrocarbons, chlorinated solvents, and dye pollutants. These approaches are generally relatively easy and inexpensive; however, long-term applications in open environmental systems are usually constrained by plasmid instability, metabolic burden, and uncontrolled gene expression (Thai et al., 2023). Hence, traditional genetic engineering is being increasingly complemented or replaced with more accurate genome-editing methods. The upcoming figure shows a schematic of traditional genetics-based bacterial bioremediation (Rafeeq et al., 2023) (Figure 2.4).

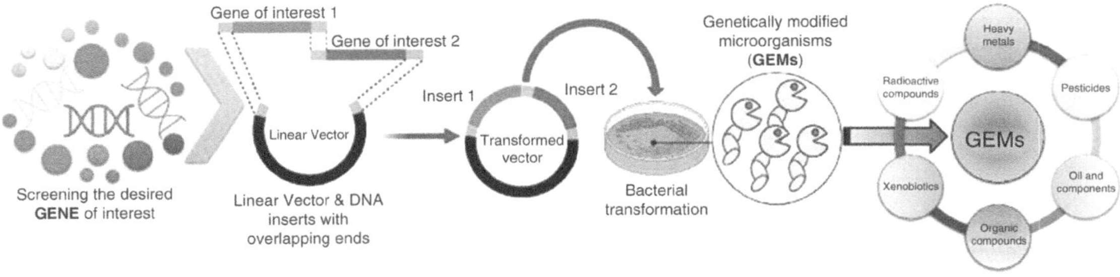

FIGURE 2.4 Genes of interest are inserted into a vector and transformed into bacteria, producing GEMs that degrade pollutants like heavy metals, pesticides, radioactive compounds, xenobiotics, oils, and organics.

2.4.2 CRISPR-Based Genome Editing and Regulation

The CRISPR-Cas systems have opened a new era in bacterial engineering by providing a site-specific, highly efficient, and stable method for genome engineering. CRISPR-Cas9 and other related systems enable precise insertion, deletion, or modification of genes associated with pollutant degradation, heavy metal resistance, and general stress tolerance independent of plasmid maintenance. In addition to editing genomes, CRISPR interference and CRISPR activation enable programmable repression or activation of target genes, thus allowing dynamic regulation of metabolic pathways in response to changing environments (Garg et al., 2025; Minhas et al., 2024).

CRISPR-based approaches have been adopted for the improvement of bacterial tolerance to heavy metals by regulating various efflux systems, intracellular sequestration pathways, and redox enzymes that perform metal transformation. CRISPR-enabled pathway optimization has also been utilized for enhancing biodegradation of recalcitrant organic pollutants, including pesticides, plastics, and PAHs. Despite their potential, CRISPR-engineered bacteria face challenges regarding environmental stability, regulatory constraints, and biosafety, especially concerning horizontal gene transfer and disturbance in the ecosystem. The following image represents the CRISPR/Cas9-mediated genome-editing workflow in engineered bacteria (Dudeja et al., 2025) (Figure 2.5).

2.4.3 Synthetic Biology and Metabolic Engineering Strategies

Synthetic biology goes beyond single-gene modification to allow the rational design of whole metabolic and regulatory networks. By assembling standardized genetic parts, synthetic circuits, and modular

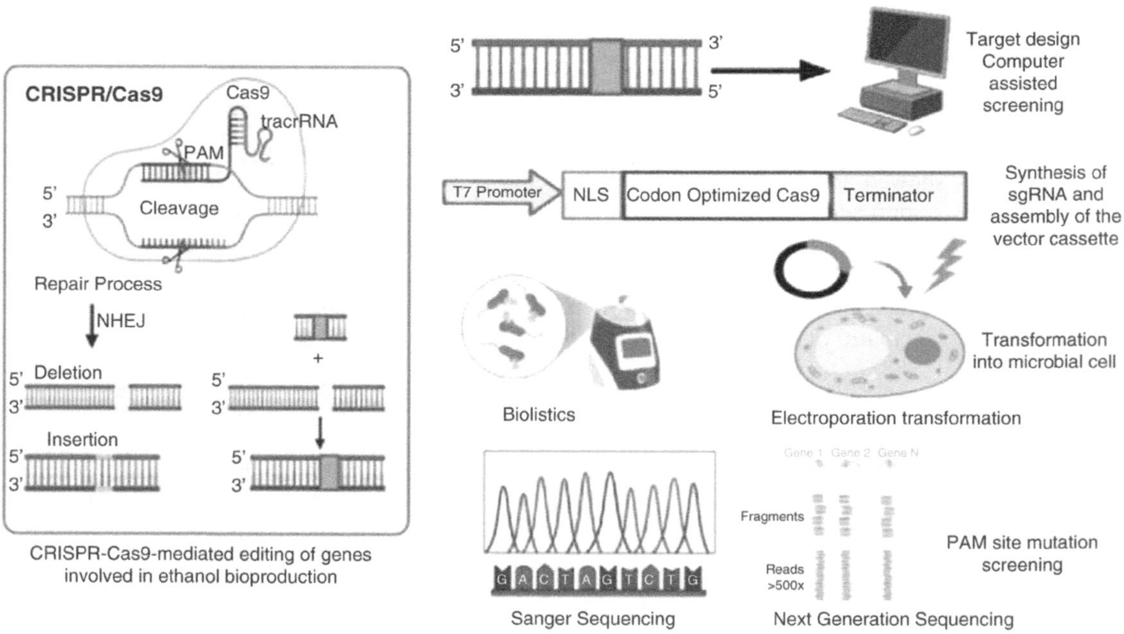

FIGURE 2.5 The process includes target design using computer-assisted screening, synthesis of sgRNA and vector assembly, microbial transformation via electroporation or biolistics, and validation through Sanger or next-generation sequencing. The left panel illustrates Cas9-mediated cleavage and Non-Homologous End Joining (NHEJ) repair resulting in insertions or deletions.

operons, bacterial systems can be programmed to act in response to pollutant presence, environmental cues, or stress signals. Such designs enable conditional expression of degradation pathways, thus minimizing metabolic burden but maximizing efficiency (Thai et al., 2023).

Metabolic engineering approaches are being developed to optimize intracellular flux distribution and enhance pollutant uptake and degradation. Examples include eliminating competing pathways, improving cofactor regeneration, and engineering transport systems for improved pollutant bioavailability. Systems biology and genome-scale metabolic models further support these approaches by the prediction of bottlenecks and the identification of optimal targets of engineering (Mohedano et al., 2022). The image depicts synthetic biology strategies for metabolic pathway engineering that include pathway construction, regulation, and spatial organization of enzymes to optimize n-butanol production (Lee et al., 2018) (Figure 2.6).

2.4.4 Engineering Stress Tolerance and Environmental Robustness

One of the most important limitations of engineered bacteria in practical bioremediation is a decrease in viability under severe environmental conditions, such as high concentrations of pollutants, high salinity, pH variation, and oxidative stress. Engineering of stress tolerance has, therefore, become a key factor in modern bioremediation. This involves membrane composition alteration, overexpression of efflux pumps, enhancement of antioxidant defences, and a shift in the programming of global transcriptional regulators to enhance robustness (Mohedano et al., 2022).

ALE, combined with genome sequencing and rational engineering, has also proved effective for developing strains more tolerant of toxic intermediates and pollutants. Collectively, these strategies enhance bacterial persistence and functional stability in contaminated aquatic environments, thereby narrowing the gap between laboratory performance and field application.

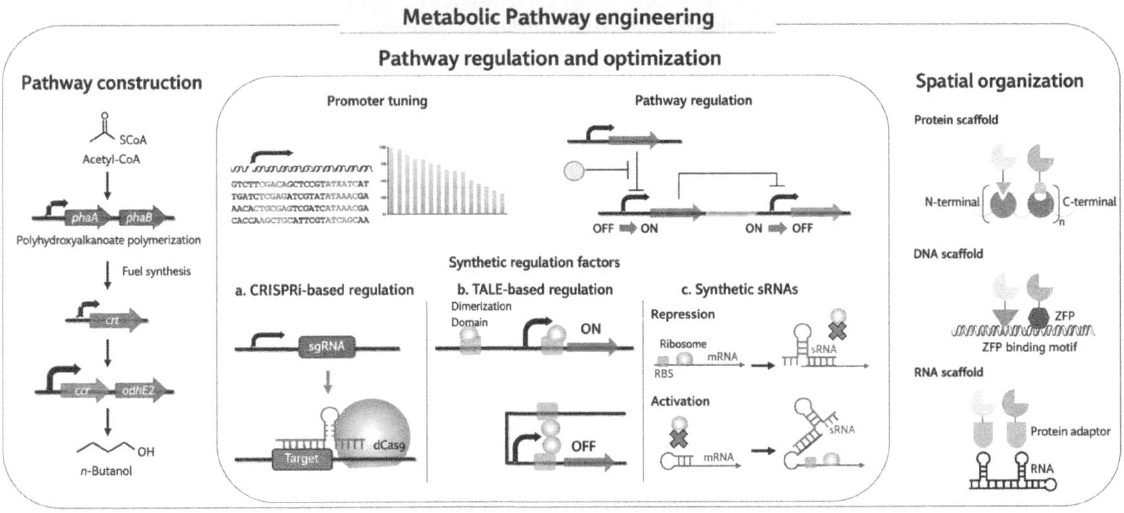

FIGURE 2.6 Metabolic pathway engineering in bacteria, showing (i) pathway construction from heterologous genes converting acetyl-CoA to n-butanol, (ii) regulation and optimization using promoter tuning, CRISPRi/TALE-based transcription factors, and synthetic sRNAs, and (iii) spatial organization of enzymes on protein, DNA, or RNA scaffolds to enhance pathway efficiency.

2.5 ENGINEERED BACTERIA FOR THE REMEDIATION OF ORGANIC WATER POLLUTANTS

Organic water contaminants are a chemically diverse and ecologically persistent class of aquatic pollutants. Such pollutants mostly come from industrial effluents, agricultural runoff, municipal wastewater discharges, and pharmaceutical manufacturing. The compounds in question will include hydrocarbons, pesticides, dyes, pharmaceuticals, endocrine disruptors, and personal care products. A significant proportion of these chemicals are highly toxic and tend to be resistant to biodegradation, thus having great potential for bioaccumulation and hence posing significant risks to aquatic environments and human health (Pandolfo et al., 2023).

Although natural degradation by indigenous microbial communities is able to partially mitigate organic pollutants in natural environments, the increasing structural complexity and concentration of emerging organic contaminants in waterbodies often exceed their inherent metabolic capacity. Engineered bacteria, therefore, have emerged as a promising strategy to improve degradation efficiency and widen substrate specificity by offering targeted removal of recalcitrant organic pollutants under realistic conditions (Chakraborty et al., 2025).

2.5.1 Engineered Bacterial Degradation of Hydrocarbons

Hydrocarbons represent one of the most studied classes of organic pollutants because of the widespread discharge from petroleum refining, oil spills, transportation activities, and industrial effluents. They can be divided into two main groups, namely aliphatic hydrocarbons and aromatic hydrocarbons, which include monocyclic aromatic hydrocarbons (BTEX) and PAHs. Hydrophobic nature and chemical stability of these compounds often limit bioavailability and decrease natural attenuation rates.

Engineered bacterial systems have been developed to overcome these limitations by enhancing hydrocarbon uptake, oxidation, and mineralization pathways. The general strategy in genetic modification typically involves the overproduction of alkane monooxygenases, dioxygenases, and ring-cleavage enzymes, which catalyse the initial activation of hydrocarbon molecules. Downstream metabolic engineering further optimizes efficient flux of intermediates into central metabolic routes such as the tricarboxylic acid cycle. Several recent studies have underlined the effectiveness of engineered strains of the genera Pseudomonas, Rhodococcus, and Mycobacterium for degrading low- and high-molecular-weight hydrocarbons both under controlled and environmental conditions (Pandolfo et al., 2023).

2.5.2 Engineered Bacteria for Pesticide and Herbicide Removal

Pesticides and herbicides are designed to be biologically active and persistent, making their removal in the aquatic environment very cumbersome. A plethora of such substances – organochlorines, organophosphates, carbamates, and neonicotinoids – are resistant to conventional WWT and thus accumulate in both surface and groundwater systems. Engineered bacterial approaches can involve the introduction or augmentation of enzymatic pathways for the hydrolysis, oxidation, or reductive dehalogenation of these compounds.

Advances in metabolism and synthetic biology have enabled the construction of bacterial strains that express pesticide-specific hydrolases, esterases, and dehalogenases, allowing for higher completeness and rates of degradation. Besides, engineering of regulatory circuits enables pesticide presence-inducible expression of pesticide-degrading enzymes, reducing metabolic burden. Altogether, these approaches enhance degradation rate, reduce the generation of toxic intermediate products, and increase the possibility of using engineered bacteria to clean up agricultural and industrial wastewater streams contaminated with pesticides (Siddiqui et al., 2022).

2.5.3 Engineered Bacterial Systems for Pharmaceutical and Personal Care Product Removal

PPCPs are considered an important class of organic micropollutants due to their continuous release, high biological activity, and partial removal by traditional treatment processes. Antibiotics, analgesics, hormones, and psychiatric drugs have been very commonly detected in surface waters at biologically relevant concentrations, creating ecological perturbation and leading to the development of antimicrobial resistance.

Engineered bacteria offer unique advantages in addressing pharmaceutical contaminants due to their ability to enable specific degradation pathways that are either missing or poorly expressed in native microbial communities. For instance, genetically enhanced strains of *Bacillus* and *Pseudomonas* have exhibited enhanced removal of pharmaceutical compounds by overexpressing oxidoreductases and conjugation enzymes concerned with drug metabolism. Experimental studies on the treatment of pharmaceutical wastewater have indeed reported significant reductions in chemical oxygen demand and the levels of active pharmaceutical residues after treatment using such engineered bacterial systems (Kehinde et al., 2022).

2.5.4 Engineered Bacteria for Dye and Synthetic Organic Pollutant Degradation

Synthetic dyes released in the environment from textile, pharmaceutical, and cosmetic industries are some of the most persistent organic colourants and chemical pollutants. Azo, anthraquinone, and triphenylmethane dyes, due to their complex aromatic structures, are often toxic, mutagenic, and recalcitrant to biodegradation. To improve the degradation of dyes, genetically engineered bacteria have been constructed that overexpress azoreductases, laccases, and peroxidases, allowing the reductive cleavage of the azo bond and subsequent mineralization of the resulting aromatic amines. Further improvement of the dye degradation is obtained by genetic optimization of the redox balance and electron transfer pathways. Such engineered systems showed superior performance compared to natural strains, especially at high dye concentrations and variable conditions found in wastewaters (Hosny et al., 2023).

2.6 ENGINEERED BACTERIA FOR INORGANIC POLLUTANT DETOXIFICATION

Heavy metals, metalloids, and radionuclides can be detoxified using engineered bacterial strains through the expression of metal-binding proteins and metal transporters and converting the xenobiotics into less toxic forms by activating enzymatic pathways with the aid of synthetic biology (Hama Aziz et al., 2023). Some of the engineered strains are biosensor-based with the ability to sense the presence of contaminants and trigger detoxification cascades, thereby increasing the effectiveness of operations in varying environmental conditions (Zhou et al., 2019). The further improvement of these microorganisms is by using systems biology and omics-based design to enhance their ability and resilience to treat industrial effluents and polluted waters (Figure 2.7).

2.6.1 Metal Reduction and Oxidation Systems

Shewanella oneidensis is a metal-reducing bacterium that is capable of converting metals like chromium, uranium, and iron into less soluble and less harmful forms through extracellular electron transfer (EET) (Zhou et al., 2019). The metals are transported by the MtrCAB/OmcA protein complex. Scientists are also

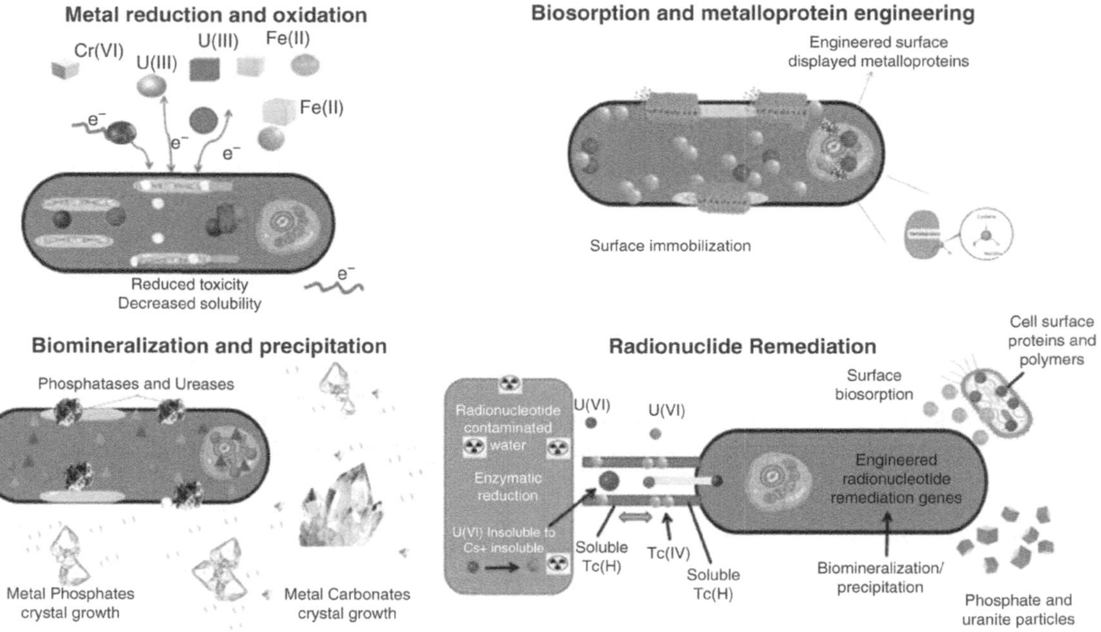

FIGURE 2.7 Microalgae-based mechanisms for remediation of metal and radionuclide contaminated wastewater. The schematic diagram illustrates key processes involved in pollutant removal, including cell surface biosorption, membrane transport mediated uptake, intracellular sequestration, enzymatic reduction of soluble radionuclides to less mobile forms, and biomineralization through precipitation. The role of engineered microorganisms expressing radionuclide-remediation genes is also shown, highlighting enhanced chelation, detoxification, and immobilization of contaminants.

focusing on the improvement of electron flow, the expression of multiheme cytochromes, and biofilm formation to enhance metal reduction in anaerobic environments (Pande et al., 2022; Domingos et al., 2015).

2.6.2 Biosorption and Metalloprotein Engineering

Bacteria can bind metals together by their cell wall components. The use of metallothionein or phytochelatin strains can be used to increase cadmium, lead, and other metal uptake (Mishra et al., 2025). Optimizing the biomass density and pH can boost biosorption by environmentally friendly mechanisms to treat wastewater.

2.6.3 Biomineralization and Precipitation Pathways

Enzymes can be used to precipitate metals as a mineral in bacteria. The carbonates are formed by urease-producing bacteria, and sulphate-reducing bacteria produce sulphides which bind up the metals like cadmium and mercury (Mishra et al., 2025; Sanusi et al., 2025). Genetic modifications enhances the quantity of enzymes or EPS, which improves the metal immobilization of soil and wastewater remediation (Tripathi et al., 2025).

2.6.4 Engineered Strains for Radionuclide Remediation

The radiation-resistant bacteria have been modified to produce uranium as insoluble phosphates through phosphatase enzymes to remove over 90% and, with improved DNA repair, are able to survive in radioactive environments (Zhou et al., 2019). Biosorption and reduction of uranium, strontium, and americium are also strong in other microbes, including *Citrobacter* and immobilized forms of *Saccharomyces cerevisiae* (Hama Aziz et al., 2023). Scientists have engineered consortia and synthetic constructs to enhance the specificity and flexibility for large-scale nuclear waste bioremediation (Khattak et al., 2025).

2.7 SYNTHETIC MICROBIAL CONSORTIA FOR MULTI-POLLUTANT REMEDIATION

Synthetic microbial consortia (SMC) refers to a system of microorganisms that are designed to stay together and perform a single function, like pollutant removal. Here, two or more strains of the same organism or more species could act together for a common cause. The strains or species of microbes that are chosen are based on the compatibility and complementary metabolic abilities between them. Using mechanisms such as metabolic DOL, cross-feeding intermediates, and quorum sensing, the removal rate and efficiency of pollutants by SMC are very high.

2.7.1 Concept and Advantages of Synthetic Consortia

The concept of SMC is inspired by natural colonies and consortia; however, major changes are made in regard to the species involved, as they are manually selected and designed in a controlled condition where all the different microbes used will co-exist and play a major role in remediation. Pollutants including hydrocarbon dyes, pesticides, pharma and nutraceuticals, and heavy metals are chemically stable and are complex to degrade by native and natural consortia, as a single type of microbe has a partial metabolic pathway that leads to partial pollutant removal and the production of toxic intermediates. This limitation is overcome with the help of SMCs by involving metabolic pathways from various strains. This phenomenon is known as metabolic DOL. Therefore, by selecting microbes that complement each other in their metabolic pathways, SMCs can be engineered to remediate target pollutants or targeted environmental conditions, making them a highly flexible tool in the field of bioremediation. SMC is suited for multi-pollutant remediation, including both organic and inorganic pollutant mixtures. The major advantage of SMC is the reduced production of toxic intermediates and by-products. This is because of the immediate use of one by-product by another microbe as a nutrient source. It also enhances process stability. The energy requirements for SMC are also very less, which makes them a better alternative for conventional physicochemical treatment methods (Adamu et al., 2023).

2.7.2 Division of Metabolic Labour and Specialization

DOL is a basic technique used in SMC, where complex metabolic functions and degradations are split up and redistributed to various strains or species of microbes. This technique helps to overcome challenges faced by monocultures such as metabolic burden that arises when microbes are expected to provide support for growth, maintenance, stress response, and expression of long metabolic pathways. The cellular performance of the microbes is reduced when the load becomes higher and is mixed with the toxic intermediates. DOL reduces such stress faced by the microbes and enhances the overall functioning of the system. Metabolic specialization refers to the degradation of specific pollutants by microbes whose native physiological strengths align with their remediation properties. Instead of a single microbe doing energetically long and demanding pathways, the activity is modularized, and each module is specialized with remediation suited to its expression. This approach reduces adenosine triphosphate (ATP) and cofactor requirements and minimizes interferences in pathways. Cross-feeding is another important mechanism in

SMC where the products produced by one microbe are consumed by another. Thus, these syntrophic interactions lead to enhanced detoxification and substrate utility that are difficult to attain using a monoculture. For example, in treating lignocellulosic biomass, cellulolytic microbes such as *Clostridium thermocellum* specialize in polymer hydrolysis, whereas their counterparts, i.e. the non-cellulolytic microbes such as *Thermoanaerobacter* spp., convert the released sugars into fuels or chemicals. Similarly, in mixed sugar fermentations, xylose or arabinose is metabolized separately by microbes such as *Escherichia coli,* which in turn enhances carbon conversion efficiency and produces acetate as a by-product. *Saccharomyces cerevisiae* uses the produced acetate for glucose fermentation. In carbon utilization, autotrophic microbes fix CO2 and methane supply and supply organic carbon to heterotrophic microbes that are engineered for product production. For example, *Synechococcus elongatus PCC 7942* is used for CO_2 fixation and sucrose secretion. *Citrobacter amalonaticus* is used in CO utilization and electron transfer, whereas *Sporomusa ovata* is used for acetate production from CO.

One of the major advantages of DOL is the reduced requirements of multiple enzymatic steps, as natural microbes need lots of enzymatic steps due to cofactor imbalance, protein misfolding, etc. DOL increases the removal efficiencies and product yield, if necessary. However, a few disadvantages of DOL are community stability, metabolite transport, and emergence of non-cooperative strains. Therefore, population dynamics and nutritional divergence should be monitored in real time (Roell et al., 2019).

2.7.3 Quorum Sensing and Engineered Communication Circuits

Quorum sensing refers to the bacterial communication process, where cells detect population density using autoinducers and signalling molecules such as acyl-homoserine lactones (AHLs) in gram-negative bacteria and peptides in gram-positive bacteria. These molecules help to activate the promoters and generally diffuse when in low densities, which forms a positive feedback loop. However, at higher densities, they accumulate and bind to receptors such as LuxR for AHL and trigger changes in gene expression for a group of behaviours such as bioluminescence, biofilm formation, or virulence. These systems are engineered by synthetic biologists using AHLs like 3OC6-HSL and autoinducer-2 (AI-2) in order to create density-dependent genetic circuits in multi-strain cultures. For example, the AI-2 systems in *E. coli* involve the uptake of transporters like LsrACDB and repressors like LsrR for cross-species signalling, but this is still limited by weak promoters and glucose repression. Therefore, orthogonal AHL variants like 3OC12-HSL/LasR are used to minimize the crosstalk, enabling complex circuits.

2.8 DEPLOYMENT STRATEGIES FOR ENGINEERED MICROORGANISMS

The engineered microorganisms are applied in practical remediation treatments in such a way that their efficiencies are enhanced and their full potential is demonstrated. Certain methods that are employed in the application of these engineered microbial systems include immobilized microbes, carrier-assisted delivery, and the design of unique bioreactors for the ex situ treatments.

2.8.1 Immobilization and Encapsulation Methods

Immobilization of microorganisms refers to the process of confining or restricting live and intact cells to a particular device or a matrix without compromising their biological functions. It is generally preferred to immobilize cells, as it helps to enhance the viability and the longevity of cells along with providing better stability, reusability, and catalytic efficiency. In bio-remediation processes, the efficiency of pollutant removal is increased significantly by immobilized microbes when compared to free microbial systems. There are a lot of ways to immobilize cells, and it includes physical and chemical methods. Physical methods based on adsorption, entrapment, encapsulation, and chemical methods include covalent bonding and cross-linking. In adsorption, live cells are immobilized by attaching them to the surface of a matrix (support material). The matrix is usually made of activated carbon (AC), glass beads, synthetic/natural polymers or natural and organic materials. A few major advantages of the adsorption method are maintenance of high cell activity, simplicity, cost-effectiveness, and the absence of a requirement for any other reagents which avoids secondary pollution. However, certain disadvantages, such as weak bond formation and the possibility of biomolecule loss, also persist. In entrapment, cells are physically trapped inside a matrix that could be a gel such as alginate, agar, or carrageenan; it forms a 3D network around the cells, and this network is known as a scaffold. The scaffold is porous and so allows for the diffusion of fluids through the scaffold and, additionally, provides mechanical support. The advantages of entrapment include the continuous operation as the cell densities are well maintained, and it also allows the transport of particles with lower molecular weight. The only disadvantage is that the entrapped cells may leak, and the number of cells that can be entrapped is limited. Encapsulation is another physical method where live cells are enclosed within a semi-permeable membrane or a microcapsule. This membrane acts as a layer of protective shield against external stress factors and sustains the cells within. The permeability of the membrane will be designed as per the requirements of the treatment process. The membrane is designed based on coacervation or interfacial polymerization. Coacervation is a technique where cell microdroplets are separated using a water-immiscible solvent. In interfacial polymerization, the interface of two droplets forms a membrane where one droplet is a hydrophobic monomer and the other is a monomer in the water-immiscible solvents. The major advantage of encapsulation is that larger molecules cannot be diffused through the membrane; however, this property may also be a limiting factor, as mass transfer could be inhibited. Covalent bonding is a type of chemical method where the cells are cross-linked to the support material. The cells are engineered in a way that the surface has functional groups, and these groups react with their corresponding groups in the matrix, thus forming a covalent bond. In short, covalent bonding can be summarized as the formation of covalent bonds between the support and amino acid side chains of the cells. Covalent bonding has the ability to maintain its integrity even at elevated temperatures, and so, immobilized cells are also protected from external heat and the binding affinity is also high, but it is not as effective as other methods. In certain cases, the immobilization is irreversible. Unlike all the above-mentioned methods, cross-linking is an irreversible method, where the cells are held without the help of a support matrix. Although diffusion of wastewater is limited, there is no leakage of cells (Najim & Radeef, 2024).

2.8.2 Biofilm-Based and Carrier-Assisted Delivery

The limitations of conventional WWT, such as less effective removal, high sludge production, and fouling, are overcome with the help of an advanced method which uses carriers including biofilms. These carriers provide additional support and nutrients needed for the growth of microbial colonies. Biofilms have the ability to accommodate more than one type of microbial system. The formation of biofilm is based on certain surface factors such as roughness, material type, specific area, and pore structure. Traditional carriers are categorized based on the nature of the material. They are inorganic and organic materials. Inorganic substances such as scoria (highly porous volcanic rocks), zeolites, ceramics, and AC are carriers for WWT, as they have stable chemical properties and a larger surface area. For instance, local scoria was used as a carrier in an aerobic submerged biofilm reactor in order to treat nitrogen and carbon that is dissolved in municipal wastewater. Zeolites are preferred for the treatment of water contaminated with nitrogen, as they have better ion-exchange properties and removal happens by adsorption. For remediation involving microbes, reactive organic materials are widely used. For example, organic pollutants are removed in up-flow anaerobic sludge blanket (UASB) reactors using Poly Vinyl Alginate (PVA) gel beads. Organic carriers are further classified into two types, and they are reactive and inert organic carrier materials. Most reactive organic materials include alginate and bamboo fibres, as they are affordable, easily available, and have good compatibility with microbial systems, as they pose no toxic effects, and they are hydrophilic (Zhao et al., 2019). Apart from providing support, reactive organic materials also act as a source of nutrients for the growth of microbes. One such example is the use of *Retinervus luffae fructus* for controlling effluent COD and denitrification. However, one major limitation is that they are rapidly degraded by microbes, and so their reusability is reduced, and they have unstable carbon release. To overcome the limitations of reactive organic materials, carriers from inert organic materials such as polyethylene (PE), polyester, polyolefin, polyurethane (PU), polypropylene (PP), polystyrene (PS), and polyvinyl chloride (PVC) were designed. The use of high-density PE in moving bed biofilm reactor (MBBR) systems led to the increased removal efficiencies of NH4–N up to 90% along with better removal of both macro- and organic micropollutants. The limitation of inert carriers is their hydrophobicity, which affects the initial biofilm formation. Even though the traditional carriers have their benefits, the disadvantages associated with them are high, as they may cause secondary pollution. So, the need for a novel carrier arose. Novel advancements in carrier-based delivery led to the design of slow-release carriers and magnetic biofilms, where slow release focuses on nutritional support and magnetic carriers focus on physiologically stimulating microbial activity by enhancing the metabolism and attachment of microbes. Further research has led to carriers with modified surfaces. This modification varies based on the requirements, such as increased bio-affinity and biofilm formation rate. Since microorganisms tend to adhere better to hydrophilic surfaces due to their higher surface energy, hydrophilic-modified surface carriers are designed, as they can form hydrogen bonds with the functional groups on the bacterial cell wall (EPS). Bacterial surfaces naturally possess a negative charge; therefore, traditional organic carriers having a negative charge face repulsion, which disrupts biofilm formation. Thus, surface modification is used to introduce positively charged content on the carrier surface. This in turn creates an electrostatic attraction with the negatively charged microorganisms. This method is known as electrophilic modified surface. Thus, over the course of pollutant removal and WWT using microbes, especially bacteria, biofilm carriers have played a major role. The design of novel biofilms that could be engineered as per the requirements of the process and bacteria has helped to overcome the limitations posed by conventional biofilms (Lago et al., 2024).

2.8.3 Bioreactor Designs for Ex Situ Treatment

Ex situ treatment refers to the process of treating wastewater away from its site of production. Wastewater remediation by treatment plants is considered as one of the ex situ treatment methods. Ex situ treatment relies on various types of bioreactors, including aerobic and anaerobic bioreactors, for the efficient removal of pollutants. Most industrial effluents are treated using aerobic bioreactors due to their easy operation and high microbial growth rates. Parameters such as temperature, pH, and oxygen can also be maintained. Continuous stirred tank reactors (CSTRs), sequencing batch reactors (SBRs), membrane sequencing batch reactors (MSBRs), MBBRs, airlift bioreactors, and fluidized bed reactors are a few of the aerated bioreactors that are widely used. Anaerobic bioreactors are used when low sludge production is necessary and for the production of methane as a by-product. Anaerobic baffled reactors (ABR) and UASB reactors are types of anaerobic bioreactors. Hybrid (anaerobic–aerobic) bioreactor systems are the upcoming trend in WWT using bioreactors. In aerobic bioreactors, CSTRs are used with acclimatized consortia, and two-stage SBRs are capable of removing 97.5% COD along with complete oil and grease removal. MBBR uses biofilms and has proved to remove 70% TPH (total petroleum hydrocarbons) when treated with high-salinity slop. Airlift bioreactors have an internal loop with immobilized bacteria such as *Sphingobium* sp. and have proved to achieve 85% TPH and 73% COD removal. In Fluidized Bed Bioreactor (FBB), bacteria are fixed on small particles that are fluidized by liquid up-flow. This provides a high immobilized biomass concentration and improves mass transfer. A three-phase FBB can achieve approximately 90% COD reduction from refinery wastewater. In anaerobic bioreactors, ABR is used to treat heavy-oil-contaminated water under high salt concentrations. USAB is also used to treat refinery effluents. Hybrid bioreactor systems are found to have better performance as they have synergistic effects and sequential degradation of recalcitrant compounds. Thus, bioreactors offer better performance in pollutant removal; however, industrial applications are still limited, as they require strict pre-treatment and skilled personnel to operate. The design of a hybrid bioreactor paves a way to overcome these limitations (Wastewater et al., 2020).

2.8.4 Discussion

The overall deployment strategies for engineered microorganisms for environmental remediation mainly focus on enhancing the stability, efficiency, and practical applicability of the engineered microbes through advanced delivery systems and reactor designs. Recent delivery techniques such as immobilization and encapsulation improve microbial viability, reusability, and resistance to environmental stress by retaining and restricting cells within matrices such as alginate, AC, synthetic polymers, or semi-permeable membranes. This increases the removal efficiency of the pollutants when compared to free-cell remediation systems. Methods, including adsorption, entrapment, covalent bonding, and cross-linking, offer specific advantages in terms of stability, mass transfer, and operational durability. However, few limitations such as diffusion constraints or irreversible binding may arise. Biofilm-based and carrier-assisted approaches overcome such limitations and further increase the performance by providing additional structural support, enhanced surface area, and improved microbial attachment. The biofilm formation is optimized by both inorganic carriers (e.g. porous minerals, ceramics) and organic materials (reactive or inert polymers) and are also used to improve the nutrient availability, and pollutant adsorption, while surface modifications such as increasing hydrophilicity or introducing positive charges promote stronger microbial adhesion and faster biofilm establishment. Precise control of operational parameters such as pH, temperature, and oxygen levels can be achieved by the use of ex situ applications of engineered microbes, especially when they are integrated with special bioreactor systems such as aerobic, anaerobic, and hybrid configurations. Reactors such as CSTRs, SBRs, MBBRs, fluidized beds, and anaerobic systems also play a major role in enhancing mass transfer, biomass retention, and sequential degradation of complex pollutants, leading to improved removal efficiencies. Overall, the combination of immobilization strategies, advanced carrier technologies, and optimized bioreactor designs represents a comprehensive framework for translating engineered microbial systems from laboratory research to scalable and efficient environmental remediation applications. Figure 2.8 gives an overall description on bacterial delivery systems in the form of encapsulation, biofilm, and covalent linking of bacterial cells.

2.9 BIOSAFETY, RISK ASSESSMENT, AND REGULATORY FRAMEWORK FOR ORGANIC AND INORGANIC POLLUTANTS IN WASTEWATER

Biosafety is one of the key approaches to risk reduction in wastewater management which poses hazards due to the presence of dangerous organic and inorganic substances. The overall aim is the safeguarding of human health, aquatic life, and downstream consumers. The basis of this effort is risk assessment, which determines what risks may occur, the degree of exposure, toxicological characterization, and, consequently, the determination of treatment standards and control-specific treatment of pollutants (Trubetskaya et al., 2025).

An example of a combination of preventive and adaptive engineering is land surveying and ground water treatment by modern WWT. There are also the continuous surveillance system, automated sensor networks, and digital control platforms that are regularly used to monitor the movement of pollutants at

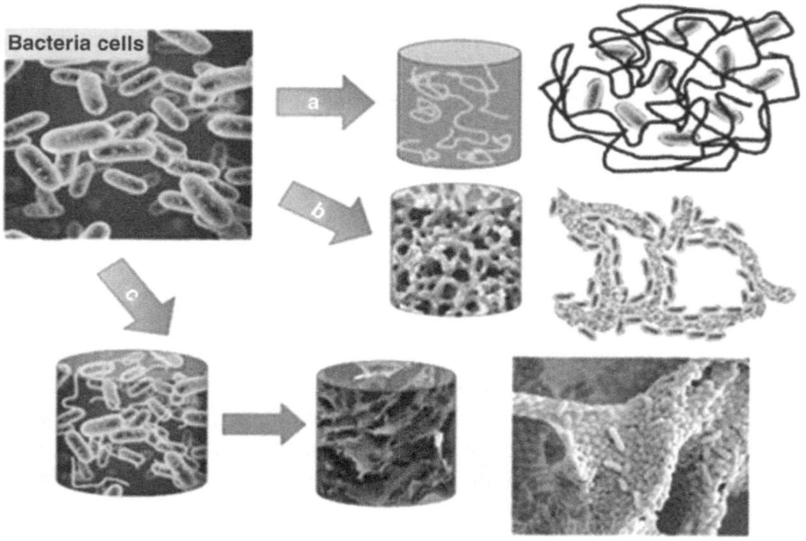

FIGURE 2.8 Strategies for bacterial deployment with the help of scaffold/polymer supports: (a) physical entrapment of bacteria cells within hydrogel structure; (b) biofilm generated on the porous surface of inorganic support; (c) macro-porous gel formation via direct covalent linkage of bacteria cells at cryo-structuration process (Berillo et al., 2021).

a high-temporal resolution. The biosafety measures go beyond chemical management to include management of sludge, spill control, and emergency response. All these measures are adjusted to international standards and national laws, which reduces environmental damage and workplace risks to a minimum (Khattak et al., 2025).

2.9.1 Environmental and Ecological Risk Considerations

Ecological risk assessment examines the effects of organic and inorganic pollutants in aquatic ecosystems, sediment action, and ecosystem functions, in general, as shown in Figure 2.3. Organic pollutants comprising pharmaceuticals, endocrine disruptors, and hydrocarbons create reproductive, growth, and behaviour distortions across taxa. Heavy metals, nitrates, and phosphates are examples of inorganic components that are involved in acute toxicity, bioaccumulation mechanisms, eutrophication, and habitat destruction (Thakur & Kumar, 2024). The common occurrence of intricate mixtures of contaminants makes it difficult to make prognostications in the long term (Figure 2.9).

Rules and regulations require full-scale environmental assessment before the discharge of wastewater and, more so, in ecologically sensitive areas. Risk governance, therefore, comprises toxicity tests, longitudinal biomonitoring programmes, and adaptive discharge licensing that responds dynamically to ecosystem stress indicators. The allies of variability of climate, the changed hydrologic programs, and the surfacing of new pollutants highlight the need to exercise precautionary and adaptive management measures.

2.9.2 Horizontal Gene Transfer and Antimicrobial Resistance in Wastewater Systems

Wastewater is a good environment where horizontal gene transfer and microbial interactions can take place, particularly with the inclusion of antibiotics, disinfectants, and heavy metals (Brauns et al., 2024).

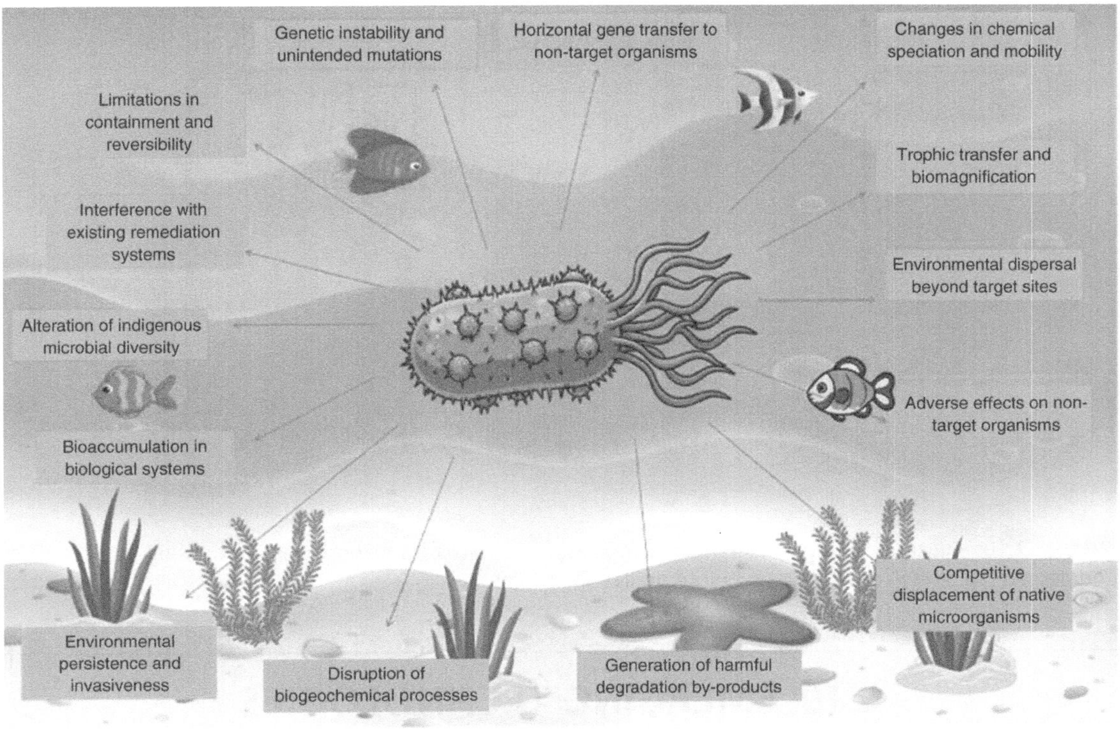

FIGURE 2.9 Potential ecological risks of engineered microorganisms in aquatic environments, including genetic instability, gene transfer, ecosystem disruption, environmental persistence, and impacts on non-target organisms.

The sharing of antimicrobial resistance genes and other adaptive information through this gene exchange increases the chances of multidrug-resistant organisms establishing themselves in surface and ground water resources (Wołowicz & Munir, 2025). The mitigation practice is based on the advanced oxidation procedure, membrane filtration, UV disinfection, and sludge stabilization methods (Olasupo et al., 2024). Monitoring resistance determinants is enabled through molecular diagnostics, such as quantitative PCR, metagenomic sequencing, and tracking of genetic elements (Zhang et al., 2023). The modern-day regulatory frameworks recognize the saliency of antimicrobial resistance as a bi-SAF issue, especially where there is the use of recycled wastewater (Shekhar et al., 2024).

2.9.3 International Guidelines and Regulatory Standards

Wastewater pollution should be controlled at the international level, which implies a combination of environmental safety, social health needs, and sustainability (Zhang et al., 2023). The World Health Organization (WHO) provides risk-based, structured reuse of wastewater and water safety planning, including the whole hydrologic cycle. The WHO Laboratory Biosafety Manual establishes internationally recognized standards of safe management of the dangerous wastewater and sludge in both scientific and practical spheres (Figure 2.10) (Khattak et al., 2025).

The regulatory thinking on environmental release of genetically engineered microorganisms and development of more advanced biotechnological interventions used in WWT has been influenced by multilateral instruments, including the Cartagena Protocol on Biosafety (Cases & de Lorenzo, 2005). Moreover, the agencies, such as the United Nations Environment Programme (UNEP), are advocates of the standardization of effluent and monitoring requirements, as well as the treatment of sludge in different geographic locations (Chemla et al., 2025).

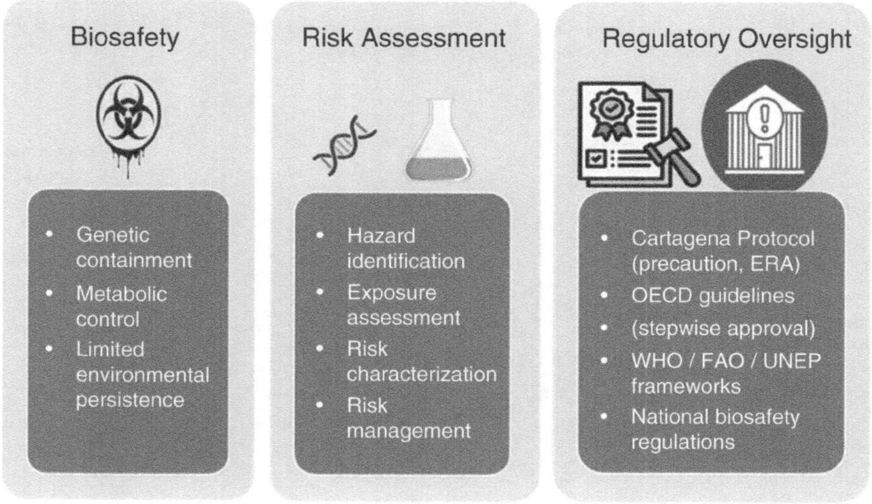

FIGURE 2.10 Biosafety, risk assessment, and regulatory oversight framework guiding the safe application of engineered bacteria in environmental bioremediation.

2.10 EMERGING TECHNOLOGIES AND FUTURE DIRECTIONS

The use of microorganisms for remediation has increased the pollutant removal efficiencies and made the remediation processes simpler. The studies on engineered microbes are ongoing and the incorporation of genetic engineering in the design of engineered microorganisms is the major outcome till now. The use of CRISPR-Cas9 for gene editing enables the accurate genetic modification of the bacteria based on needs, including enhancing specific pollutant degradation pathways, which has in turn increased the pollutant removal exponentially. The genetic design of microbes to degrade heavy metals and targeted pollutants could lead to more effective remediation.

Synthetic biology is the field of science where entirely new genomes are built from scratch.

Microbes with a built genome could increase the removal rates and efficiencies that are difficult for natural strains. The role of bioinformatics will play a major role as it helps to identify specific traits and the genes corresponding to them, which can be used in other genetic engineering methods for the development of better engineered microbes. By combining engineered microbes with conventional methods, a synergistic approach towards remediation could be achieved. Targeted pollutant removal can be enhanced by combining engineered microbes and nanotechnology. Nanoparticles can be used to deliver the microbes to the site of contamination. Due to the complexity of contaminants, interdisciplinary collaboration in upcoming research is needed. Different fields like microbiology and environmental science and engineering should work together to deal with the complexity of pollutants. Further research should be conducted to design systems that improve effective delivery and efficiency of engineered microbes by integrating techniques such as automated monitoring, optimized conditions for microbial activities, and decision-making by policymakers for the safe and responsible use of engineered microbes (Akintola, 2024). For certain engineered microbes, such as *Bacillus subtilis* and its spores, programmable biofilms can be designed. The responsiveness, pollutant removal, self-regeneration of films and sensing environmental signals can be achieved with the advancements in synthetic biology and metabolic engineering (Mohsin et al., 2021). The development of material science, biodegradable and smart living materials, is enhanced by interdisciplinary integration. By combining machine learning, synthetic consortia, and genetic engineering tools, the degradation of persistent pollutants such as PET can also be achieved. Challenges such as stability, scalability, and economic feasibility hinder the industrial applications of engineered microbial systems. Therefore, future research should focus more on rational consortium design using computational models, multi-omics with integrated high-throughput screening, and AI-driven analyses (Zhou et al., 2025). Thus, focusing on interdisciplinary integration and studies on the combination of microbes is important for advanced research on engineered microbes. Innovative and sustainable solutions can be developed when expertise from other fields is implemented for complex environmental challenges, which ultimately contribute to a healthier ecosystem and a cleaner environment.

2.11 CONCLUSION

One of the most concerning environmental issues of these days are wastewater pollutants, as they have increased complexity due to the mixing of organic and inorganic pollutants. Rapid industrialization, enhanced agricultural practices, and increased population has led to the excessive generation of wastewater that is highly contaminated. The conventional treatment processes were designed for pollutants with simple complexity in mind and so are not suitable for the remediation of current days' pollutants. The limitations of conventional treatments also include the inability to degrade chemicals that are highly stable, has higher complexity, and are present in trace amounts. Higher operational costs, energy demands, and secondary waste generation limit the long-term use of traditional treatment methods. These challenges have led to the search for new alternatives that involves biological components that are capable of degrading pollutants using their metabolic pathways, even in trace concentrations. Microbial remediation is also considered environmentally compatible and adaptable alternative to transform, immobilize, or detoxify a wide range of contaminants under monitored conditions. Single strain of microbes has certain limitations including their narrow metabolic scope, sensitivity towards toxicity, and instability in fluctuating conditions. Therefore, to overcome these restrictions, microbes work as communities in wastewater remediation and synthetically or genetically engineered microbes were designed. The use of advanced molecular biology, genetic engineering, and metabolic pathway analyses have helped in achieving the design of engineered microbes that are specific to pollutant degradation. By DOL, the metabolic tasks are divided among different strains so that the cellular burden and accumulation of toxic materials are minimized. This approach is advantageous for the treatment of complex, multi-pollutant wastewater streams. Immobilization, biofilms, and advanced ex situ bioreactors are other techniques that are equally important to DOL. These methods enhance the microbial viability, retention, and long-term operations. The gap between lab scale and industrial applications can be closed with the help of these methods. However, few limitations still exist for the upscaling and industrial applications, and this includes mass transfer, complete protection against environmental stress, and enabling reuse of the microbes, scalability, operational complexity, economic feasibility, and long-term system stability.

Interdisciplinary integration leads to the future of wastewater remediation. The combined use of synthetic biology, computational modelling, and multi-omics analysis, along with material science and AI tools, offer future scope and opportunities to design more efficient, smarter, and more resilient remediation systems. For the safe use of technology, regulatory policies are important to maintain public safety and trust. Biosafety and regulatory considerations are therefore central to the future of engineered microbial remediation. Careful risk assessments, regulatory frameworks, and international guidelines are mandatory if any genetically engineered microbes are to be used in industrial scale. Detailed study on how they affect the environment, possible effects on the ecosystems, and unintended outcomes is also necessary. Thus, engineered bacterial consortia and systems form a transformative solution for modern water pollution challenges. Even though hurdles remain regarding safety and scalability, continued innovation along with interdisciplinary support has the potential to redefine sustainable wastewater management.

REFERENCES

Adamu, K. S., Bichi, Y. H., Nasiru, A. Y., Babangida, A. M., Umar, M. M., Usman, G., & Muhammad, R. (2023). Synthetic microbial consortia in bioremediation and biodegradation. *International Journal of Research and Innovation in Applied Science, 8,* 232–241.

Ahamad, A., Madhav, S., Singh, A.K., Kumar, A., Singh, P. (2020). Types of Water Pollutants: Conventional and Emerging. In: Pooja, D., Kumar, P., Singh, P., Patil, S. (eds.), *Sensors in Water Pollutants Monitoring: Role of Material. Advanced Functional Materials and Sensors.* Springer, Singapore. https://doi.org/10.1007/978-981-15-0671-0_3

Ahmad, N. B., Jaafaru, M. S., Isa, Z., Abdulhamid, Y., Kakudi, R. A., Ugya, A. Y., & Meguellati, K. (2024). High pollution loads engineer oxygen dynamics, ecological niches, and pathogenicity shifts in freshwater environments. *Journal of Hazardous Materials Advances, 14,* 100425. https://doi.org/10.1016/j.hazadv.2024.100425

Akintola, A. A. (2024). Advancements in microbial bioremediation: Harnessing genetically engineered microbes for enhanced pollutant. *International Research Journal of Modernization in Engineering Technology and Science, 10,* 488–501. https://www.doi.org/10.56726/IRJMETS62065

Bala, S., Garg, A., Kumar, R., & Singh, D. (2022). Recent strategies for bioremediation of emerging pollutants: A review. *Environmental Science and Pollution Research, 29,* 1–20.

Berillo, D., Al-jwaid, A., & Caplin, J. (2021). Polymeric materials used for immobilisation of bacteria for the bioremediation of contaminants in water. *Polymers, 13,* 1073. https://doi.org/10.3390/polym13071073

Brauns, B., Chandra, S., Civil, W., Lapworth, D. J., MacDonald, A. M., McKenzie, A. A., Read, D. S., Sekhar, M., Singer, A. C., Thankachan, A., & Tipper, H. J. (2024). Presence of emerging organic contaminants and microbial indicators in surface water and groundwater in urban India. *Environmental Pollution, 362,* 124983. https://doi.org/10.1016/j.envpol.2024.124983

Cases, I., & de Lorenzo, V. (2005). Genetically modified organisms for the environment: stories of success and failure and what we have learned from them. *International Microbiology: The Official Journal of the Spanish Society for Microbiology, 8*(3), 213–222.

Chakraborty, P., Mohapatra, S., Verma, S., & Singh, R. (2025). Emerging trends in engineered microbial systems for environmental remediation. *Discover Environment, 3,* 17. https://doi.org/10.1007/s44274-025-00217-7

Chemla, Y., Sweeney, C. J., Wozniak, C. A., & Voigt, C. A. (2025). Design and regulation of engineered bacteria for environmental release. *Nature Microbiology, 10*(2), 281–300. https://doi.org/10.1038/s41564-024-01918-0

Choudhary, A. K., Sonwani, R. K., & Panjiara, D. (2025). *Challenge in Degradation of Micropollutants by Biological Wastewater Treatment Processes* (pp. 657–682). https://doi.org/10.1007/978-3-032-03530-1_27

de Sousa, D. N. R., Mozeto, A. A., Carneiro, R. L., & Fadini, P. S. (2014). Electrical conductivity and emerging contaminant as markers of surface freshwater contamination by wastewater. *Science of the Total Environment, 484,* 19–26. https://doi.org/10.1016/j.scitotenv.2014.02.135

Domingos, R. F., Gélabert, A., Carreira, S., Cordeiro, A., Sivry, Y., & Benedetti, M. F. (2015). Metals in the aquatic environment—Interactions and implications for the speciation and bioavailability: A critical overview. *Aquatic Geochemistry, 21*(2–4), 231–257. https://doi.org/10.1007/s10498-014-9251-x

Dudeja, C., Mishra, A., Ali, A., Singh, P. P., & Jaiswal, A. K. (2025). Microbial genome editing with CRISPR–Cas9: Recent advances and emerging applications across sectors. *Fermentation, 11*(7), 1–20. https://doi.org/10.3390/fermentation11070410

Elyamine, A. M., Kan, J., Meng, S., Tao, P., Wang, H., & Hu, Z. (2021). Aerobic and anaerobic bacterial and fungal degradation of pyrene: Mechanism pathway including biochemical reaction and catabolic genes. *International Journal of Molecular Sciences, 22*(15), 8202. https://doi.org/10.3390/ijms22158202

Garg, A., Minhas, P. S., Kumar, A., & Singh, R. (2025). CRISPR-based genome engineering of bacteria for environmental remediation. *Frontiers in Bioengineering and Biotechnology, 11,* 1178680. https://doi.org/10.3389/fbioe.2023.1178680

Grzegorzek, M., Wartalska, K., & Kowalik, R. (2024). Occurrence and sources of hormones in water resources—Environmental and health impact. *Environmental Science and Pollution Research, 31*(26), 37907–37922. https://doi.org/10.1007/s11356-024-33713-z

Hama Aziz, K. H., Mustafa, F. S., Omer, K. M., Hama, S., Hamarawf, R. F., & Rahman, K. O. (2023). Heavy metal pollution in the aquatic environment: Efficient and low-cost removal approaches to eliminate their toxicity: A review. *RSC Advances, 13*(26), 17595–17610. https://doi.org/10.1039/D3RA00723E

He, Y., Song, K., Yang, C., Li, Y., He, W., & Xu, F. (2021). Suspended particulate matter (SPM)-bound polycyclic aromatic hydrocarbons (PAHs) in lakes and reservoirs across a large geographical scale. *Science of The Total Environment, 752,* 142863. https://doi.org/10.1016/j.scitotenv.2020.142863

Hosny, R., Abdel-Gawad, F. K., & El-Sayed, W. S. (2023). Engineered bacterial systems for azo dye degradation under wastewater *conditions. Scientific Reports, 13*, 41106. https://doi.org/10.1038/s41598-023-41106-4

Jee, H. K., Kim, H. N., & Park, J. H. (2024). Immobilization of manganese in solution and soil contaminated with trivalent chromium using biochars. *Applied Biological Chemistry*, 67(1), 118. https://doi.org/10.1186/s13765-024-00969-5

Jones, A. M., Griffin, P. J., & Waite, T. D. (2015). Ferrous iron oxidation by molecular oxygen under acidic conditions: The effect of citrate, EDTA and fulvic acid. *Geochimica et Cosmochimica Acta, 160*, 117–131. https://doi.org/10.1016/j.gca.2015.03.026

Kehinde, A. O., Aziz, H. A., & Yusuf, R. O. (2022). Bioremediation of wastewater from pharmaceutical industries using bacteria: Challenges and prospects. *Water, 15*(2), 375. https://doi.org/10.3390/w15020375

Khattak, W. A., Anas, M., Hakki, E. E., Iqbal, J., Abbasi, B. A., & Iqbal, R. (2025). Regulatory framework and quality assurance in biomanufacturing. In Ayush Madan, et al. (ed.), *Industrial applications for bioprocessing and biomanufacturing* (pp. 385–414). IGI Global Scientific Publishing. https://doi.org/10.4018/979-8-3373-2873-7.ch013

Lago, A., Rocha, V., Barros, O., Silva, B., & Tavares, T. (2024). Bacterial biofilm attachment to sustainable carriers as a clean-up strategy for wastewater treatment: A review. *Journal of Water Process Engineering, 63*, 105368.

Lednicka, B., Kubacka, M., Freda, W., Haule, K., Dembska, G., Galer-Tatarowicz, K., & Pazikowska-Sapota, G. (2022). Water turbidity and suspended particulate matter concentration at dredged material dumping sites in the southern Baltic. *Sensors, 22*(20), 8049. https://doi.org/10.3390/s22208049

Lee, H. M., Vo, P. N. L., & Na, D. (2018). Advancement of metabolic engineering assisted by synthetic biology. *Catalysts, 8*(12), 619. https://doi.org/10.3390/catal8120619

Mahanty, S., Sarkar, A., Chaudhuri, P., & Krishna Darbha, G. (2023). Mycosynthesized magnetic iron-oxide nanoparticles for the remediation of heavy metals – An insight into the mechanism of adsorption, process optimization using algorithmic approach and its application for the treatment of groundwater. *Environmental Nanotechnology, Monitoring & Management, 20*, 100854. https://doi.org/10.1016/j.enmm.2023.100854

Maria, M. P., Torres, N. H., Nascimento, V. R. S., Chagas, T. S. A., Saratale, G. D., Mulla, S. I., Bharagava, R. N., Cavalcanti, E. B., & Romanholo Ferreira, L. F. (2023). Current advances in the brewery wastewater treatment from anaerobic digestion for biogas production: A systematic review. *Environmental Advances, 13*, 100394. https://doi.org/10.1016/j.envadv.2023.100394

Minhas, P. S., Garg, A., & Singh, R. (2024). Programmable CRISPR interference and activation systems for microbial environmental applications. *Trends in Biotechnology, 42*(3), 295–308. https://doi.org/10.1016/j.tibtech.2023.11.004

Miller, D. H., LaLone, C. A., Villeneuve, D. L., & Ankley, G. T. (2024). Projection of Interspecific Competition (PIC) matrices: A conceptual framework for inclusion in population risk assessments. *Environmental Toxicology and Chemistry, 43*(6), 1406–1422. https://doi.org/10.1002/etc.5867

Mishra, T., Tiwari, P. B., Kanchan, S., & Kesheri, M. (2025). Advances in microbial bioremediation for effective wastewater treatment. *Water, 17*(22), 3196. https://doi.org/10.3390/w17223196

Mohedano, A. F., Costa, R. H. R., & Gutiérrez, J. L. (2022). Systems biology and metabolic engineering approaches for wastewater bioremediation. *Science of the Total Environment, 807*, 150834. https://doi.org/10.1016/j.scitotenv.2021.150834

Mohsin, M. Z., Omer, R., Huang, J., Mohsin, A., Guo, M., Qian, J., & Zhuang, Y. (2021). Advances in engineered Bacillus subtilis biofilms and spores, and their applications in bioremediation, biocatalysis, and biomaterials. *Synthetic and Systems Biotechnology*, 6(July), 180–191.

Morin-Crini, N., Lichtfouse, E., Liu, G., et al. (2022). Worldwide cases of water pollution by emerging contaminants: A review. *Environmental Chemistry Letters, 20*, 2311–2347. https://doi.org/10.1007/s10311-022-01447-4

Najim, A. A., & Radeef, A. Y. (2024). *Immobilization: The promising technique to protect and increase the efficiency of microorganisms to remove contaminants*. February. https://doi.org/10.1002/jctb.7638

Narayanan, M. S. S., Pitchaimani, V. S., Sivakumar, M., Dinesh Kumar, T., Abishek, S. R., & Karuppannan, S. (2025). Spatial assessment of heavy metal contamination in groundwater in the Kadaladi region, Tamil Nadu, India. *Scientific Reports, 15*(1), 27704. https://doi.org/10.1038/s41598-025-12120-5

Nayak, S., Sahoo, G., Das, I. I., Mohanty, A. K., Kumar, R., Sahoo, L., & Sundaray, J. K. (2023). Poly- and Perfluoroalkyl Substances (PFAS): Do they matter to aquatic ecosystems? *Toxics, 11*(6), 543. https://doi.org/10.3390/toxics11060543

Niu, L., Gärtner, A. A. E., König, M., Krauss, M., Spahr, S., & Escher, B. I. (2025). Role of suspended particulate matter for the transport and risks of organic micropollutant, mixtures in rivers: A comparison between baseflow and high discharge conditions. *Environmental Science & Technology, 59*(10), 4857–4867. https://doi.org/10.1021/acs.est.4c13378

Olasupo, A., Corbin, D. R., & Shiflett, M. B. (2024). Trends in low temperature and non-thermal technologies for the degradation of persistent organic pollutants. *Journal of Hazardous Materials, 468,* 133830. https://doi.org/10.1016/j.jhazmat.2024.133830

Pande, V., Pandey, S. C., Sati, D., Bhatt, P., & Samant, M. (2022). Microbial interventions in bioremediation of heavy metal contaminants in agroecosystem. *Frontiers in Microbiology, 13.* 824084. https://doi.org/10.3389/fmicb.2022.824084

Pandolfo, E., Barra Caracciolo, A., & Rolando, L. (2023). Recent advances in bacterial degradation of hydrocarbons. *Water, 15*(2), 375. https://doi.org/10.3390/w15020375

Porras-Rivera, G., Górski, K., & Colin, N. (2024). Behavioral biomarkers in fishes: A non-lethal approach to assess the effects of chemical pollution on freshwater ecosystems. *Environmental Research, 260*, 119607. https://doi.org/10.1016/j.envres.2024.119607

Rafeeq, H., Afsheen, N., Rafique, S., Arshad, A., Intisar, M., Hussain, A., Bilal, M., & Iqbal, H. M. N. (2023). Genetically engineered microorganisms for environmental remediation. *Chemosphere, 310*(September), 136751. https://doi.org/10.1016/j.chemosphere.2022.136751

Roell, G. W., Zha, J., Carr, R. R., Koffas, M. A., Fong, S. S., & Tang, Y. J. (2019). Engineering microbial consortia by division of labor. *Microbial Cell Factories, 18*, 35. https://doi.org/10.1186/s12934-019-1083-3

Sanusi, I. O., Adepoju, A. A., & Abdulrahman, B. D. (2025). Heavy metal contamination, human health impact, and remediation techniques in water bodies: a review. *Discover Environment, 3*(1), 268. https://doi.org/10.1007/s44274-025-00475-5

Shah, M. P., & Gandhi, N. (2021). Conventional and advanced wastewater treatment technologies: Limitations and opportunities. In Shahid-ul-Islam (ed.), *Advances in wastewater treatment I* (pp. 1–28). Springer.

Sharma, S., Pathania, S., Bhagta, S., Kaushal, N., Bhardwaj, S., Bhatia, R. K., & Walia, A. (2024). Microbial remediation of polluted environment by using recombinant E. coli: A review. *Biotechnology for the Environment, 1*(1), 1–21. https://doi.org/10.1186/s44314-024-00008-z

Shekhar, C., Khosya, R., Sharma, A. K., Thakur, K., Mahajan, D., Kumar, R., Kumar, S., & Sharma, A. K. (2024). A systematic review on health risks of water pollutants: Classification, effects and innovative solutions for conservation. *Toxicology Research, 14*(1), tfaf014. https://doi.org/10.1093/toxres/tfaf014

Siddiqui, S., Naushad, M., & Khan, A. (2022). Microbial degradation of pesticides: Advances in engineered bacterial systems. *Journal of Environmental Chemical Engineering, 10*(6), 108405. https://doi.org/10.1016/j.jece.2022.108405

Srinivasan, R., Kumar, M., & Singh, P. (2025). Microbial mediators in heavy metal biotransformation and detoxification. *Toxics, 13*(1), 484.

Thakur, A., & Kumar, A. (2024). Unraveling the multifaceted mechanisms and untapped potential of activated carbon in remediation of emerging pollutants: A comprehensive review and critical appraisal of advanced techniques. *Chemosphere, 346*, 140608. https://doi.org/10.1016/j.chemosphere.2023.140608

Thai, T. Q., Bui, H. M., & Tran, L. T. (2023). Synthetic biology approaches for environmental bioremediation. *Biotechnology Advances, 62*, 108072.

Tripathi, P., Varma, K., Srivastava, V., Dlamini, N. S., & Jha, P. K. (2025). Heavy metal contamination and associated health risk assessment in groundwater of Mirzapur, Uttar Pradesh. *Frontiers in Water*, *7*, 1694963. https://doi.org/10.3389/frwa.2025.1694963

Trubetskaya, A., Hennessy, A., Ryan, I., Stockil, K., & Gaskin, C. (2025). Industrial assessment tools of risks associated with wastewater and water technologies. *ACS ES&T Water*, 5(10), 6131–6151. https://doi.org/10.1021/acsestwater.5c00839

Wang, F., Xiang, L., Sze-Yin Leung, K., Elsner, M., Zhang, Y., Guo, Y., Pan, B., Sun, H., An, T., Ying, G., Brooks, B. W., Hou, D., Helbling, D. E., Sun, J., Qiu, H., Vogel, T. M., Zhang, W., Gao, Y., Simpson, M. J., … Tiedje, J. M. (2024). Emerging contaminants: A One Health perspective. *The Innovation, 5*(4), 100612. https://doi.org/10.1016/j.xinn.2024.100612

Wastewater, H., Kuyukina, M. S., & Krivoruchko, A. V. (2020). Applied sciences advanced bioreactor treatments of hydrocarbon-containing wastewater. *Applied Sciences*, *10*, 831. https://doi.org/10.3390/app10030831.

Wołowicz, A., & Munir, H. M. S. (2025). Emerging organic micropollutants as serious environmental problem: A comprehensive review. *Science of The Total Environment, 958*, 177948. https://doi.org/10.1016/j.scitotenv.2024.177948

Xiang, N., Jiang, C., Huang, W., Nordhaus, I., Zhou, H., Drews, M., & Diao, X. (2019). The impact of acute benzo(a) pyrene on antioxidant enzyme and stress-related genes in tropical stony corals (Acropora spp.). *Science of The Total Environment, 694*, 133474. https://doi.org/10.1016/j.scitotenv.2019.07.280

Xu, Z., & Tsang, D. C. W. (2022). Redox-induced transformation of potentially toxic elements with organic carbon in soil. *Carbon Research, 1*(1), 9. https://doi.org/10.1007/s44246-022-00010-8

US Environmental Protection Agency (US EPA). (2012). *Emerging contaminants in water: Definitions and policy framework*. Washington, DC, USA.

Zhang, S., Liang, C., Xiao, M., Chui, C., Wang, N., Ji, Y., Wang, Z., Shi, J., & Liu, L. (2023). Metagenomic characterization of the enhanced performance of multicomponent synergistic thermophilic anaerobic co-digestion of food waste utilizing kitchen waste or garden waste as co-substrate. *Water Research, 244*, 120457. https://doi.org/10.1016/j.watres.2023.120457

Zhao, Y., Liu, D., Huang, W., Yang, Y., Ji, M., Nghiem, D. L., & Trinh, Q. T. (2019). Insights into biofilm carriers for biological wastewater treatment processes: Current state-of-the-art, challenges, and opportunities. *Bioresource Technology, 288*, 121619. https://doi.org/10.1016/j.biortech.2019.121619. Epub 2019 Jun 7. PMID: 31202712.

Zhou, Y., Meng, J., Zhang, M., Chen, S., He, B., Zhao, H., Li, Q., Zhang, S., & Wang, T. (2019). Which type of pollutants need to be controlled with priority in wastewater treatment plants: Traditional or emerging pollutants? *Environment International, 131*, 104982. https://doi.org/10.1016/j.envint.2019.104982

Zhou, Y., Zeeshan, M., & Haq, U. (2025). Engineering of synthetic microbial consortia for sustainable management of wastewater and polyethylene terephthalate: A comprehensive review. *International Journal of Molecular Sciences, 2024*, 1–22.

3 Low-Cost and Advanced Carbon-Based Adsorbents for Heavy Metal Removal in Decentralized Water Treatment

Abubakar M. Bilyaminu, Nimish Khandelwal, Shishir Kumar Behera, Seyed Morteza Zamir, Anina James, Eldon R. Rene, and Damian Kasperczyk

3.1 INTRODUCTION

Water contamination by heavy metals remains a serious environmental and public health concern because they are persistent, non-biodegradable, and capable of accumulating in aquatic systems, soils, and food chains. Industrial discharges, mining, electroplating, metallurgy, battery production, tanneries, and agricultural runoff are among the major pathways by which toxic heavy metals such as lead, cadmium, chromium, mercury, and arsenic enter water bodies (Duan et al., 2020; Song et al., 2024). Even at relatively low concentrations, these contaminants can cause chronic toxic effects, and their long-term presence in drinking-water sources is especially problematic for decentralized communities with limited advanced treatment infrastructure (Song et al., 2024).

Conventional methods for heavy metal removal include chemical precipitation, coagulation–flocculation, ion exchange, membrane filtration, and electrochemical treatment. Although these techniques are effective, they often result in high operational costs, sludge generation, membrane fouling, or

DOI: 10.1201/9781003741626-3

decreased efficiency at low metal concentrations (Duan et al., 2020). Thus, adsorption has become one of the most promising alternatives due to its operational simplicity, flexible design, and compatibility with both household- and community-scale water treatment systems. Carbon-based adsorbents have garnered a lot of attention due to their high surface area, pore structure, and the presence of oxygen-, nitrogen-, and sulfur-containing functional groups that facilitate metal binding through electrostatic attraction, ion exchange, complexation, and redox-assisted immobilization (Yang et al., 2019; Song et al., 2024).

Within the broad class of carbon adsorbents, the field has progressed from low-cost biomass-derived materials toward increasingly engineered adsorbents with improved surface chemistry and higher selectivity. Activated carbon (AC) and biochar remain central because of their availability and practical relevance, whereas modified and advanced carbon materials, including carbon nanotubes (CNTs), graphene-based materials, and hybrid nanocomposites, offer enhanced adsorption kinetics and greater control over the interaction between the adsorbent surface and dissolved metal species (Yang et al., 2019; Liu et al., 2019). This transition is important for decentralized water treatment because small-scale systems require materials that combine high removal efficiency with compact design, low chemical demand, regeneration potential, and robustness under variable water quality conditions (Song et al., 2024; Schmidt et al., 2023).

Accordingly, this chapter aims to: (i) examine the use of low-cost natural carbonaceous materials for heavy metal removal from water; (ii) evaluate how physical and chemical modification helps in improving their adsorption capacity, selectivity, and stability; and (iii) discuss the role of these adsorbents, including CNTs, graphene-derived materials, and hybrid carbon nanomaterials, in achieving high-performance heavy metal removal.

3.2 LOW-COST NATURAL CARBON-BASED MATERIALS FOR HEAVY METAL REMOVAL

Low-cost natural materials primarily serve as biosorbents, enabling household- or community-level removal of heavy metals through simple adsorption-based units rather than centralized, high-cost treatment plants. They support decentralized water safety by being locally available, inexpensive, and compatible with low-energy, low-chemical designs and applications. Additionally, using biomass residues as biosorbents supports organic waste management and aligns with sustainable and climate-resilient water-use strategies, particularly in rural areas. The seeds, leaves, wood, bark, and roots of plants, including fruit and vegetable peels, rinds, and residues, have all been used to remove metals from water. The cellulose, hemi-cellulose, and lignin components include nucleophilic functional groups such as hydroxyl, carboxylic, amino, thiol, and phosphate groups, and they can either chemically or physically adsorb metal ions (Li et al., 2025). The non-toxicity and environmental friendliness encourage the use of natural and renewable materials for metal removal from water.

A study examined the removal of heavy metals from drinking water using a gravity-driven membrane (GDM) filtration system, comparing wood membranes with pore sizes of 4–7 nm and polymer membranes with a pore size of 0.45 μm (Liu et al., 2023). The results showed that wood membranes were more effective at removing iron, copper, and manganese. Unlike polymer membranes, which have a cobweb-like structure, wood membranes have a sponge-like fouling layer that prolongs the retention period of heavy metals. For wood membranes, the concentration of carboxylic groups (–COOH) in the fouling layer was higher than for polymer membranes, enabling better adsorption. Furthermore, the microbial analysis showed that, compared to polymer membranes, the wood membranes had a greater abundance of bacteria known to remediate heavy metals. Many studies report the use of ground plant seeds as sorbents to remove heavy metals. *Allium cepa* seed biomass (ACSB), at 90–120 min of contact duration, neutral pH, and 50 mg/L of sorbent, showed maximum removal for lead, copper, zinc, cadmium, and chromium to be 1.75, 1.03, 1.68, 1.50, and 0.67 mg/L, respectively (Sheikh et al., 2021). Maity et al. (2022) showed that *Artocarpus heterophyllus* seed powder could remove different heavy metals from water. The adsorption study was conducted in batch mode with 10 μg/L arsenic and 10 mg/L cadmium, chromium, nickel, zinc, and copper as the initial concentrations. Arsenic showed the highest removal (61.22%), followed by chromium (54.63%) and cadmium (50.3%). Copper, nickel, and zinc showed low removal effectiveness, with removal values of 10.27%, 18.68%, and 15.07%, respectively. The Langmuir adsorption model and pseudo-second-order kinetics were consistent with endothermic, spontaneous adsorption kinetics (Maity et al., 2022). Table 3.1 lists the studies reporting the use of native carbon-based materials for removing heavy metals from water.

While natural materials offer simplified strategies for removing heavy metals from water, certain physicochemical parameters must be considered to achieve optimal results. The chemistry of both the adsorbent and the adsorbate can be strongly influenced by the solution pH. It can change the charge on the adsorbate, interfering species, and the adsorbent surface properties. Peels from fruits and vegetables are high in aldehydes, polyphenols, and polyhydroxy ketones. Their adsorption mechanism and reaction kinetics will be impacted by the pH shift. Differences in adsorption sites can alter the adsorption process by protonation or deprotonation of the adsorbent or adsorbate (Tariq et al., 2024). However, it is crucial to remember that the reduction potentials of heavy metals vary and are strongly influenced by pH. Wattanakornsiri et al. (2022) demonstrated the pH-dependent removal of lead and cadmium ions from water at an initial concentration of 100 mg/L using peels of dragon fruit, rambutan, and passion fruit. The untreated fruit peels showed lead and cadmium removal ranging from 50% to 85%. The highest heavy metal removal was observed at pH 4.0. They proposed that competitive adsorption between H^+ and metal ions occurred at low pH and that the functional groups were not deprotonated. Furthermore, the initial concentration of heavy metals determines the extent of adsorption. The interaction of adsorbent active sites increases in direct proportion to the initial concentration of heavy metals. This direct association holds for a given amount of adsorbent up to a specific concentration, beyond which a drop is observed. At 50 mg/L, onion peels had an adsorption efficiency of 98.29% for chromium in water. Beyond this concentration, the adsorbent's surface area-to-volume ratio dropped since there weren't enough binding sites (Prokopov et al., 2021). The adsorbent dose can also significantly affect the removal of heavy metal ions, as increasing the dose provides more active sites, resulting in a higher removal efficiencies. However, empty active sites have a detrimental impact on adsorption capacity after a certain point. For instance, the removal efficiency of the pomegranate peel adsorbent for chromium increased with increasing adsorbent dose, while the adsorption capacity decreased (Chen et al., 2023). The maximum adsorption capacity of the pomegranate peel adsorbent was achieved at pH 4.0, where chromium adsorption was 3.75 mg/g, corresponding to a removal efficiency of 78.17%. Interestingly, the removal efficiency of chromium on pomegranate peel adsorbent increased from 63.13% to 81.35%, while the adsorption capacity of chromium demonstrated a marked reduction from 7.58 mg/g to 3.25 mg/g when the dosage of pomegranate peel was increased from 2.5 to 7.5 g. Hence, for optimal application of biomass residues as adsorbents for heavy metals in water, operational parameters such as pH, initial metal concentration, and adsorbent dosage are critical.

The adsorbents are subjected to further adsorption and elution tests to assess their reusability. Reusable adsorbents are advantageous for both cost and environmental reasons (Li et al., 2024a). For chromium ions, walnut shell-derived adsorbents showed remarkable regeneration and reusability potential (Garg et al., 2023). When comparing natural walnut shell powder to its base and acid-treated equivalents, the highest desorption efficiency was found in the native form. After ten cycles of testing, these adsorbents showed only a slight decline in adsorption potential. Another striking advantage of biomass-based adsorbents is their low cost. Tolkou et al. (2024) conducted a cost analysis for the fabrication of adsorbents from fruit peels. For a dosage of 6.0 g/L, the cost of adsorbent made from banana peels was 0.196 €/L, from orange peels 0.129 €/L, and from pomegranate peels 0.277 €/L. Kurniawan et al. (2006), while summarizing the critical parameters of heavy metal adsorption by agricultural wastes such as coconut shell, also included the raw material's retail price of 0.25 US$/kg. There are few studies that compare the actual prices of adsorbents made from biomass with those made from synthetic materials for heavy metal removal from drinking

TABLE 3.1 Removal of heavy metals from water by natural carbon-based materials

CARBON MATERIAL	*TARGET HEAVY METAL*	*EXPERIMENTAL CONDITIONS*	*ADSORPTION CAPACITY /REMOVAL EFFECIENCY*	*CRITICAL INSIGHTS*	*REFERENCE*
Walnut shell, chestnut shell, pinewood, and burned pinewood	Pb^{2+}, Ni^{2+}, Cr^{6+}	pH 3.0–7.5; initial concentration: 200 mg/L; temperature: 20°C	Pb^{2+} >90%, Ni^{2+} >90%, Cr^{6+} >90%	Adsorption is closely linked to pH; heavy metals can be categorized into two groups: first, the adsorption increases with pH before precipitation occurs (Ni^{2+}, Pb^{2+}); second, accumulation is not linked to precipitation (Cr^{6+})	Cruz-Lopes et al. (2021)
Corncob	Ni^{2+}	pH 8.0; initial concentration: 10 mg/L; adsorbent dose: 1.5 g/50 mL; contact time: 1 h	74.93%	A high carbon content and low ash level of the corncob allowed for the adsorption of Ni^{2+}	Chaudhari and Patkar (2022)
Grape seeds	Cr^{6+}	pH 5.5; initial concentration: 10 mg/L; adsorbent dose: 15 g/L; contact time: 1 h; temperature: 20°C	91.7%	For the high maximum sorption capacity (q_{max} = 208.3 mg/g), a combination of chemical and physical sorption with endothermic and spontaneous thermodynamic characteristics	Sánchez-Ponce et al. (2023)
Watermelon rind	Pb^{2+},	pH 5.0; initial concentration: 100–1,500 mg/L; adsorbent dose: 1.0 g/L; contact time: 8–10 h	230.5 mg/g	Adsorption occurred by ion exchange, electrostatic attraction, and microprecipitation; interactions between Pb^{2+} and binding sites depended on the carboxyl functional group	Wang et al. (2022a)
Mango seed	Pb^{2+}, Cd^{2+}	pH 5, 7.5; initial concentration: 500 mg/L; adsorbent dose: 10 mg; temperature: 25°C	263.4 mg-Pb/g 93.7 mg-Cd/g	Binding sites provided by carboxyl, hydroxyl, amine, and ether groups; adsorption was attributed to electrostatic attraction, microprecipitation, complexation, and ion exchange	Wang et al. (2022b)
Water hyacinth roots	Pb^{2+}	pH 5.0; initial concentration: 100 mg/L; adsorbent dose: 0.1 g; contact time: 90 min	~50 mg Pb^{2+} per g of dried roots	Chemisorption	Jahangiri et al. (2021)

water; however, the above values indicate that biomass-derived adsorbents would be cheaper than synthetic ones, as they are predominantly waste. Despite these promising developments, a few problems need to be fixed before the inexpensive adsorbents can be widely employed in commercial water filtration systems. Scaling up processes and creating durable, reusable materials that function well in real-world scenarios are the primary challenges. For example, scaling up adsorption processes necessitates meticulous attention to mass transfer limitations, hydrodynamics, and adsorbent properties to guarantee performance uniformity between laboratory and industrial scales. For this purpose, Cruz-Olivares et al. (2022) suggested applying models such as the Thomas model, Adams–Bohart model, Yoon–Nelson model, dose–response model, and Bed Depth Service Time (BDST) model to the data obtained from continuous adsorption studies. Additionally, fixed-bed reactor design (Espina de Franco et al., 2017), regeneration efficiency, and economic feasibility (Bayuo et al., 2024) must be optimized to enable sustainable and cost-effective large-scale operation. Thus, as outlined in this section, systems-level integration—comprising of optimal column design, regeneration cycles, and process coupling—guarantees long-term heavy metal removal performance, resource efficiency, and scalability under fluctuating operating conditions.

3.3 MODIFIED CARBON-BASED MATERIALS FOR HEAVY METAL REMOVAL

Modified carbon-based materials are essential for decentralized drinking-water treatment, primarily because they are sustainable, cost-effective, and widely available (Gupta et al., 2020; Mariana et al., 2021; Ullah and Rahman, 2024). Since decentralized areas often lack large-scale infrastructure, the ability to produce high-performance adsorbents from local biomass (e.g., agricultural residue, forest waste) is an advantage (Hamda et al., 2025; Joshi and Ragupathy, 2025). While raw carbon materials often lack sufficient active sites required for efficient trace-level heavy metal removal, surface modification strategies enable the engineering of "designer" materials with enhanced selectivity and capacity (Liang et al., 2022; Wang et al., 2019). These materials are often engineered through chemical functionalization, heteroatom doping, or composite formation (Rahmani et al., 2022; Sultana et al., 2022).

3.3.1 Modified Activated Carbons

Among the various classes of biomass-derived adsorbents, modified ACs are the most technologically mature. Generated through the thermochemical processing of carbon-rich precursors such as coconut shells, pistachio shells, and/or fruit stones, these materials exhibit high mechanical strength and a well-developed internal pore network (Mariana et al., 2021; Sultana et al., 2022). Recent synthesis trends favor the chemical activation of adsorbent materials using potassium hydroxide or phosphoric acid, which generates extensive microporosity and increases the specific surface area, often exceeding 1,000 m^2/g (Gul et al., 2022; Sultana et al., 2022). By adding groups such as amino groups or 1,4-butane sultone, researchers can create acidic or basic sites on AC that greatly enhance its capacity to adsorb lead, arsenic, and cadmium (Nejadshafiee and Islami, 2019; Sultana et al., 2022). These modifications help to form stable coordination complexes, which are vital for removing low levels of heavy metals that still exceed the safety limits in groundwater.

3.3.2 Modified Biochars for Decentralized Applications

Modified biochars (BCs) represent another critical material class, generally produced at lower pyrolysis temperatures (300–700°C) than ACs (Gupta et al., 2020; Ullah and Rahman, 2024; Wang et al., 2019). Biochar is particularly attractive for decentralized applications because it can be produced using simple, low-cost technologies from diverse feedstocks such as banana stems, corn stalks, or water hyacinth (Joshi and Ragupathy, 2025; Ullah and Rahman, 2024). Modification of biochar often involves incorporating inorganic metal oxides or nanoscale zero-valent iron (nZVI) to overcome its low inherent porosity (Liang et al., 2022; Wang et al., 2019). For example, loading biochar with magnesium oxide, manganese dioxide, or zinc oxide enhances surface complexation and introduces new active sites for metal binding (Li et al., 2018; Ullah and Rahman, 2024). Synthesizing magnetic biochar enables the rapid recovery of used materials with magnets, thereby simplifying operations and preventing secondary turbidity (Rahmani et al., 2022; Ullah and Rahman, 2024).

3.3.3 Biopolymer-Derived Carbons and Hybrid Composites

Biopolymer-derived carbons and composites, such as those based on chitosan, alginate, or carrageenan, utilize natural structural matrices to provide abundant nitrogen- and oxygen-rich functional groups (Rahmani et al., 2022; Shen et al., 2020). Adding nitrogen, sulfur, or phosphorus (heteroatom doping) fine-tunes/alters the chemical structure and the hydrophilicity of carbon materials (Fang et al., 2022; Liang et al., 2022). These nitrogen-doped or sulfonated materials exhibit high electronegativity, which favors the adsorption of cationic species through strong electrostatic attraction and coordinate covalent bonding (Fang et al., 2022; Rahmani et al., 2022). The integration of carbon dots or CNTs into these biopolymer matrices further enhances the specific surface area and structural stability, thereby creating hybrid materials capable of treating multi-contaminant water sources with high selectivity (Deng et al., 2019; Rahmani et al., 2022; Shen et al., 2020).

3.3.4 Physical and Chemical Removal Mechanisms

The effectiveness of these materials depends on a combination of physical and chemical processes. Surface complexation is frequently the dominant pathway, involving the formation of covalent bonds between metal ions and functional groups such as carboxyl (–COOH), hydroxyl (–OH), and amine ($-NH_2$) moieties (Duan et al., 2020; Liang et al., 2022; Yang et al., 2019). Ion exchange also plays a significant role, with heavy metal cations replacing inherent competing ions, such as calcium, magnesium, or sodium, present on the modified carbon surface (Duan et al., 2020; Joshi and Ragupathy, 2025; Liang et al., 2022). To treat toxic hexavalent chromium, modified carbons act as electron donors, converting it into the safer form Cr(III), which is then trapped on the material's surface (Liang et al., 2022; Ullah and Rahman, 2024; Wang et al., 2019) (Figure 3.1).

The adsorption performance is highly sensitive to operating conditions, with solution/water pH serving as the primary control variable. Most studies indicate that optimal removal of divalent cations occurs at pH 4.0–7.0 (Duan et al., 2020; Hamda et al., 2025; Rahmani et al., 2022). Efficiency drops at low pH because hydronium ions (H_3O^+) compete for active sites and cause electrostatic repulsion (Duan et al., 2020; Liang et al., 2022; Rahmani et al., 2022). Conversely, excessive alkaline conditions may lead to the formation of insoluble metal hydroxides, complicating the interpretation of true adsorptive removal (Rahmani et al., 2022). Furthermore, the presence of competing ions commonly found in natural water matrices (e.g., calcium, magnesium, and chloride) can influence the selectivity and overall removal of target metals (Hamda et al., 2025; Liang et al., 2022). In high-saline water, high chloride levels can hinder mercury removal because mercury binds to chloride rather than attaching to the carbon surface (Krishna et al., 2023).

Despite these competitive effects, modified carbon-based materials demonstrate robust performance across various aqueous environments, achieving high removal efficiencies while maintaining potential for regeneration. Quantitative details regarding the maximum adsorption capacities (q_{max}) and specific experimental conditions (pH, concentration, and contact time) for the materials discussed in this section are summarized in Table 3.2. Purposeful functionalization creates a sustainable, practical platform for providing safe drinking water to decentralized communities by significantly improving the carbon based adsorbents performance.

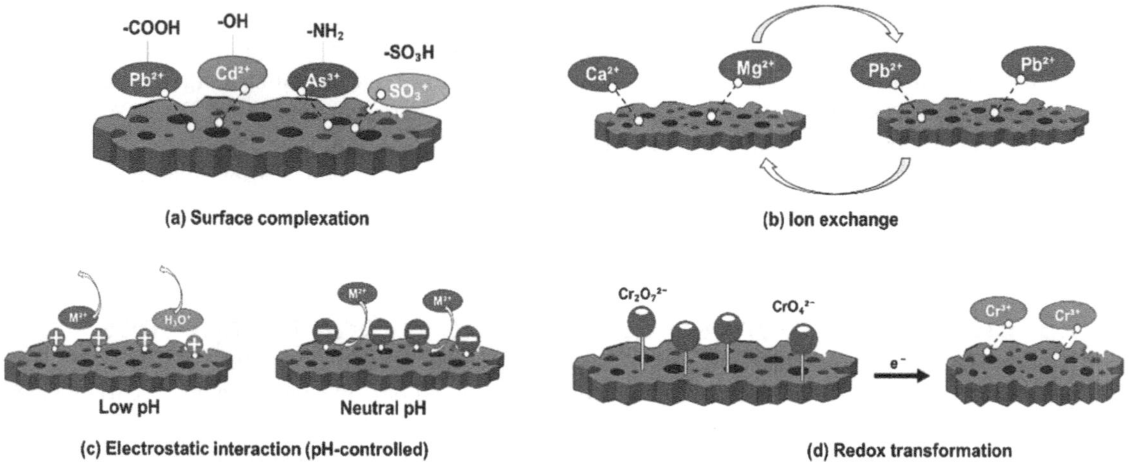

FIGURE 3.1 Predominant mechanisms for heavy metal removal using modified carbon-based materials (Duan et al., 2020; Liang et al., 2022; Yang et al., 2019; Joshi and Ragupathy, 2025; Rahmani et al., 2022; Hamda et al., 2025; Ullah and Rahman, 2024; Wang et al., 2019).

TABLE 3.2 Adsorption capacities and mechanisms for the removal of heavy metals by modified carbon-based materials

MODIFIED CARBON MATERIAL	*TARGET HEAVY METAL*	*EXPERIMENTAL CONDITIONS*	*ADSORPTION CAPACITY (mg/g)*	*ADSORPTION/ REMOVAL MECHANISM*	*REFERENCE*
AO-g-MWCNTs[1]	U(VI)	pH 5.0; initial concentration: 50–300 mg/L; temperature: 298 K	145.00	Surface complexation between metal ions and oxygen/ nitrogen atoms in amidoxime groups	Wang et al. (2014)
SWCNTs-SH[2]	Hg(II)	pH 5.0; initial concentration: 40 mg/L; adsorbent dose: 0.25 mg/mL; contact time: 1 h	131.58	Chemisorption via stable complexation between Hg(II) and mercapto groups	Bandaru et al. (2013)
GO[3]–MgO nanohybrid	Pb(II)	pH 6.5; adsorbent dose: 0.4 g/L; temperature: 303 K	190.00	Synergistic ion exchange (Mg(II) for Pb(II)) and surface complexation with oxygenated groups	Mohan et al. (2017)
Fe_3O_4 NPs@AC@SO_3H [4]	As(III)	pH 5.5; adsorbent dose: 0.4 g/L; contact time: 10 min; ultrasound-assisted	151.51	Electrostatic interaction driven by deprotonated surface sulfonic and carboxyl groups	Nejadshafiee and Islami (2019)

(Continued)

TABLE 3.2 (Continued)

MODIFIED CARBON MATERIAL	*TARGET HEAVY METAL*	*EXPERIMENTAL CONDITIONS*	*ADSORPTION CAPACITY (mg/g)*	*ADSORPTION/ REMOVAL MECHANISM*	*REFERENCE*
MWCNT-PEI[5]/PAN membrane	Pb(II)	pH 6.0; adsorbent dose: 10 mg; contact time: 120 min; temperature: 298 K	232.70	Chemisorption via coordination interaction between metal ions and PEI amino ligands	Deng et al. (2019)
BN[6] char	Pb(II)	pH 5.0; initial concentration: 200 ppm; adsorbent dose: 0.5 g/L; contact time: 180 min	252.46	Surface complexation with carboxyl/ hydroxyl groups and electrostatic attraction	Joshi and Ragupathy (2025)
Nitrogen-doped magnetic carbon nanoparticles	Cr(III)	pH 8.0; contact time: 10 min; temperature: 298 K	638.56	Chemical adsorption enhanced by high nitrogen group density on the graphitic core	Shin et al. (2011)
MMCG[7] with N-doped carbon dots	Pb(II)	pH 6.0; initial concentration: 200 mg/L; adsorbent dose: 1.0 g/L; contact time: 6 h	114.60	Complexation and chelation involving amine, hydroxyl, and sulfate ($-OSO_3^-$) groups	Rahmani et al. (2022)
nZVI/BC[8]	Pb(II)	pH 5.0; initial concentration: 10–300 mg/L; contact time: 120 min	295.30	Synergistic effect of surface complexation, precipitation, and reduction to Pb^0	Li et al. (2020)
EDTA-GO[9]	Pb(II)	pH 6.8; initial concentration: 100 mg/L; temperature: 298 K	525.00	Strong coordination interaction between metal ions and polycarboxylic acid groups	Madadrang et al. (2012)
Graphene oxide–$MnFe_2O_4$ magnetic nanohybrids	As(V)	pH 4.0; contact time: 10 min; temperature: 298 K	207.00	Oxidation of As(III) to As(V) followed by complexation with manganese oxide sites	Kumar et al. (2014)
Amidoxime-functionalized hollow carbon spheres	U(VI)	pH 5.0; contact time: 30 min; temperature: 293 K	355.60	Surface complexation with amidoxime moieties on high-surface-area hollow structures	Xia et al. (2019)

Note: 1 = amidoxime-grafted multi-walled carbon nanotubes; 2 = thiol-derivatized single-walled carbon nanotubes; 3 = graphene oxide; 4 = sultone-modified magnetic activated carbon; 5 = polyethyleneimine; 6 = banana stem; 7 = magnetically modified κ-carrageenan; 8 = porous biochar-supported nanoscale zero-valent iron; 9 = ethylenediaminetetraacetic acid-modified graphene oxide

3.4 ADVANCED CARBON-BASED MATERIALS FOR HEAVY METAL REMOVAL

Advanced carbon-based materials represent the highly engineered end of the adsorbent spectrum and include CNTs, graphene and graphene oxide derivatives, magnetic carbon nanohybrids, carbon nanofibers, and membrane-supported carbon composites. Figure 3.2 illustrates the application of several carbon-based adsorbents in fixed-bed column systems for the removal of heavy metals from contaminated water. It highlights the ability of these materials to effectively adsorb heavy metal ions during continuous water treatment and provides a means for recovering heavy metals from the adsorbent. The extraction of heavy metals from the adsorbents is generally accomplished using one or more desorption methods, wherein chemical solutions like acids, bases, or chelating agents are added to the adsorbent to release the adsorbed metal ions for subsequent collection and reutilization.

Compared with conventional AC and many (un)modified biochars, the carbon-based materials offer greater control over pore accessibility, defect density, surface chemistry, and interfacial reactivity, all of which are critical for trace-level heavy metal removal from complex water matrices (Liu et al., 2019; Song et al., 2024). Their high specific surface area and tunable functional groups make them especially attractive for decentralized treatment technologies where compactness, fast kinetics, and regeneration are important design requirements (Song et al., 2024). Representative advanced carbon-based adsorbents, together with their dominant removal mechanisms, operational strengths, and practical limitations, are summarized in Table 3.3. The main distinction among these materials is not only adsorption performance but also their suitability for structured deployment, recoverability, and continuous-flow operation.

3.4.1 Carbon Nanotubes and Nanofiber Membranes

CNTs have received considerable attention because of their hollow tubular morphology, strong mechanical properties, and accessible external surface. However, pristine CNTs generally exhibit limited affinity for dissolved metal ions due to their hydrophobic and chemically inert graphitic structure. For this reason, surface functionalization is often necessary to introduce binding sites such as carboxyl, hydroxyl, amine, and thiol groups. Bandaru et al. (2013) demonstrated that thiol-derivatized single-wall CNTs exhibited enhanced mercury adsorption because sulfur-containing groups have a strong affinity for Hg(II). Likewise, Deng

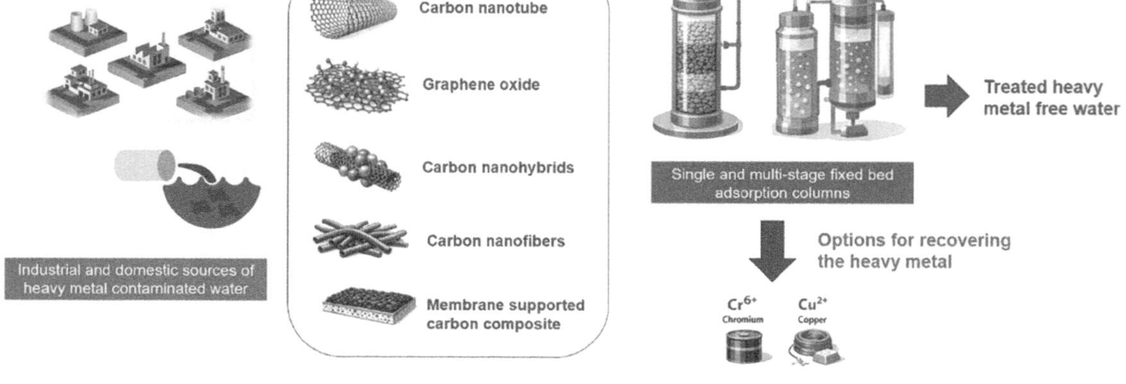

FIGURE 3.2 An illustration of the application of different carbon-based adsorbents in fixed-bed columns for the removal of heavy metals from contaminated water.

TABLE 3.3 Representative advanced carbon-based materials for metal removal from water

MATERIAL CLASS	*REPRESENTATIVE DESIGN*	*DOMINANT MECHANISMS*	*MAIN STRENGTHS*	*MAIN LIMITATIONS*	*ILLUSTRATIVE REFERENCES*
Graphene oxide/reduced graphene oxide composites	GO, magnetic GO, rGO-oxide hybrids, rGO@titanate	Electrostatic attraction, surface complexation, ion exchange, cation–π interactions	High tunability, abundant oxygenated groups, strong affinity for multiple metals	Restacking of sheets, recovery challenges for fine powders unless structured	GO performance depends strongly on oxygenated functionality and can be intensified under magnetic assistance (Melchor-Durán et al., 2024); rGO-THs combined high Pb/Cd/Cu uptake with fixed-bed operation (Yang et al., 2025)
Functionalized CNTs	Aminated, oxidized, polymer-grafted, oxide-decorated MWCNTs/SWCNTs	Coordination/complexation, electrostatic attraction, ionic interactions	Fast interfacial transport, high surface accessibility, and chemically tailorable	Aggregation, high cost, regeneration burden, unresolved nanotoxicity at scale	Amino-functionalized MWCNTs reached 92.8% Pb(II) removal under optimized conditions (Zhao et al., 2025)
Carbon/graphene aerogels and foams	rGO aerogels, magnetic carbon aerogels, biomass-derived carbon aerogels	Surface complexation, electrostatic attraction, pore-assisted transport	The 3D porous structure reduces agglomeration and enables easy handling and the formation of reusable monoliths	Fabrication complexity, possible fragility, cost depends on drying/functionalization route	Cysteamine-modified rGO aerogel removed Ni^{2+} effectively (Madduri & Kommalapati, 2024); $MnFe_2O_4$-loaded bamboo-pulp carbon aerogel removed Pb^{2+}, Cu^{2+}, and Cd^{2+} (Jing et al., 2024)
Carbon-based nanofibrous membranes/structured hybrids	Electrospun membranes containing GO and bio-derived additives	Adsorption plus permeation, surface complexation, and electrostatic interactions	Suited for flow-through treatment, easier recovery, lower solids-handling burden	Fouling, fabrication control, and membrane stability over long runs	GO/CNC/sericin electrospun membrane showed high Pb/Cr/Cu rejection with high flux (Hazarika et al., 2025)

Note: GO: graphene oxide, TH: titanate hybrids. The table emphasizes reactor-relevant functionality rather than equilibrium capacity alone, because recent studies suggest that structured recoverability and continuous-flow compatibility are increasingly decisive in practical water treatment (Hazarika et al., 2025; Yang et al., 2025; Zhao et al., 2025).

et al. (2019) incorporated polyethylenimine-modified multi-walled CNTs into polyacrylonitrile nanofiber membranes and reported improved removal of Pb(II) and Cu(II), demonstrating that CNT performance can be substantially enhanced when adsorption sites are integrated with membrane-support architecture.

3.4.2 Graphene and Graphene Oxide Derivatives

Graphene-based materials form another important class of advanced adsorbents. Graphene oxide and its derivatives contain abundant oxygenated functional groups that enhance metal uptake via electrostatic interactions, ion exchange, and surface complexation (Liu et al., 2019). Nevertheless, GO sheets may restack or agglomerate in aqueous media, thereby lowering the effective surface area and reducing the adsorption efficiency. To address this limitation, several studies have focused on hybridization and chemical modification. Madadrang et al. (2012), for example, showed that EDTA-functionalized graphene oxide markedly improved Pb(II) adsorption by providing additional coordination sites. Membrane-embedded CNTs, magnetic graphene-based composites, and structured graphenic filter media helps to overcome these limitations by improving handling and simplifying regeneration. Schmidt et al. (2023) developed regeneratable graphene-based household water filters that removed substantially more lead, cadmium, and mercury from tap water than activated charcoal and could be regenerated using mild treatments. Such findings indicate that the value of advanced carbon materials lies not only in their high adsorption capacity but also in their ability to serve as engineered platforms for point-of-use and point-of-entry treatment devices (Deng et al., 2019; Schmidt et al., 2023).

3.4.3 Magnetic Carbon Nanohybrids

The magnetic carbon nanohybrids serve as efficient adsorbent materials, characterized by a high surface area and magnetic characteristics, facilitating the effective removal of heavy metals from water and the convenient separation of the adsorbent via an external magnetic field. Mohan et al. (2017) reported effective lead removal using a graphene oxide–MgO nanohybrid, while Kumar et al. (2014) developed graphene oxide–$MnFe_2O_4$ magnetic nanohybrids that demonstrated efficient removal of lead and arsenic, as well as easier post-treatment separation. A key advantage of these advanced carbon materials is that they can be integrated into practical treatment configurations rather than being used only as dispersed powders. This is highly relevant to decentralized water treatment because unsupported nanomaterials can pose challenges for recovery, and repeated reuse. Almotairy (2026) synthesized and assessed a mesoporous nitrogen-doped carbon–barium ferrite nanomagnetic composite ($BaFe_2O_4$–NC) for the removal of mercury from contaminated water. The author emphasized the following significant findings: (i) the experimental data followed the pseudo-second-order kinetic model, while equilibrium studies conformed to the Langmuir isotherm, indicating a maximum adsorption capacity of ~487 mg/g; (ii) in adsorption–desorption studies, the adsorbent maintained approximately 96% of its original adsorption capacity and efficiency after five regeneration cycles; and (iii) the interaction between $BaFe_2O_4$ magnetic nanoparticles and the nitrogen-doped carbon matrix resulted in numerous active sites, favorable/enhanced mass transfer characteristics, and a strong affinity for binding mercury ions. Kumar and Bellare (2025) studied Fe_3O_4/MWCNTs–COOH (magnetite/carboxylated multi-walled CNTs) nanohybrids and combined them with polysulfone to produce modified hollow fiber membranes (HFMs) designed for the removal of arsenic and cadmium from contaminated water (both synthetic and lake water). The authors demonstrated that incorporating the Fe_3O_4/MWCNTs–COOH nanohybrid into HFMs facilitated the separation of arsenic and cadmium ions via mild acid wash. The adsorption capacities of the modified membrane were determined to be 54 ± 4.13 mg/g for arsenic and 85.06 ± 7.12 mg/g for cadmium. Long-term studies indicated that the adsorbent reached saturation after 23 h for cadmium ions and 26 h for arsenic ions.

3.4.4 Mechanism of Pollutant/Heavy Metal Removal

The removal mechanisms associated with these advanced materials are broadly similar to those reported for other carbon adsorbents, but they are often intensified by deliberate control of surface chemistry and nanostructure. Surface complexation and chelation are commonly dominant because functional groups on CNTs and graphene derivatives can form stable coordination interactions with metal cations (Yang et al., 2019). Electrostatic attraction also contributes significantly, particularly when solution pH favors deprotonation of the adsorbent surface, whereas ion exchange and surface precipitation may become important depending on the metal species and the composition of the hybrid material (Liu et al., 2019; Song et al., 2024). For redox-sensitive contaminants, engineered carbon materials may also support the reduction-coupled immobilization pathways, further broadening their applicability in polluted water treatment (Song et al., 2024).

Despite their excellent adsorption performance, advanced carbon-based materials still face several constraints that limit large-scale implementation. Agglomeration of nanoparticles, difficult recovery from treated water, relatively high synthesis cost, chemical-intensive functionalization routes, and uncertainties regarding long-term stability and end-of-life management remain significant concerns (Liu et al., 2019; Song et al., 2024). For decentralized treatment, these issues are particularly important because material replacement, regeneration, and disposal must remain technically simple and economically feasible. Future work should therefore prioritize immobilized or magnetically recoverable adsorbents, low-intensity regeneration methods, and performance evaluation under realistic multi-ion water conditions rather than idealized single-metal treating laboratory systems. In this regard, advanced carbon-based materials are best understood not as a replacement for all low-cost adsorbents, but as a complementary class of high-performance materials that can be selectively deployed where higher efficiency, selectivity, and compact system design are required.

3.5 CONTINUOUS ADSORPTION SYSTEMS AND COLUMN (FIXED-BED) REACTORS FOR HEAVY METAL REMOVAL

In continuous column reactors or fixed-bed experiments, the use of carbon-based adsorbents provides high adsorption capacity, enables the use of various materials (e.g., AC, biochar, CNTs, graphene-based adsorbents, and surface-modified or functionalized low-cost, bio-based materials), and demonstrates excellent adsorbent regeneration capacity. Hussain et al. (2025) employed raw oyster (RO) shell, oyster shell biochar (OB), and ginkgo-leaf biochar (LB) as biosorbents in two hybrid columns, achieving good removal efficiencies of a mixture of heavy metal ions: (i) primary column: 71% cadmium, 93% copper, 6% manganese, 52% iron, and 11% zinc; and (ii) secondary column: 99.5% cadmium, 100% copper, 55.2% manganese, 100% iron, and 74.03% zinc. The authors advised that the operation of the columns utilizing acid mine drainage (AMD) wastewater may be impeded by clogging from the metal precipitates. Besides, the authors noted that sourcing oyster shells and ginkgo leaves locally could reduce transportation costs and alleviate regional solid waste issues. Senaut et al. (2025) assert that carbon-based nanotubes are effective adsorbents for heavy metal removal. Nevertheless, their inherent physical and chemical properties limit their adsorption capacity, necessitating functionalization to introduce desired functional groups such as hydroxyl (–OH), carboxyl (–COOH), and amide (–CONH). According to Senaut et al. (2025), the following functionalization methods have been documented in the literature: oxidation using H_2O_2, oxidation using HNO_3, oxidation using $KMnO_4$ and H_2SO_4, reduction of graphene oxide by citric acid, modified Hummer's method, alkaline hydrothermal treatment, reduction of graphene oxide by gallic acid and sodium ascorbate, and the application of Al_2O_3, TiO_2, Fe_2O_3, and Cu_2O_3. The efficacy of carbon-based adsorbents depends on the production methods, surface area, pore size, and functional groups. Khan et al. (2025) emphasized that the adsorptive removal of heavy metals in batch and continuous operations depends on factors such as pH, temperature, reaction/contact time, adsorbent dosage, adsorbent type, functional groups present in the adsorbent, and the metal ion concentration. While utilizing vegetable and fruit-derived adsorbents for heavy metal removal, Tariq et al. (2024) identified the predominant adsorption mechanisms as physical adsorption, chemisorption, electrostatic interactions, hydrogen bonding, π–π interactions, and precipitation.

Carbon-based nanosheets have garnered considerable interest as highly effective adsorbents for removing heavy metals from polluted water. These nanosheets (graphene, graphene oxide, reduced graphene oxide, and carbon nanosheet composites) exhibit distinctive structural and physicochemical characteristics, including high specific surface area, diverse surface functional groups, and good mechanical stability. Seifpour et al. (2025) developed sulfur-doped boron carbon nitride (S-BCN) nanosheets as an efficient solid-phase extraction adsorbent for the concurrent preconcentration of chromium, copper, cadmium, and lead in aqueous environments and wastewater. The authors reported detection limits of 0.40–0.90 μg/L, preconcentration factors >125, a breakthrough volume of 500 mL, and good precision (relative standard deviation <4.5%). Syeda et al. (2025) examined the removal of lead, zinc, and copper in single- and multi-metal systems using functionalized biopolymer-based aerogel beads in a fixed-bed (column) reactor. The authors evaluated the impact of influent flow rate, bed depth, and initial metal ion concentration, and reported maximum adsorption capacities of 193.50, 51.18, and 54.30 mg/g for lead, zinc, and copper, respectively. In another study, Rahman et al. (2025) employed a fixed-bed column adsorption technique using rice husk biochar (FRHBC) and optimized the effects of process variables, including concentration, flow rate, and bed height, on lead removal efficiency. The highest removal efficiency was 85.01%, with a maximum adsorption capacity of 179.77 mg/g. The authors employed column/fixed-bed adsorption models, including the Thomas, Adams–Bohart, and Yoon–Nelson models, to analyze breakthrough curves and identify the reaction mechanism and adsorption isotherm. These models elucidate the adsorption kinetics, estimates the breakthrough time, and reveals information about the adsorption capacity under continuous-flow conditions. The Thomas model predicts the adsorption column's efficiency, while the Adams–Bohart model elucidates the initial phase/segment of the breakthrough curve. The Yoon–Nelson model simplifies predicting the time required to achieve 50% adsorbate breakthrough. These models are useful for the design engineers to: (a) optimize the column design, (b) scale up the adsorption column, and (c) enhance the long-term operational efficiency for heavy metal removal. Some recent studies that have used various types of fixed-bed columns to remove heavy metals and other pollutants from water are summarized in Table 3.4. Evidently, different adsorbents possess distinct functional groups on their surfaces, which facilitate the binding of heavy metal ions in fixed-bed column systems through varied interactions. The breakthrough models facilitate the prediction of adsorption efficiency and the estimation of bed exhaustion/breakthrough. Comprehending these characteristics, in conjunction with techno-economic assessment, facilitates effective adsorption column design, cost optimization, and the scale-up of adsorption columns from lab to semi-pilot and full-scale systems.

Li et al. (2026a) employed waste straw-derived N-doped magnetic AC for the adsorptive removal of chromium and lead and reported maximum adsorption capacities of 283.3 mg/g and 197.5 mg/g, respectively. The breakthrough curve was analyzed using the Thomas, Yoon–Nelson, and Adams–Bohart models, revealing that the bed saturation rate of the novel adsorbent was significantly influenced by the flow rate and initial concentration. Villar da Gama et al. (2026) investigated the removal of cadmium and copper

TABLE 3.4 Different types of fixed-bed adsorption studies for removing heavy metals and other pollutants from water

ADSORBENT	*TARGET POLLUTANTS*	*FUNCTIONAL GROUPS AND MECHANISMS*	*PERFORMANCE ANALYSIS (ADSORPTION CAPACITY/ REMOVAL)*	*ECONOMIC AND OPERATIONAL DATA*	*REFERENCES*
Raw oyster shell and biochar	AMD: Cd, Cu, Fe, and Mn	Carbonate, hydroxyl, phosphate, and amine groups	High sorption capacity confirmed by the Thomas and Yoon–Nelson models	Metal uptake is enhanced by increasing hydraulic retention time	Hussain et al. (2025)
SCBB	CIP and heavy metals (Pb, Cd, Zn)	Ion exchange, complexation, and pore diffusion	105.5 mg/g (Pb), 100.3 mg/g (Cd), 82.2 mg/g (Zn), 55.7 mg/g (CIP)	$0.31/kg preparation cost; 0.88-year payback; 10 mL/min flow rate; 25.68 cm bed height	El-Aswar et al. (2024)

(Continued)

TABLE 3.4 (Continued)

ADSORBENT	*TARGET POLLUTANTS*	*FUNCTIONAL GROUPS AND MECHANISMS*	*PERFORMANCE ANALYSIS (ADSORPTION CAPACITY/ REMOVAL)*	*ECONOMIC AND OPERATIONAL DATA*	*REFERENCES*
FRHBC	Lead (Pb^{2+}) and MB dye	2D honeycomb-like mesoporous structure; active sites: –OH, –C–O–C–, and phosphate groups; synergistic mono/multilayer physisorption and chemisorption	179.77 mg/g (Pb) and 173.01 mg/g (MB); removal efficiencies of 85.01% (Pb) and 87.38% (MB)	Flow rate: 2.5–4.5 mL/min; bed height: 1.0–1.5 cm; optimal performance at pH 4.0	Rahman et al. (2025)
Calcined natural clay (heated to 550°C)	Pb(II), Cr(VI), and Cd(II)	Two-stage mechanism: surface adsorption followed by inner layer diffusion; performance fits the Thomas and Advection Diffusion models	39.4 mg/g (Pb), 29.95 mg/g (Cd), and 21.65 mg/g (Cr)	Flow rate: 3–9 mL/min; bed height: 2–6 cm; pH 6.0	Khalfa et al. (2021)
3DG/A-MWCNTs	NPX and various pharmaceutical products	Simultaneous adsorption and PMS activation; generates reactive oxygen species (SO_4^{-}, •OH, $^1O_2$) and promotes electron transfer	56.2% removal after 1,150 min; maintains removal above 66% after six reuse cycles via in situ regeneration	Flow rate: 2.3 mL/min; bed height: 20–40 mm; PMS concentration: 0.2 g/L	Liu et al. (2024)
PHW mixed with sand	Pb, Cu, Co, and Ni	Functional groups: hydroxyl, carboxyl, and amine; high surface area (4.0218 m^2/g); metal precipitation via cation exchange (K, Ca)	Exhaustion time doubled from ~400 to 800 min; batch removal up to 88% (Ni) and 87% (Pb)	5 g PHW/kg sand ratio; flow rate: 5–15 mL/min; sand prevents clogging and color issues	Ganji et al. (2024)
Fe/Mg-WHBC	Imidacloprid (pesticide) and other micropollutants	Surface complexation (major), hydrogen bonding, π–π interaction, and pore-filling	114 mg/g; fixed-bed penetration time reduced only 27.6% after five cycles	Flow rate: 3.5–7.5 mL/min; filler mass: 0.3–0.9 g; safe metal leaching levels between pH 2.0 and 10.0	Ji et al. (2024)
HA-HFO-D201	Cu(II), Cd(II), and Pb(II)	Ligand exchange; HFOs provide capture sites while HA coating adds functional groups and restricts Fe leaching at low pH	101.4 mg/g (Pb), 88.96 mg/g (Cu), and 65.67 mg/g (Cd); treated 410 bed volumes for Pb	Flow rate: 25 mL/h; Na_2EDTA regeneration allows for repeated use with >90% efficiency	Hao et al. (2022)

Note: 3DG/A-MWCNTs: 3D graphene-doped acidified multi-walled carbon nanotubes; AMD: acid mine drainage; CIP: ciprofloxacin; Fe/Mg-WHBC: Fe–Mg co-modified water hyacinth-based biochar; FRHBC: functionalized rice husk biochar; HA-HFO-D201: humic acid-coated hydrated ferric oxides in resin; MB: methylene blue; NPX: naproxen; PHW: pistachio hull waste; PMS: peroxymonosulfate; SCBB: sugarcane bagasse modified with bentonite.

ions in a fixed-bed column using a peanut shell adsorbent, reporting adsorption efficiencies >85% for both metal ions. They presented a comprehensive scientific analysis regarding the total adsorptive capacity of the adsorbent, stating that: (i) the adsorption capacity increases with increasing flow rates, which correlates with a diminished mass transfer zone, resulting in reduced diffusional resistance and enhanced adsorption, and (ii) the mass transfer zone for copper ions is larger than that for cadmium ions, as copper ions possess superior access to the adsorbent pores and demonstrate enhanced electronic stability.

Heavy metal removal from polluted water is effective and cheap with olive stone waste, a biomass byproduct. The porous structure and functional groups improve the metal ion binding capacity. Thus, using agricultural waste reduces the disposal issues and treatment costs and provides an eco-friendly alternative to conventional adsorbents for water filtration. Awad (2024) employed Jordanian olive stone waste as a biosorbent for lead ion adsorption, with breakthrough results indicating over 90% removal at an initial concentration of 18 mg/L, achieved in ~110 h. Zuo et al. (2025) employed a fixed-bed column packed with functionalized AC to remove lead from water, thereby purifying approximately 127 L of simulated drinking water containing 150 μg/L lead, with a maximum adsorption capacity of 20.34 mg/g. Signori-Iamin et al. (2025) reported that cationic cellulose nanofibers are effective for chromium adsorption, achieving adsorption capacities >100 mg/g. The authors developed two models (A and B) employing random forests to forecast/predict the metal ion (VI and total) concentrations based on the operational/process data (Input parameters for model A: flow rate and concentration, adsorbent mass, pH, and volume; Input parameters for model B: color changes and conductivity). Artificial intelligence (AI) is increasingly used to predict the adsorption breakthrough in fixed-bed systems (Karishma et al., 2026; Shanmughan et al., 2025). Machine learning methods such as artificial neural networks (ANNs), random forests, and support vector machines evaluate experimental process factors, including flow rate, input concentration, temperature, pressure, and adsorbent characteristics. These models capture intricate nonlinear linkages and precisely predict the breakthrough time and adsorption capacity without relying exclusively on conventional mathematical models. The application of AI tools and techniques reduces the experimental time, enhances column design, and increases the efficiency of fixed-bed adsorption columns.

3.6 FUTURE RESEARCH PERSPECTIVES

The rapid development of natural and modified carbon-based materials (Figure 3.3) has significantly expanded the available tools for metal(loid) removal in decentralized water treatment. However, these adsorbents also have their own limitations (as mentioned in Figure 3.3), such as the degradation of the adsorbent (e.g., natural) over prolonged periods of use, the high production costs of carbon-based nanotubes, and the fate of carbon-based nanomaterials in the environment. The adsorption mechanisms and material synthesis strategies are now relatively well understood; hence, future research should focus on addressing issues such as comparability, chemical complexity, operational durability, and constraints in real-world applications. These challenges are particularly critical in decentralized settings, where treatment systems operate under variable water chemistry and limited process control.

One of the most continuing limitations is the lack of standardized experimental protocols for evaluating the adsorption performance. Reported adsorption capacities for similar modified carbon materials often differ by orders of magnitude due to inconsistent influent concentrations, pH control strategies, different electrolytes, and evaluation endpoints. This issue has been repeatedly highlighted in recent critical reviews as a major barrier to translating laboratory findings into engineering practice (Putra et al., 2025). Future research should emphasize comparative testing, in which newly developed adsorbents are evaluated alongside reference materials (e.g., commercial AC or unmodified biochar) under identical, environmentally relevant conditions (Li et al., 2026b). Importantly, adsorption performance should be assessed at trace-level metal(loid) concentrations, as a representative of contaminated groundwaters rather than at artificially elevated concentrations commonly used to maximize the adsorbents capacity. Such standardization will allow engineers to select materials based on performance reliability rather than isolated laboratory metrics.

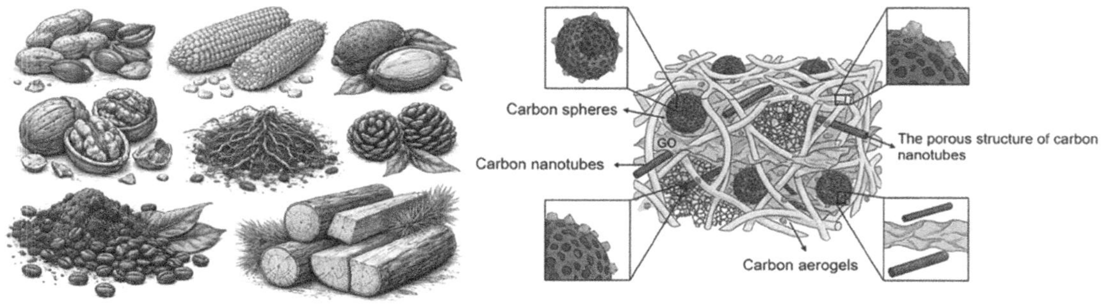

Limitations of natural adsorbents

- **Reduced adsorption capacity** - Attributable to less surface area and a scarcity of active sites
- **Low adsorption kinetics** - Necessitate extended contact time as the diffusion of metal ions into pores occurs at a reduced rate
- **Inadequate selectivity** - When several metal ions coexist in the water, the treatment efficiency reduces
- **Structural instability** - Prolonged operation may result in material degradation and the leaching of organic matter into water

Limitations of advanced carbon-based adsorbents

- **High production costs** - Due to the utilization of costly precursors, specialized equipment's, and energy-demanding procedures
- **Nanomaterial aggregation** - Occurs in aqueous environments, leading to the formation of agglomerates that reduces the surface area and efficiency
- **Challenges in recovery** - Necessitates filtering or centrifugation processes, which are difficult to implement at an industrial scale
- **Fate of the adsorbent** - Raises toxicity concerns if discharged into the environment, necessitating meticulous handling and disposal protocols

FIGURE 3.3 A schematic overview and the limitations of the different natural and carbon-based adsorbents used for heavy metal removal from contaminated water.

Despite growing recognition of the complexity of the water matrix, most adsorption studies still focus on single-metal systems. In decentralized water sources, multiple competing ions coexist. These species can suppress adsorption through electrostatic screening, site competition, and disruption of ion-exchange equilibrium (Joshi and Ragupathy, 2025). Future research must systematically investigate multi-metal and multi-ion systems, moving beyond qualitative observations toward quantitative analyses of selectivity and competition. This is particularly important for modified carbon materials designed to address both cationic metals and oxyanions, where functional groups or embedded metal oxides may preferentially bind one species over others (Dhokpande et al., 2024). Without such data, adsorption performance under real decentralized conditions will be overestimated.

Many advanced carbon-based materials rely on coupled adsorption–redox mechanisms, particularly for redox-sensitive contaminants such as chromium and arsenic species. While these mechanisms can significantly enhance the removal efficiency, they also introduce uncertainty regarding the stability of immobilized transformation products. Recent studies emphasized that reduced species may be remobilized under changing pH or redox conditions, which are common in decentralized systems with intermittent operation (Seder-Colomina et al., 2022). Future work should obviously distinguish the adsorption from transformation by combining speciation analysis with surface characterization. Long-term experiments to check the fate of immobilized metal(loid)s under fluctuating water chemistry are essential for assessing the true treatment robustness and environmental safety.

Decentralized water sources increasingly contain organic micropollutants and natural organic matter alongside metal(loid)s. Carbon-based adsorbents exhibit strong affinity for organic compounds, which may compete with metals for adsorption sites or modify the surface charge and hydrophobicity. Recent studies show that the presence of natural organic matter can significantly suppress metal adsorption, particularly at low metal concentrations, where site competition becomes critical (Zhang et al., 2023). Future research should examine metal(loid) adsorption in the presence of representative organic contaminants and different fractions of natural organic matter. Knowledge about these interactions is essential before proposing carbon-based adsorbents as multifunctional treatment media for decentralized systems.

Although many studies discuss the regenerability of carbon-based adsorbents, systematic assessments for the regeneration remain limited. Chemical regeneration can degrade the surface functional groups, whereas thermal regeneration is often impractical for decentralized applications. Moreover, regeneration processes may result in a secondary contamination stream due to concentrated metal(loid) release (Wibowo, 2025). Future research should prioritize low-intensity regeneration strategies compatible with decentralized operation and evaluate the performance over multiple cycles. The stability and disposal pathways of spent adsorbents must also be assessed to ensure that adsorption does not merely transfer contamination from water to solid waste.

Moreover, a major research gap exists between laboratory-based adsorption studies and the performance of continuous-flow systems. Batch adsorption capacities rarely explain directly what is happening in the column or fixed-bed systems due to mass transfer limitations, fouling, and influent variability. Recent studies emphasize the need to integrate adsorption data into predictive column models relevant to decentralized operation. Future research should therefore link material behavior to parameters such as breakthrough curves, empty bed contact time, and operational resilience. This transition from material optimization to system integration is essential for realizing the full potential of carbon-based adsorption technologies in decentralized water treatment.

3.7 CONCLUSIONS

Low-cost, carbon-based adsorbents provide an effective and sustainable option for removing heavy metals from contaminated water. Biochar and AC offer high surface area and functional groups, which improve the adsorption performance. Surface modification (by physical or chemical methods) can enhance selectivity, stability, and adsorption capacity when removing a mixture of heavy metals. Fixed-bed systems are particularly well suited for continuous water/wastewater treatment and for scalable operations in practice. The effective selection of adsorption column parameters, their optimization, and an overview of breakthrough behavior are crucial for sustaining long-term adsorption column performance. The regeneration and reuse of adsorbents facilitate cost savings and enhance sustainability. Thus, the incorporation of low-cost, modified carbon-based adsorbents with optimized fixed-bed systems presents intriguing opportunities for sustainable and cost-effective removal of heavy metals from water.

ACKNOWLEDGMENTS

The authors extend their gratitude to their respective universities, institutes, and departments for providing staff time and infrastructural support that enabled the completion of this chapter. This acknowledgment is a testament to our appreciation for the collaborative environment and resources that have enabled the completion of this chapter, thereby promoting the collaborative excellence of both research and learning.

REFERENCES

Almotairy, A.R.Z. (2026). Wastewater treatment using a novel magnetic mesoporous adsorbent composed of nitrogen-doped carbon for the removal of highly toxic metal ions. *Diamond and Related Materials*, *164*, 113503. https://doi.org/10.1016/j.diamond.2026.113503.

Awad, A. (2024). Removal efficiency, metal uptake, and breakthrough curve of aqueous lead ions removal using olive stone waste. *Results in Engineering*, *22*, 102311. https://doi.org/10.1016/j.rineng.2024.102311.

Bandaru, N.M., Reta, N., Dalal, H., Ellis, A.V., Shapter, J., Voelcker, N.H. (2013). Enhanced adsorption of mercury ions on thiol derivatized single wall carbon nanotubes. *Journal of Hazardous Materials*, *261*, 534–541. https://doi.org/10.1016/j.jhazmat.2013.07.076.

Bayuo, J., Rwiza, M.J., Choi, J.W., Mtei, K.M., Hosseini-Bandegharaei, A., Sillanpää, M. (2024). Adsorption and desorption processes of toxic heavy metals, regeneration and reusability of spent adsorbents: Economic and environmental sustainability approach. *Advances in Colloid and Interface Science*, *329*, 103196. https://doi.org/10.1016/j.cis.2024.103196.

Chaudhari, V., Patkar, M. (2022). Removal of nickel from aqueous solution by using corncob as adsorbent. *Materials Today: Proceedings*, *61*, Part 2, 307–314, https://doi.org/10.1016/j.matpr.2021.09.458.

Chen, Y., Yang, J., Abbas, A. (2023). Enhanced chromium (VI) adsorption onto waste pomegranate-peel-derived biochar for wastewater treatment: performance and mechanism. *Toxics*, *11*, 440. https://doi.org/10.3390/toxics11050440.

Cruz-Lopes, L.P., Macena, M., Esteves, B., Guiné, R.P.F. (2021). Ideal pH for the adsorption of metal ions Cr^{6+}, Ni^{2+}, Pb^{2+} in aqueous solution with different adsorbent materials. *Open Agriculture*, *6*(1), 115–123. https://doi.org/10.1515/opag-2021-0225.

Cruz-Olivares, J., Pérez-Alonso, C., Martínez-Barrera, G., Roa-Morales, G., López-Téllez, G., del Campo-López, E.M. (2022). Modeling and scaling up of the Cr(VI) adsorption process by using mexicalcite natural mineral in a packed bed column. *Results in Engineering*, *16*, 100687. https://doi.org/10.1016/j.rineng.2022.100687.

Deng, S., Liu, X., Liao, J., Lin, H., Liu, F. (2019). PEI modified multiwalled carbon nanotube as a novel additive in PAN nanofiber membrane for enhanced removal of heavy metal ions. *Chemical Engineering Journal*, *375*, 122086. https://doi.org/10.1016/j.cej.2019.122086.

Dhokpande, S.R., Deshmukh, S.M., Khandekar, A.R., Sankhe, A.A. (2024). A review of carbon-based adsorbents for the removal of organic and inorganic components. *Reviews in Inorganic Chemistry*, *44*, 655–670. https://doi.org/10.1515/revic-2024-0027.

Duan, C., Ma, T., Wang, J., Zhou, Y. (2020). Removal of heavy metals from aqueous solution using carbon-based adsorbents: A review. *Journal of Water Process Engineering*, *37*, 101339. https://doi.org/10.1016/j.jwpe.2020.101339.

El-Aswar, E.I., Ibrahim, S.S., Abdallah, Y.R., Elsharkawy, K. (2024). Removal of ciprofloxacin and heavy metals from water by bentonite/activated carbon composite: Kinetic, isotherm, thermodynamic and breakthrough curve modeling studies. *Journal of Molecular Liquids*, *403*, 124821, https://doi.org/10.1016/j.molliq.2024.124821.

Espina de Franco, M.A., de Carvalho, C.B., Bonetto, M.M., de Pelegrini Soares, R., Féris, L.A. (2017). Removal of amoxicillin from water by adsorption onto activated carbon in batch process and fixed bed column: Kinetics, isotherms, experimental design and breakthrough curves modelling. *Journal of Cleaner Production*, *161*, 947–956. https://doi.org/10.1016/j.jclepro.2017.05.197.

Fang, Y., Liu, L., Xiang, H., Wang, Y., Sun, X. (2022). Biomass-based carbon microspheres for removing heavy metals from the environment: A review. *Materials Today Sustainability*, *18*, 100136. https://doi.org/10.1016/j.mtsust.2022.100136

Ganji, H., Taghavijeloudar, M., Khodashenas, S.R. (2024). A sustainable approach for heavy metals removal from surface water by sand column amendment with bio-adsorbent of pistachio hull waste (PHW): Batch and fixed-bed column adsorption. *Journal of Water Process Engineering*, *67*, 106061, https://doi.org/10.1016/j.jwpe.2024.106061.

Garg, R., Garg, R., Sillanpää, M., Alimuddin, Khan, M.A., Mubarak, N.M., Tan, Y.H. (2023). Rapid adsorptive removal of chromium from wastewater using walnut-derived biosorbents. *Scientific Reports*, *13*, 6859. https://doi.org/10.1038/s41598-023-33843-3.

Gul, A., Ma'amor, A., Khaligh, N.G., Julkapli, N.M. (2022). Recent advancements in the applications of activated carbon for the heavy metals and dyes removal. *Chemical Engineering Research and Design*, *186*, 276–299. https://doi.org/10.1016/j.cherd.2022.07.051

Gupta, S., Sireesha, S., Sreedhar, I., Patel, C.M., Anitha, K.L. (2020). Latest trends in heavy metal removal from wastewater by biochar based sorbents. *Journal of Water Process Engineering*, *38*, 101561. https://doi.org/10.1016/j.jwpe.2020.101561

Hamda, A.S., Areti, H.A., Gudeta, R.L., Abo, L.D., Jayakumar, M. (2025). Carbon-based nanomaterials for water treatment: A comprehensive review of recent advances and mechanisms. *Chemical Engineering Journal Advances*, *23*, 100834. https://doi.org/10.1016/j.ceja.2025.100834

Hao, L., Li, L., Yu, S., Liu, J. (2022). Humic acid-coated hydrated ferric oxides-polymer nanocomposites for heavy metal removal in water. *Science of the Total Environment*, *834*, 155427. https://doi.org/10.1016/j.scitotenv.2022.155427.

Hazarika, P., Duarah, R., Goswami, R., Rajguru, P., Baruah, S., Goswami, P.K., Feng, X., Hazarika, S. (2025). Electrospun nanofibrous adsorptive membranes for removal of heavy metals from wastewater via a flow-through adsorption/permeation process. *Chemical Engineering Journal*, *519*, 164877. https://doi.org/10.1016/j.cej.2025.164877.

Hussain, F., Kim, L.H., Oh, S-E., Kim, S. (2025). Neutralization of pH and removal of heavy metals from acid mine water by using low-cost biosorbents in batch and column studies. *Groundwater for Sustainable Development*, *31*, 101506. https://doi.org/10.1016/j.gsd.2025.101506.

Jahangiri, F.M., Moutushi, H.T., Moniruzzaman, M., Hoque, S., Hossain, M.E. (2021). Removal of lead from aqueous solutions and wastewaters using water hyacinth (*Eichhornia crassipes*) roots. *Water Practice and Technology*, *16*(2), 404–419. doi: https://doi.org/10.2166/wpt.2021.005.

Ji, X., Liu, Y., Gao, Z., Lin, H., Xu, X., Zhang, Y., Zhu, K., Zhang, Y., Sun, H., Duan, J. (2024). Efficiency and mechanism of adsorption for imidacloprid removal from water by Fe-Mg co-modified water hyacinth-based biochar: Batch adsorption, fixed-bed adsorption, and DFT calculation. *Separation and Purification Technology*, *330*, 125235. https://doi.org/10.1016/j.seppur.2023.125235.

Jing, W., Yang, C., Lin, X., Tang, M., Lian, D., Yu, Y., Liu, D. (2024). $MnFe_2O_4$-loaded bamboo pulp carbon-based aerogel composite: Synthesis, characterization and adsorption behavior study for heavy metals removal. *RSC Advances*, *14*, 39995–40005. https://doi.org/10.1039/D4RA06363E.

Joshi, A.A., Ragupathy, G. (2025). Performance and mechanisms of waste-based carbon adsorbents in heavy metal removal – An experimental and theoretical approach. *RSC Advances*, *15*(42), 34609–34621. https://doi.org/10.1039/D5RA03863D.

Karishma, S., Kamalesh, R., Saravanan, A., Yaashikaa, P.R., Vickram, A.S. (2026). AI-driven neural time series network forecasting and cost analysis for dye removal prediction in packed bed adsorption using ultrasonic biomass composites for sustainable wastewater management. *Environmental Research*, *289*, 123396. https://doi.org/10.1016/j.envres.2025.123396.

Khan, K., Shafi, M., Raza, J., Ahmed, I.A., Zada, A., Narasimharao, K., Sun, X. (2025). Mechanistic breakthroughs in affordable adsorbents for heavy metal remediation: An in-depth exploration of next-generation sustainable water purification technologies. *Journal of Hazardous Materials Advances*, *19*, 100847. https://doi.org/10.1016/j.hazadv.2025.100847.

Khalfa, L., Sdiri, A., Bagane, M., Cervera, M.L. (2021). A calcined clay fixed bed adsorption studies for the removal of heavy metals from aqueous solutions. *Journal of Cleaner Production*, *278*, 123935. https://doi.org/10.1016/j.jclepro.2020.123935.

Krishna, R.H., Chandraprabha, M.N., Samrat, K., Murthy, T.P.K., Manjunatha, C., Kumar, S.G. (2023). Carbon nanotubes and graphene-based materials for adsorptive removal of metal ions – A review on surface functionalization and related adsorption mechanism. *Applied Surface Science Advances*, *16*, 100431. https://doi.org/10.1016/j.apsadv.2023.100431.

Kumar, V., Bellare, J. (2025). Heavy metals removal from contaminated water with Psf hollow fiber membrane blended with Fe_3O_4/MWCNTs –COOH nanohybrid. *Journal of Water Process Engineering*, *69*, 106880. https://doi.org/10.1016/j.jwpe.2024.106880.

Kumar, S., Nair, R.R., Pillai, P.B., Gupta, S.N., Iyengar, M.A.R., Sood, A.K. (2014). Graphene oxide–$MnFe_2O_4$ magnetic nanohybrids for efficient removal of lead and arsenic from water. *ACS Applied Materials & Interfaces*, *6*(20), 17426–17436. https://doi.org/10.1021/am504826q.

Kurniawan, T.A., Chan, G.Y.S., Lo, W.-H., Babel, S. (2006). Comparisons of low-cost adsorbents for treating wastewaters laden with heavy metals. *Science of The Total Environment*, *366*(2–3), 409–426. https://doi.org/10.1016/j.scitotenv.2005.10.001.

Li, C., Zhang, L., Gao, Y., Li, A. (2018). Facile synthesis of nano ZnO/ZnS modified biochar by directly pyrolyzing zinc-contaminated corn stover for Pb(II), Cu(II) and Cr(VI) removals. *Waste Management*, *79*, 625–637. https://doi.org/10.1016/j.wasman.2018.08.035.

Li, G., Du, R., Cao, Z., Li, C., Xue, J., Ma, X., Wang, S. (2024a). Research progress in graphene-based adsorbents for wastewater treatment: Preparation, adsorption properties and mechanisms for inorganic and organic pollutants. *C*, *10*(3), 78. https://doi.org/10.3390/c10030078.

Li, L., Hu, Z., Guo, W., Xia, H., Wang, Y., Wang, S., Chen, G., Qiu, M., Hu, B. (2024b). Recent advances in various adsorbents for the extraction of uranium from saline lakes: A review. *Journal of Molecular Liquids*, *395*, 123862. https://doi.org/10.1016/J.MOLLIQ.2023.123862.

Li, C., Wang, Q., Zeng, L., Zhang, K., Mu, J. (2026a). Waste straw-derived N-doped magnetic activated carbon toward efficient adsorption of anionic/cationic heavy metal. *Colloids and Surfaces A: Physicochemical and Engineering Aspects*, *735*, 139565. https://doi.org/10.1016/j.colsurfa.2026.139565.

Li, S., Liu, X., Sun, Y., Zhang, M., Wu, L., Guo, Z., Tang, Y. (2026b). The fortune of biochar: A comprehensive dissection from preparation and modification to multipollutants removal and engineering application. *Separation and Purification Technology*, *382*, 135862. https://doi.org/10.1016/j.seppur.2025.135862.

Li, S., Yang, F., Li, J., Cheng, K. (2020). Porous biochar-nanoscale zero-valent iron composites: Synthesis, characterization and application for lead ion removal. *Science of the Total Environment*, *746*, 141037. https://doi.org/10.1016/j.scitotenv.2020.141037.

Li, Z., Lu, G., Du, D., Zhao, D. (2025). Harnessing low-cost adsorbents for removal of heavy metals and metalloids in contaminated water: Progress in the past decade and future perspectives. *Journal of Cleaner Production*, *518*, 145845. https://doi.org/10.1016/j.jclepro.2025.145845.

Liang, W., Wang, G., Peng, C., Tan, J., Wan, J., Sun, P., Li, Q., Ji, X., Zhang, Q., Wu, Y., Zhang, W. (2022). Recent advances of carbon-based nano zero valent iron for heavy metals remediation in soil and water: A critical review. *Journal of Hazardous Materials*, *426*, 127993. https://doi.org/10.1016/j.jhazmat.2021.127993.

Liu, M., Lu, Q., Yu, W. (2023). The improvement of heavy metals removal by wood membrane in drinking water treatment: Comparison with polymer membrane and associated mechanism. *Chemosphere*, *324*, 138297, https://doi.org/10.1016/j.chemosphere.2023.138297.

Liu, Z., Liu, L., Lu, J., Kong, W., Wen, C., Zhang, Y. (2024). A fixed-bed reaction system packed with a graphene/multiwalled carbon nanotube scaffold for the oxidative removal of naproxen. *ACS Applied Nano Materials*, *7*(8), 8869–8879. https://doi.org/10.1021/acsanm.4c00383.

Liu, X., Ma, R., Wang, X., Ma, Y., Yang, Y., Zhuang, L., Zhang, S., Jehan, R., Chen, J., Wang, X. (2019). Graphene oxide-based materials for efficient removal of heavy metal ions from aqueous solution: A review. *Environmental Pollution, 252*, 62–73. https://doi.org/10.1016/j.envpol.2019.05.050.

Madadrang, C.J., Kim, H.Y., Gao, G., Wang, N., Zhu, J., Feng, H., Gorring, M., Kasner, M.L., Hou, S. (2012). Adsorption behavior of EDTA-graphene oxide for Pb (II) removal. *ACS Applied Materials & Interfaces*, *4*(3), 1186–1193. https://doi.org/10.1021/am201645g.

Madduri, S.B., Kommalapati, R.R. (2024). Harnessing novel reduced graphene oxide-based aerogel for efficient organic contaminant and heavy metal removal in aqueous environments. *Nanomaterials, 14*(21), 1708. https://doi.org/10.3390/nano14211708.

Maity, S., Patil, P.B., SenSharma, S., Sarkar, A. (2022). Bioremediation of heavy metals from the aqueous environment using *Artocarpus heterophyllus* (jackfruit) seed as a novel biosorbent. *Chemosphere*, *307*(Part 4), 136115. https://doi.org/10.1016/j.chemosphere.2022.136115.

Mariana, M., Khalil, A.H.P.S., Mistar, E.M., Yahya, E.B., Alfatah, T., Danish, M., Amayreh, M. (2021). Recent advances in activated carbon modification techniques for enhanced heavy metal adsorption. *Journal of Water Process Engineering*, *43*, 102221. https://doi.org/10.1016/j.jwpe.2021.102221.

Melchor-Durán, A.P., Moreno-Virgen, M.R., Bonilla-Petriciolet, A., Reynel-Ávila, H.E., Lucio Ortiz, E., González-Vázquez, O.F. (2024). Heavy metal removal from water using graphene oxide in magnetic-assisted adsorption systems: Characterization, adsorption properties, and modelling. *Separations, 11*(10), 294. https://doi.org/10.3390/separations11100294.

Mohan, S., Kumar, V., Singh, D.K., Hasan, S.H. (2017). Effective removal of lead ions using graphene oxide-MgO nanohybrid from aqueous solution: Isotherm, kinetic and thermodynamic modeling of adsorption. *Journal of Environmental Chemical Engineering*, *5*(3), 2259–2273. https://doi.org/10.1016/j.jece.2017.03.031.

Nejadshafiee, V., Islami, M.R. (2019). Adsorption capacity of heavy metal ions using sultone-modified magnetic activated carbon as a bio-adsorbent. *Materials Science & Engineering C*, *101*, 42–52. https://doi.org/10.1016/j.msec.2019.03.081.

Prokopov, T.V., Nikolova, M.I., Taneva, D.S., Petkova, N.T. (2021). Removal of chromium (VI) from aqueous solution by extracted onion processing waste. *Scientific Study and Research*, *22*(4), 399–410.

Putra, N.R., Zaini, M.A.A., Kusuma, H.S., Darmokoesoemo, H., Faizal, A.N.F. (2025). Advances in chromium removal using biomass-derived activated carbon: A comprehensive review and bibliometric analysis. *Environmental Progress & Sustainable Energy*, *44*, e14598. https://doi.org/10.1002/ep.14598.

Rahman, M.M., Maniruzzaman, M., Ealam, N., Mahmud, P., Khatun, S., Hossain, M.K., Hossain, M.I., Hasanuzzaman, M., Alam, M.A., Al-amin, M., Ghos, B.C. (2025). Adsorptive removal of toxic heavy metal and dyes from wastewater by rice husk (lignocellulosic biomass) derived activated biochar: A fixed-bed column adsorption study. *Carbohydrate Polymer Technologies and Applications*, *9*, 100698. https://doi.org/10.1016/j.carpta.2025.100698.

Rahmani, Z., Ghaemy, M., Olad, A. (2022). Removal of heavy metals from polluted water using magnetic adsorbent based on κ-carrageenan and N-doped carbon dots. *Hydrometallurgy*, *213*, 105915. https://doi.org/10.1016/j.hydromet.2022.105915.

Sánchez-Ponce, L., Granado-Castro, M.D., Casanueva-Marenco, M.J., Galindo-Riaño, M.D., Díaz-de-Alba, M. (2023). Sherry wine industry by-product as a potential biosorbent for the removal of Cr(VI) from aqueous medium. *Biomass Conversion and Biorefinery*, *13*, 12489–12507. https://doi.org/10.1007/s13399-021-02053-0.

Schmidt, S.J., Dou, W., Sydlik, S.A. (2023). Regeneratable graphene-based water filters for heavy metal removal at home. *ACS ES&T Water, 3*(8), 2179–2185. https://doi.org/10.1021/acsestwater.3c00010.

Seder-Colomina, M., Mangeret, A., Bauda, P., Brest, J., Stetten, L., Merrot, P., Julien, A., Diez, O., Barker, E., Billoir, E., Poupin, P., Thouvenot, A., Cazala, C., Morin, G. (2022). Influence of microorganisms on uranium release from mining-impacted lake sediments under various oxygenation conditions. *Environmental Science: Processes and Impacts*, *24*, 1830–1843. https://doi.org/10.1039/d2em00104g.

Seifpour, A., Ghanemi, K., Larki, A. (2025). Sulfur-doped boron carbon nitride nanosheets: A high-performance adsorbent for simultaneous preconcentration of heavy metal ions in water and wastewater. *Microchemical Journal*, *219*, 115953, https://doi.org/10.1016/j.microc.2025.115953.

Senaut, A.F., Mubarak, N.M., Ruslan, K.N., Tan, Y.H., Karri, R.R., Prasad, D.M.R. (2025). Mechanistic insights into carbon-nano-based adsorbents for heavy metal removal: A comprehensive review. *Groundwater for Sustainable Development*, *30*, 101498. https://doi.org/10.1016/j.gsd.2025.101498.

Shanmughan, B., Nighojkar, A., Kandsubramanian, B. (2025). Advancements in characterization Techniques, empirical Models, and Artificial intelligence for comprehensive understanding of heavy metal adsorption on sewage sludge biochar. *Waste Management Bulletin*, *3(1)*, 193–206. https://doi.org/10.1016/j.wmb.2025.01.003.

Sheikh, Z., Amin, M., Khan, N., Khan, M.N., Sami, S.K., Khan, S.B., Hafeez, I., Khan, S.A., Bakhsh, E.M., Cheng, C.K. (2021). Potential application of *Allium Cepa* seeds as a novel biosorbent for efficient biosorption of heavy metal ions from aqueous solution. *Chemosphere*, *279*, 130545, https://doi.org/10.1016/j.chemosphere.2021.130545.

Shen, W., An, Q.-D., Xiao, Z.-Y., Zhai, S.-R., Hao, J.-A., Tong, Y. (2020). Alginate modified graphitic carbon nitride composite hydrogels for efficient removal of Pb(II), Ni(II) and Cu(II) from water. *International Journal of Biological Macromolecules*, *148*, 1298–1306. https://doi.org/10.1016/j.ijbiomac.2019.10.105.

Shin, K.Y., Hong, J.Y., Jang, J. (2011). Heavy metal ion adsorption behavior in nitrogen-doped magnetic carbon nanoparticles: Isotherms and kinetic study. *Journal of Hazardous Materials*, *190*(1–3), 36–44. https://doi.org/10.1016/j.jhazmat.2010.12.102.

Signori-Iamin, G., Lombardo, S., Fiol, N., Carrasco, F., Aguado, R.J., Santos, A.F., Delgado-Aguilar, M. (2025). Cr(VI) removal by cationic cellulose nanofiber aerogels: batch and fixed-bed adsorption, reduction mechanism, and machine learning breakthrough predictions. *Chemical Engineering Journal Advances*, *24*, 100919. https://doi.org/10.1016/j.ceja.2025.100919.

Song, Y., Liu, Z., Zhang, Q. (2024). Engineering the future: Unveiling novel paths in heavy metal wastewater remediation with advanced carbon-based nanomaterials-Beyond performance comparison, tackling challenges, and exploring opportunities. *Chemosphere*, *366*, 143477. https://doi.org/10.1016/j.chemosphere.2024.143477.

Sultana, M., Rownok, M.H., Sabrin, M., Rahaman, M.H., Nur Alam, S.M. (2022). A review on experimental chemically modified activated carbon to enhance dye and heavy metals adsorption. *Cleaner Engineering and Technology*, *6*, 100382. https://doi.org/10.1016/j.clet.2021.100382.

Syeda, H.I., Muthukumaran, S., Baskaran, K. (2025). Dynamic adsorption of heavy metals on functionalized and regeneratable biopolymeric aerogels: Fixed-bed column reactor modelling and dual functionality elution technique. *Separation and Purification Technology*, *363*(Part 1), 131861. https://doi.org/10.1016/j.seppur.2025.131861.

Tariq, A., Yahaya, N., Sajid, M. (2024). Low cost adsorbents derived from vegetables and fruits: Synthesis, properties, and applications in removal of heavy metals from water. *Desalination and Water Treatment*, *320*, 100626, https://doi.org/10.1016/j.dwt.2024.100626.

Tolkou, A.K., Tsoutsa, E.K., Kyzas, G.Z., & Katsoyiannis, I.A. (2024). Sustainable use of low-cost adsorbents prepared from waste fruit peels for the removal of selected reactive and basic dyes found in wastewaters. *Environmental Science and Pollution Research*, *31*, 14662–14689. https://doi.org/10.1007/s11356-024-31868-3.

Ullah, M.H., Rahman, M.J. (2024). Adsorptive removal of toxic heavy metals from wastewater using water hyacinth and its biochar: A review. *Heliyon*, *10*, e36869. https://doi.org/10.1016/j.heliyon.2024.e36869

Villar da Gama, B.M., Campos, N.F., Sales, D.C.S., Rodriguez-Díaz, J.M., de Menezes Barbosa, C.M.B., Duarte, M.M.M.B. (2026). Sustainable removal of cadmium and copper ions using peanut shell-based adsorbent: fixed-bed column performance and phenomenologic insights. *Chemical Engineering Science*, *320*(Part B), 122511. https://doi.org/10.1016/j.ces.2025.122511

Wang, L., Wang, Y., Ma, F., Tankpa, V., Bai, S., Guo, X., Wang, X. (2019). Mechanisms and reutilization of modified biochar used for removal of heavy metals from wastewater: A review. *Science of the Total Environment*, *668*, 1298–1309. https://doi.org/10.1016/j.scitotenv.2019.03.011

Wang, Q., Wang, Y., Tang, J., Yang, Z., Zhang, L., Huang, X. (2022a) New insights into the interactions between Pb(II) and fruit waste biosorbent. *Chemosphere*, *303*(Part 1), 135048, https://doi.org/10.1016/j.chemosphere.2022.135048.

Wang, Q., Wang, Y., Yang, Z., Han, W., Yuan, L., Zhang, L., Huang, X. (2022b) Efficient removal of Pb(II) and Cd(II) from aqueous solutions by mango seed biosorbent. *Chemical Engineering Journal Advances*, *11*, 100295, https://doi.org/10.1016/j.ceja.2022.100295.

Wang, Y., Gu, Z., Yang, J., Liao, J., Yang, Y., Liu, N., Tang, J. (2014). Amidoxime-grafted multiwalled carbon nanotubes by plasma techniques for efficient removal of uranium (VI). *Applied Surface Science*, *320*, 10–20. https://doi.org/10.1016/j.apsusc.2014.08.182

Wattanakornsiri, A., Rattanawan, P., Sanmueng, T., Satchawan, S., Jamnongkan, T., Phuengphai, P. (2022). Local fruit peel biosorbents for lead(II) and cadmium(II) ion removal from waste aqueous solution: A kinetic and equilibrium study. *South African Journal of Chemical Engineering*, *42*, 306–317, https://doi.org/10.1016/j.sajce.2022.09.008.

Wibowo, Y.G. (2025). Carbon-Based adsorbents for heavy metal removal in Wastewater: Green technology or environmental burden? *Environmental Nanotechnology, Monitoring & Management*, *23*, 101051. https://doi.org/10.1016/j.enmm.2025.101051

Xia, L., Jin, W., Zhang, S., Ding, C., Sheng, G., Alsaedi, A., Hayat, T., Li, J., Song, Y. (2019). Amidoxime-functionalized hollow carbon spheres for efficient removal of uranium from wastewater. *ACS Sustainable Chemistry & Engineering*, *7*(12), 10800–10807. https://doi.org/10.1021/acssuschemeng.9b01616.

Yang, X., Liu, P., Yu, H. (2025). Adsorption of heavy metals from wastewater using reduced graphene oxide@titanate hybrids in batch and fixed bed systems. *BMC Chemistry, 19*(1), 72. https://doi.org/10.1186/s13065-025-01443-z.

Yang, X., Wan, Y., Zheng, Y., He, F., Yu, Z., Huang, J., Wang, H., Ok, Y.S., Jiang, Y., Gao, B. (2019). Surface functional groups of carbon-based adsorbents and their roles in the removal of heavy metals from aqueous solutions: A critical review. *Chemical Engineering Journal*, *366*, 608–621. https://doi.org/10.1016/j.cej.2019.02.119.

Zhang, X., Xu, H., Xi, M., Jiang, Z. (2023). Removal/adsorption mechanisms of Cr(VI) and natural organic matter by nanoscale zero-valent iron-loaded biochar in their coexisting system. *Journal of Environmental Chemical Engineering*, *11*, 109860. https://doi.org/10.1016/j.jece.2023.109860.

Zhao, D., Zhang, Y., Wu, C. (2025). Adsorption of Pb(II) ions by functionalized multi-walled carbon nanotubes MWCNTs-NH2. *Journal of Hazardous Materials Advances, 19*, 100764. https://doi.org/10.1016/j.hazadv.2025.100764.

Zuo, Q., Zheng, H., Zhang, P., Zhang, Y., Zhang, J. (2025). Adsorptive removal of trace lead from drinking water in a functionalized activated carbon fiber fixed-bed column: Experimental and modeling study. *Journal of Molecular Liquids*, *438*(Part A), 128659. https://doi.org/10.1016/j.molliq.2025.128659.

4 Fluorescent and Optical Sensors for Real-Time Detection of Waterborne Contaminants

Arumugam Thangamani, Kumarasamy Chandra Prakash, Pandian Bothi Raja, and Kuppusamy Kanagaraj

4.1 INTRODUCTION

Water contaminants pose an increasingly increasing risk to the health of people, the environment, and industries around the globe. The existence of these pollutants in the water can be disastrous whether they are microbial pathogens like bacteria, viruses, and protozoa or chemical pollutants like heavy metals, pesticides, and industrial chemicals. The prevalence of water contamination, be it through industrial processes, agricultural drainage, or natural catastrophes, requires stable and swift detection techniques to protect the quality of water and avert the damage to human beings and the environment [1]. Traditionally, the monitoring of water contaminants has been based on the laboratory-based testing techniques, which, despite their effectiveness, are also associated with a number of limitations. These old methods are usually characterized by lengthy turn-around times, expensive laboratory equipment and reagents, and sophisticated sample preparation. Moreover, they do not necessarily give real-time information, and it is difficult to intervene in water treatment and the protection of human health in a timely manner. Accordingly, there has been a growing requirement of sophisticated, on-site detection technologies which are capable of providing real-time, in situ outputs [2].

Fluorescence-based and optical sensors are an exciting alternative to the drawbacks of the conventional water quality monitoring techniques. The sensors can offer real-time and continuous data about the presence of contaminants in water, which can be responded to in a quicker and more effective manner [3]. Fluorescent sensors are based on the fact that some contaminants (e.g., microbial pathogens or toxic chemicals) have fluorescent properties, which the sensor identifies or can be induced to fluoresce by specialized probes. However, optical sensors require measurement of the interaction of light with the target

DOI: 10.1201/9781003741626-4

contaminants, such as alterations in light absorption, scattering, or reflection. These sensors are sensitive and specific and can detect contaminants at low levels—often parts per billion (ppb) or parts per trillion (ppt)—essential to detecting waterborne contaminants before it will inflict major damage [4]. The fact that these sensors can sense pollutants at very low levels and the fact that real-time monitoring is possible makes such sensors a valuable asset in the water quality management system, especially in situations that require quick responses.

Fluorescence-based and optical sensors have benefits that are not limited to sensitivity and real-time measurements. These sensors can be easily carried around, unlike traditional laboratory-based systems, so on-site measurements can be done in a broad spectrum of degrading environments, both in remote locations and in an emergency situation, such as when water pollution is a possible issue. This portability enables instant monitoring of water quality, and therefore, the local authorities or water treatment facilities will find it easier to act immediately, whether by change in the treatment process or by public health alert message. Moreover, the sensors are cost-efficient in the long run [5]. Although the upfront cost of installing some of these technologies might be high, they are usually less costly to operate continuously compared to the traditional ways, where laboratory tests, transportation of samples, and reagent cost may be expensive. Fluorescent and optical sensors also do not need much preparation of the sample, and less time and effort are needed to perform testing. Consequently, these sensors not only facilitate faster decision-making but also contribute to the cost savings of the total monitoring, which makes them an interesting choice to consider when monitoring water quality in the long term.

However, challenges do not mean that fluorescent and optical sensors cannot be integrated into real-time water quality monitoring systems despite the obvious merits of their utilization. The interference of other substances that may be in the water is one of the major obstacles. As an example, sensor readings can be influenced by high turbidity, the presence of dissolved organic matter, or other fluorescent substances and cause a false positive or loss of data accuracy. In addition to this, environmental conditions like temperature, salinity, and pH may impact the operation of these sensors, and this means that they need routine calibration and maintenance to give an accurate reading [6]. Moreover, although these sensors are highly sensitive with regard to the detection of some types of contaminants, often times, their detection capabilities are limited, especially with regard to some pollutants, such as complex organic substances or viruses, which do not interact with the sensor in the same manner as other contaminants do. Sensors also

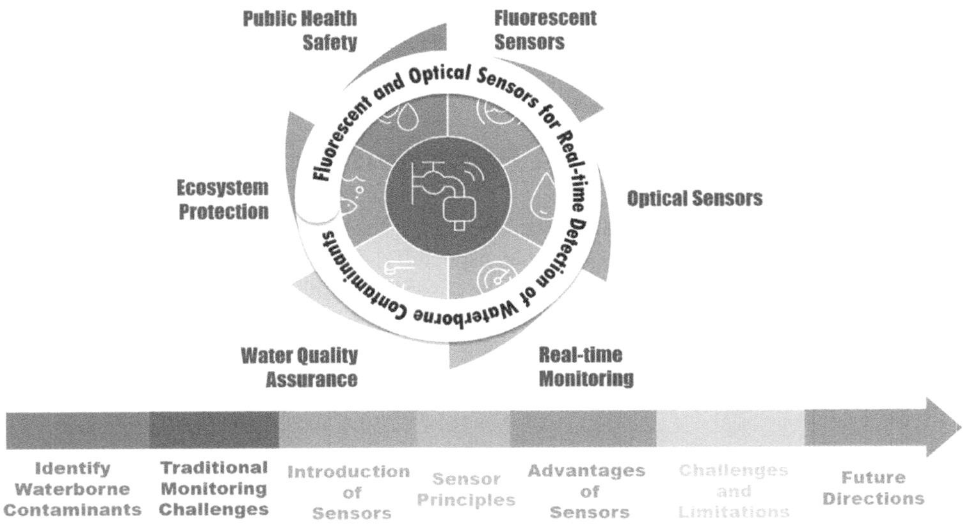

FIGURE 4.1 An overview of the chapter structure is presented, mapping the key topics from ecosystem protection and real-time monitoring challenges to the introduction and future directions of hyperspectral and optical sensors.

may degenerate or foul with use in harsh or dynamic environments over the long term, thereby affecting the reliability and life cycle of these devices [7]. These issues highlight the need to continue research and development to maximize the performance of fluorescent and optical sensors, in particular by incorporating nanomaterials, artificial intelligence (AI), and wireless capabilities, which can improve sensitivity, minimize interference, and automate the processing of data to improve decision-making.

In this chapter, the role of the fluorescent and optical sensors in the real-time detection of waterborne contaminants will be examined, covering the principles underlying such technologies, their use in different sectors, and the benefits of such technologies over conventional detection methods (Figure 4.1). We will also discuss the challenges and limitations that exist, including interference of environmental factors, degradation of sensors, and detection limits. Lastly, we will discuss the future of these technologies, especially regarding their adoption in smart water monitoring systems and IoT (Internet of Things) networks, which could revolutionize how the quality of water is monitored and managed across the world. With the ever-increasing demand for safer and more reliable management systems of water quality, the introduction and adoption of the real-time sensor technologies will be instrumental in the sustainability of water resources and the security of human health [8].

4.2 THE NEED FOR REAL-TIME DETECTION OF WATERBORNE CONTAMINANTS

The water is a basic essential to life, as it is an essential resource in drinking, sanitation, agriculture, and industry. Nevertheless, different contaminants can sometimes ruin the quality of water, and they can be very dangerous to human health, environmental sustainability, and economic development. Waterborne contaminations can broadly be categorized into three (shown in Figure 4.2), such as microbial pathogens, chemical pollutants, and physical contaminants. All categories are distinguished in terms of their features and pose certain difficulties of detection and management. The necessity of real-time monitoring of these contaminants has been of growing importance, especially in areas that are vulnerable to natural calamities and industrial pollution, as well as fast urbanization [9, 10]. This chapter discusses why real-time detection systems are necessary, particularly with reference to microbial pathogens, chemical contaminants, and physical contaminants, and how the timely, accurate detection can aid in reducing health hazards, environmental protection, and regulatory compliance.

4.2.1 Microbial Pathogens: A Persistent Threat to Public Health

Among the most common causes of waterborne diseases in the world are microbial pathogens in water, which have serious health consequences on people, especially in developing countries. These are bacteria, viruses, and protozoa, which can cause very diverse diseases. Only a few of such diseases are cholera, dysentery, giardiasis, and cryptosporidiosis, which are associated with contamination by microbes. The World Health Organization (WHO) says that waterborne diseases are a major cause of preventable deaths worldwide, with an estimated 2 billion people having no access to safely managed drinking water [11].

4.2.1.1 Bacterial Contaminants

Escherichia coli (*E. coli*), *Salmonella*, and *Vibrio cholerae* are bacteria that are very common in fecal content and may be contaminated in water bodies and in places where there are poor sanitation facilities. Fecal coliforms (usually *E. coli*) are commonly considered as a marker of possible bacterial contamination. Once these bacteria get into the drinking water systems, they are capable of triggering serious gastrointestinal diseases and even death in worst-case scenarios. Poor hygiene habits, waste management,

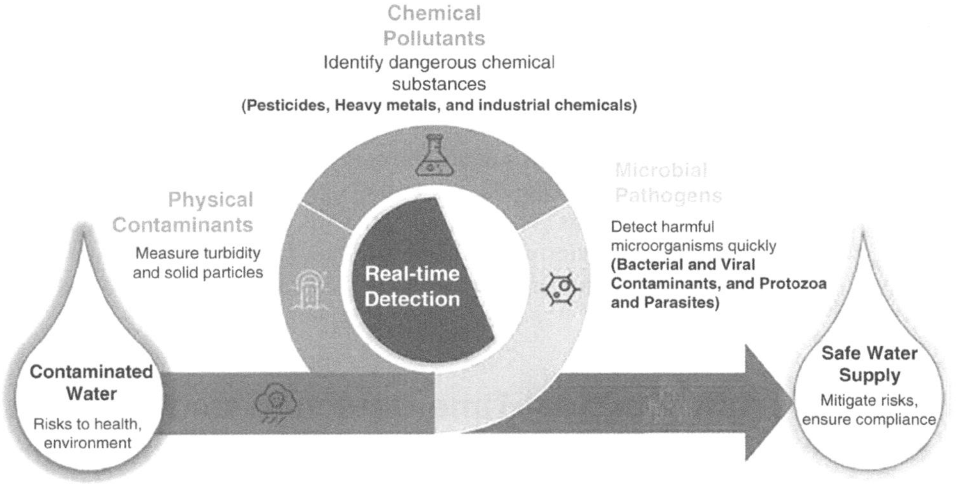

FIGURE 4.2 Outline of the critical need for real-time detection of waterborne contaminants, categorizing threats from chemical, physical, and microbial sources to mitigate risks and ensure a safe water supply.

and water filtration systems also compound the proliferation of bacterial pathogens in developing countries. An example of this is a cholera outbreak that can spread fast in flood-prone regions or other areas where there is a lack of water treatment. Early identification of these bacterial pathogens in real time is very essential in preventing large-scale epidemics. Conventional water testing techniques, whereby the samples are cultured in a laboratory, may require several hours or even days; by that time, it might be too late to contain the outbreak [12–15]. The necessity of quick on-site detection of bacterial contaminants is hence a prerequisite of successful interventions in the area of public health.

4.2.2 Chemical Pollutants: Long-Term Health and Environmental Risks

The chemical contaminants of water constitute an extensive variety of organic and inorganic compounds, including pesticides, heavy metals, and industrial chemicals that may have drastic long-term health and environmental impacts. The sources of these pollutants are agricultural runoffs, industrial effluents, and urban wastewater. In contrast to microbial contaminants, which usually induce immediate illnesses, chemical pollutants may build up in the environment over a long period, thus causing chronic health problems, including cancer, nervous system disorders, and endocrine disorders. Also, heavy metals such as lead, mercury, and arsenic may be toxic to aquatic organisms, which upset ecosystems and food chains [16–19].

4.2.3 Physical Contaminants: Indicators of Pollution

Indirect indicators of water quality are often referred to as physical contaminants, e.g., turbidity and suspended solids. Turbidity is normally linked with pollution, and it shows the presence of particulate matter in water. This substance can be organic or inorganic, a microorganism, or runoffs such as construction area, erosion, and agricultural runoffs. Even though turbidity is not an immediate health hazard, it is frequently an indication of other pollution sources, including pathogens or chemicals, and may impair the filtration process of water, which is why it is a significant parameter to use in real-time measurement [20].

4.2.3.1 Turbidity and Suspended Solids

Turbidity measurement is based on light scattering and absorption of light through a water sample. When there is a high level of turbidity, the efficacy of the water treatment process decreases and, in effect, the cost of filtration and disinfection increases. Moreover, the suspended solids may hamper the provision of clear, aesthetically pleasing water by the water utilities. Turbidity sensors are becoming popular in real time to measure water treatment plants and natural water bodies to give information on the concentration of particles and water clarity. These sensors are based on optical technologies, including nephelometry, or measure the scattering of light by particles in the water. Constant monitoring of the turbidity of water is crucial to guarantee that the water treatment processes are functioning effectively, and there are no incidences of possible contamination [21].

4.2.4 The Importance of Real-Time Detection for Public Health, Environmental Protection, and Regulatory Compliance

Through new technology, waterborne contaminants can be detected promptly and efficiently in order to protect the environment, safeguard human health, and ensure that the water is treated according to the water quality requirements. Conventional water testing procedures, though effective, are usually slow, taking several hours to days to produce results. This time lag can give the contaminants time to propagate without being detected, resulting in disease outbreaks, destruction of the ecology, and failure to meet the set standards [22, 23].

4.2.4.1 Public Health Protection

By being able to detect microbial pathogens (bacteria, viruses, and protozoa) in real time, outbreaks of waterborne diseases can be prevented because the authorities learn about contamination events immediately. The real-time identification of pathogens is necessary in areas that experience frequent floods like coastal zones or low river basins where flood runoff can easily contaminate water supplies, leading to a large number of sick people [24].

4.2.4.2 Environmental Protection

In the context of environmental protection, real-time analysis of chemical pollutants is vital in the monitoring of the well-being of aquatic ecosystems. Heavy metals, pesticides, and industrial pollutants, among others, can be devastating to aquatic life and the overall food chain. In real time, the presence of these pollutants can be tracked to avoid their build-up in water bodies to guarantee the sustainability of water resources and biodiversity.

4.2.4.3 Regulatory Compliance

Government and environment agencies have come up with water quality standards that are designed to maintain the safety of drinking water and safeguard aquatic ecosystems. Such standards control the tolerable concentrations of different contaminants, which include the microbial pathogens, chemical pollutants, and physical parameters. Real-time detection enables water utilities and regulatory authorities to continuously monitor the quality of water to ensure that it is managed in accordance with these standards and to provide the data to facilitate decision-making. It is urgent because the threat of contaminated water by microbial pathogens, chemical pollutants, and physical contaminants is larger, and as a result, there is a need to obtain real-time intelligence on the presence of these contaminants [25, 26]. The potential of rapid detection technologies is to safeguard the health of the population, maintain the quality levels of the environment, and control compliance with the regulations. The real-time monitoring of the water quality in flood-prone regions, industrial areas, and cities where the pollution can be introduced rather quickly is not only an option but a must. More sophisticated, precise, and affordable real-time detection systems will be essential in the future of water quality management and will be instrumental in preventing disease outbreaks, reducing environmental harm, and preventing drinking water from becoming unsafe in the future.

Fluorescent and optical sensors have become potent tools to detect and monitor the presence of contaminants on different surfaces, including water. The sensors have a high capacity to detect minute changes in their environment compared to normal techniques; they are fast and can give real-time data. The technologies can detect a broad spectrum of contaminants, such as microbial pathogens, chemical pollutants, and physical contaminants by examining the interaction of light and the target substances in water. This chapter discusses the principles of these two categories of sensors—fluorescent sensors and optical sensors—their components, the mechanism of action, and the uses of such sensors as applied in the monitoring of water quality [27].

4.2.5 Fluorescence-Based Sensors: Harnessing Light for Detection

Fluorescence is a physical process where a substance takes in the light of a particular wavelength and gives it back at a different longer wavelength. The technique is especially applicable in detection of contaminants in water since it can be very sensitive and selective. Fluorescent sensors can be used to detect a wide range of substances, including contaminants that either inherently have a fluorescent effect or can be made to have a fluorescent effect by using special reagents. These sensors can provide real-time and quantitative information about water quality, which is why they can be of great value in monitoring the environment [28–33].

4.2.6 Optical Sensing: Detecting Contaminants through Light Interaction

This is in contrast to a fluorescence-based sensor, which uses the emission of light by the target substance to detect contaminants, and an optical sensor detects contaminants by observing variations in the characteristics of light as it passes through the sample (Figure 4.3). These sensors are capable of detecting a broad spectrum of contaminants—organic and inorganic—based on optical characteristics of absorption, reflection, scattering, and polarization. Application of optical sensors in water monitoring for turbidity, chemical contamination, and physical contamination is popular in the use of optical sensors.

4.2.6.1 Working Principle of Optical Sensors

Optical sensing works on the basis that the passage of light through a substance makes the light properties change. It is possible to track the presence of contaminants in the water sample with the help of these interactions. The most widely used light interactions in optical sensing are as follows:

- **Absorption**: As light is passing through a substance, it is absorbed by the substance absorbing some of its energy. Absorption Spectrum: An absorption spectrum may be characteristic of a material and can be used to identify and quantify certain contaminants [34].
- **Reflection**: Light reflecting off the surface of a contaminant can reflect back to the sensor. The reflected amount of light may give data on the concentration of the contaminant.
- **Scattering**: The scattering process takes place when the light is interacted with the particles in the water, like suspended solids or microorganisms. The angle and intensity of the scattered light may give us details on the concentration of the particles and the size of the particles in the water.

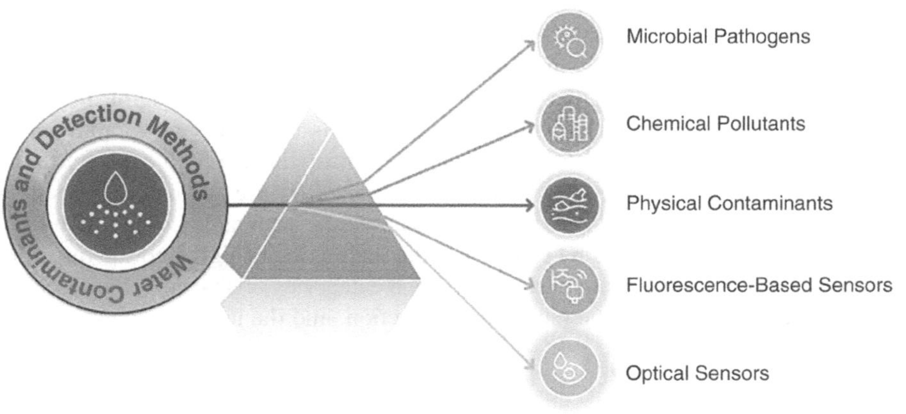

FIGURE 4.3 An overview of water quality monitoring targets (pathogens, chemicals, physical contaminants) and the optical detection methodologies used for their identification and analysis.

- **Polarization**: Polarized light is light which vibrates in a certain direction. The polarization has been measured following the interaction of light with a contaminant and using the difference in the polarizations to form information about the properties of the contaminant [35–37].

4.3 APPLICATIONS OF FLUORESCENT AND OPTICAL SENSORS IN WATERBORNE CONTAMINANT DETECTION

The quality of water is vital in public health, environmental sustainability, and economic development. Contaminants that are waterborne, regardless of being microbial, chemical, or physical, are very dangerous, such as disease spread, degradation, and biodiversity loss. Water quality is important in getting safe drinking water, the proper treatment of wastewater, and the health of the aquatic life. Fluorescent and optical sensors have become important instruments in the detection and monitoring of these contaminants in real time [38–40]. These technologies not only enable highly sensitive, rapid, and on-site analysis but also enable the invaluable data in environmental management as well as in protecting the health of the people (Figure 4.4).

This chapter discusses possible uses of fluorescent and optical sensors in the measurement of waterborne pollutants (microbial pathogens, chemical pollutants, turbidity, pH, and dissolved oxygen) (Figure 4.4). These sensors have superiority over conventional approaches such as high sensitivity, quicker responses, and the capability to monitor water quality in real time.

4.3.1 Microbial Pathogen Detection

Waterborne diseases, such as cholera, dysentery, and giardiasis, are one of the primary causes of waterborne diseases worldwide caused by microbial pathogens in water. The presence of microbial pathogens like *E. coli*, *Vibrio cholerae*, and *Cryptosporidium* is important in terms of drinking water safety and preventing waterborne disease outbreaks. The use of fluorescent sensors as a method of microbial pathogen detection has become a common phenomenon because of their sensitivity and capability to deliver real-time data [41].

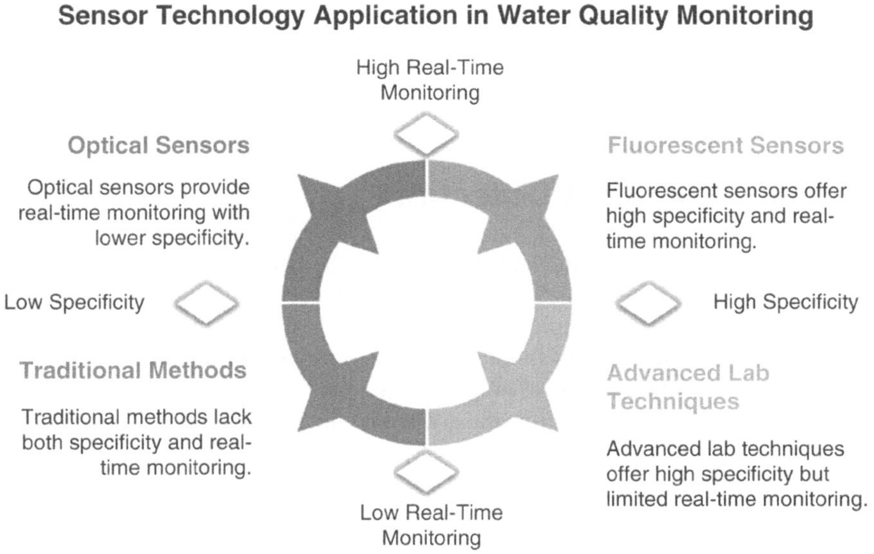

FIGURE 4.4 A comparison of various water quality monitoring methods, highlighting the trade-offs between specificity and real-time data acquisition across different sensor technologies.

4.3.1.1 Fluorescence for Microbial Detection

Fluorescent microbial pathogen detector sensors are based on the principles of fluorescent probes that have the potential to bind to the pathogen or identify the inherent fluorescence of microbial components. The main benefit of fluorescence sensors in pathogen detection is that they can give in situ and real-time readings of microbial concentrations. When light hits the target pathogens, the fluorescence emission of the target pathogens can be directly correlated with that concentration of the microbial species in the sample [42].

4.3.1.2 Detection of E. coli

E. coli is an organism that is among the frequent indicators of fecal contamination, and its existence in water may indicate the possibility of other dangerous pathogens to be found in water. The most common approach of developing fluorescent sensors that detect *E. coli* involves the utilization of antibodies or nucleic acid probes that specifically attach to *E. coli* cells. These probes are frequently conjugated with fluorescent dyes, which give off light on excitation by a given wavelength of light. Fluorescence is directly proportional to bacteria concentration in the sampled water, and thus, quantitative real-time detection is achievable.

An example of a common fluorescent sensor-based *E. coli* detection system is to label antibodies or nucleic acid probes that specifically bind to *E. coli* surface proteins or the genetic material of *E. coli*. The binding of these probes to the bacteria results in fluorescence whose values can be detected using a photodetector. This can be used to instantly identify and measure *E. coli* in water samples, which is important in the monitoring of water and safeguarding the health of the population.

4.3.1.3 Other Pathogen Detection

Fluorescent sensors can also be used to identify a broad variety of other pathogenic microorganisms such as *Vibrio cholerae* (causative agent of cholera) and *Cryptosporidium* (a protozoan parasite that causes gastrointestinal disease) in addition to *E. coli*. The pathogens may be readily visualized due to their natural fluorescence or by the combination of a particular probe with their DNA, proteins, or other biomolecules. Fluorescent sensors have been used to detect such harmful microorganisms at a field level promptly, and this is most useful especially in those areas that have limited access to laboratory facilities. Fluorescent sensors can also distinguish between living and dead microorganisms; this is essential to comprehend the health hazard of contamination. This capability can be accomplished by applying fluorescent viability stains that label live cells in contrast to dead cells, enabling a better evaluation of the presence of microbial contamination in water [42].

4.3.2 Chemical Pollutant Detection

Water has the chemical pollutants of the water, which are the heavy metals, pesticides, and industrial chemicals that are harmful to human health and the environment. Conventional techniques used to analyze these contaminants, including atomic absorption spectroscopy (AAS) or high-performance liquid chromatography (HPLC), involve costly laboratory tools as well as take a lot of time to analyze the samples. Another manner of detecting chemical pollutants in a rapid and on-site fashion is using optical and fluorescent sensors, which can offer real-time data, which could be used to take immediate action [43].

4.3.2.1 Heavy Metal Detection

Lead, mercury, cadmium, and arsenic are heavy metal elements and are very toxic and may accumulate in the water bodies through the discharge of industries, agricultural runoffs, and mining. These metals are not biodegradable and may take long before they are cleared in the environment, thus creating long-term health hazards to human beings and aquatic life. These metals are of great importance, as they need to be detected to provide safe water conditions and avoid exposure to such dangerous substances.

Heavy metal sensors, which are optical sensors, usually operate on the principle of measuring the absorption spectra of the metals. Different metals absorb different colors, and their characteristics enable the identification and determination of their concentration in a sample. Also, we can selectively bind metal ions with fluorescent chelating agents, and this will cause the fluorescence to increase in direct proportion to the concentration of the metal within the sample. These sensors are highly sensitive and are able to detect metal ions at very low levels suitable to environmental monitoring and regulatory compliance [44].

As one example, mercury-detecting sensors are built to utilize mercury-specific chelating agents, which combine with mercury ions and fluoresce when subjected to light of certain wavelengths. The fluorescence is directly proportional to the concentration of mercury in the solution, and thus real-time monitoring of mercury pollution in water is possible.

4.3.2.2 Pesticide Detection

Agricultural practices may cause water to be contaminated by the leaching of pesticides into water bodies, and because of water runoff, drinking water sources and aquatic ecosystems will be polluted. The residues of pesticides may be detrimental to human health, inducing cancer and disturbing hormones and brain injuries. Pesticide residue in water is also being measured with fluorescent and optical sensors, which offer a quick and effective way of measuring water quality.

Pesticide sensors based on optical detection are based on the light absorption or scattering characteristics, which may change as the pesticide reacts with light. Specific pesticide molecules can also be selectively bound with fluorescent probes to produce a signal in terms of fluorescence proportional to the presence of the pesticide. The advantage of these sensors is that they can analyze various pesticide remnants at the same time; this gives a more detailed analysis of water quality [45].

4.3.3 Turbidity and Suspended Solids

Turbidity is used to describe the haze or mistiness of a fluid that is due to suspended solids, particles, microorganisms, and organic matter. Turbidity is thought to be a characteristic of pollution, and high levels of turbidity may mean the presence of pollutants like microbial pathogens, heavy metals, or organic matter. The turbidity also has an impact on the effectiveness of the water treatment process since it may disrupt the process of filtration and disinfection [46,47].

4.3.4 pH and Dissolved Oxygen Measurement

Two of the most significant parameters that are strictly observed in the analysis of water quality are pH and dissolved oxygen (DO). pH has an impact on the solubility and toxicity of most contaminants, and DO is a crucial parameter in the analysis of aquatic organisms and the well-being of the aquatic environment. The use of fluorescent and optical sensors is an efficient method of measuring such parameters in real time, which can help in monitoring the water quality of both natural and artificial water bodies [48].

4.4 ADVANTAGES OF FLUORESCENT AND OPTICAL SENSORS

The introduction of fluorescent and optical sensors has transformed the water quality monitoring process by providing some of the main benefits over the traditional ones. With the world being more worried about water safety and environmental sustainability than ever before, the need to have quicker, more precise, and cost-effective water quality monitoring solutions is never high (Figure 4.5). These sensors have high sensitivity to detect waterborne contaminants in real time, which offers useful information to inform decision-making in water management, environmental protection, and the health of the people. The key benefits of including fluorescent and optical sensors in the water quality monitoring systems that we will discuss in this chapter include real-time monitoring, high sensitivity and specificity, low cost, portability, and low sample preparation [6].

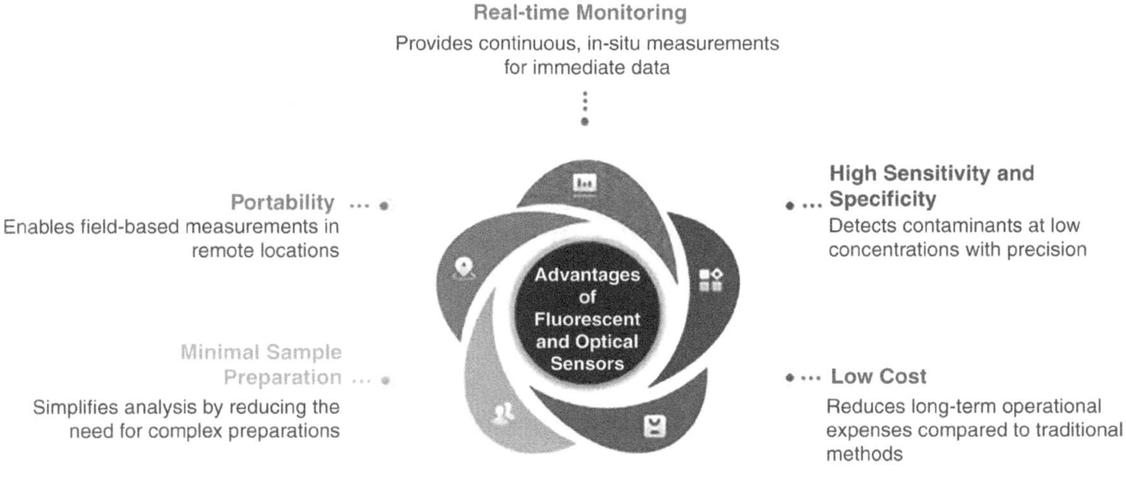

FIGURE 4.5 Key advantages of modern water monitoring technologies, including real-time data, portability, high sensitivity, minimal sample preparation, and cost-effectiveness.

4.4.1 Real-Time Monitoring

The possibility of real-time monitoring of water quality is one of the greatest benefits of fluorescent and optical sensors. Conventional water testing techniques, e.g., microbiological culture, laboratory-based chemical testing, take a lot of time to collect samples, transport samples, process, and analyze. Such approaches usually require hours or even days before giving results, thus complicating the detection of contamination events and response to them.

Conversely, fluorescent and optical sensors offer continuous and in situ measurements and allow collecting water quality data on a real-time basis. This enables water quality professionals, researchers, and those in charge of public health to keep a check on the contaminants as they are introduced into a water system so that corrective actions can be taken promptly.

As an example, with microbial contamination, like *E. coli* or *Cryptosporidium*, that was detected, real-time sensors can notify authorities about the presence of dangerous microorganisms as soon as they are detected [49,50]. This real-time response is essential in averting the outbreak of waterborne diseases and ensuring healthy people.

Furthermore, real-time monitoring may also be essential in dynamic conditions like in the wastewater treatment plant or in the natural water bodies. With wastewater treatment, say, operators have to continually modify processes in order to maximize the elimination of contaminants. Real-time monitoring enables the operators to identify changes in the quality of water when they occur, thus enabling the treatment processes to be adjusted to the changed water quality and the water quality to be maintained [51].

4.4.2 High Sensitivity and Specificity

Fluorescent and optical sensors are characterized by high sensitivity and high specificity, which makes them priceless tools in detecting contaminants at very low concentrations. This is particularly significant with respect to fluorescence-based sensors, since they are capable of detecting contaminants at parts per billion (ppb) or even parts per trillion (ppt) levels, which is frequently far below the detection limits of conventional detection techniques.

As an example, in the process of detecting heavy metal such as mercury or lead, the traditional laboratory means might need a high volume of samples and extensive analysis. Fluorescent and optical sensors, however, are capable of detecting these metals with a minimum concentration as low as a few parts per trillion, giving early warning of the existence of toxic materials in water. This is particularly important in the setting where minute quantities of pollutants can be a serious hazard to human health or aquatic life [5].

Fluorescent and optical sensors are also highly specific in the detection of individual contaminants in addition to its sensitivity. Fluorescent probes may be trained to specifically bind to a particular pathogen, chemical, or pollutant, preventing the chances of cross-interference by other compounds present in the sample. This particularity plays a key role in the case of complex water matrices, as there can be several contaminants at once. An example is that fluorescent sensors can be engineered to be pesticide- and pharmaceutical-selective, which can allow reliable and focused monitoring [52].

4.4.3 Low Cost

In comparison to the traditional water quality testing methods, the initial installation cost of fluorescent and optical sensors may be expensive, but the overall cost of their operation is generally lower in the long run. Conventional techniques, like microbiological culture or chemical testing, are also resource-intensive in both terms of special reagents, laboratory instruments, and skilled manpower. Not to mention, these techniques may require transportation and processing of water samples, which can be very expensive logistically.

By comparison, fluorescent and optical sensors need briefer maintenance intervals and can often prove quite cost-effective in the long term. These sensors are built to be used continuously, and once fitted, it can be used to check the water quality over long durations without any calibration or expensive consumables. This gives them an appealing choice in applications that involve constant monitoring, including municipal water treatment plants, industrial facilities, or environmental monitoring programs [52].

In addition, the capability of performing on-site monitoring without necessarily transporting samples to a laboratory also saves money by removing transportation, sample preparation, and laboratory analysis charges. This is especially helpful in remote or inaccessible locations where there is a shortage of resources to monitor the quality of the water.

The cost saving of using fluorescent and optical sensors in most cases can cover the initial investment, especially when one considers the possibility of early detection of contaminants and avoidance of waterborne disease epidemics or environmental pollution.

4.4.4 Portability

The ease of use of fluorescent and optical sensors is another major strength that increases the use of the sensors in various water quality monitoring systems. Most of these sensors are small, portable, and geared toward field measurements, and therefore, they can be applied in distant areas or dynamic conditions where the conventional laboratory testing would not be feasible.

Portable fluorescence sensors can be used as an example; they can be placed in a natural water body, like a river, lake, or coastal waters, to determine the presence of any microbial pathogens or chemical pollution. These sensors have the capacity to give real-time information on the water quality even in regions where laboratory facilities are not readily available or even absent. Likewise, turbidity or suspended solids in water bodies associated with industrial effluents, agricultural runoffs, or city effluents can be monitored using portable optical sensors [53].

These sensors can be transported directly to the site of monitoring, which minimizes the expenses associated with the transportation of other samples, which can be expensive and time-consuming. Moreover, it can be used to sample more often, guaranteeing the continuous updating of water quality data and prompting the reaction to the contamination incidents.

4.4.5 Minimal Sample Preparation

Conventional methods of water quality testing may necessitate significant sample treatment, which may include filtration, chemical treatment, or reagent treatment, which further complicates and increases the cost and time of analysis. Fluorescent and optical sensors, in contrast, do not frequently require elaborate preparations to be able to carry out analyses on raw water samples [54].

Considering the example of turbidity sensors, the sensors observe the light scattering on the water sample, and it is not necessary to filter or chemically treat the sample. Likewise, fluorescent sensors to measure microbial pathogens, heavy metals, or organic contaminants can scan through the sample of water in its natural form, eliminating time-sensitive sample processing.

This capability of analyzing the samples of water directly on the ground without the use of sophisticated laboratory facilities does not only make the monitoring exercise easy but also enhances the effectiveness of water quality tests. Real-time results can be obtained by water quality professionals, and therefore decisions can be made faster and water resources can be managed more efficiently.

4.5 CHALLENGES AND LIMITATIONS

Although fluorescent and optical sensors have immense benefits in water quality monitoring, they have not escaped without challenges and limitations. Like any technology, the implementation of these sensors in the physical world may be faced with problems that can impact their performance, reliability, and usability in general. In this chapter, we shall discuss the major problems and limitations of fluorescent and optical sensors, which are interference with other substances, environmental factors, limitations of the detection distance, and lifetime concerns. It is important to understand these limitations in order to fully optimize the effectiveness of these sensors and make sure that they are implemented properly in water quality monitoring systems [55,56].

4.6 FUTURE DIRECTIONS

The future of the fluorescent and optical sensors for water contaminant detection is in a number of evolutions:

- **Multi-Parameter Sensing**: The capabilities of real-time water quality monitoring systems will be widened by creating sensors that have the capacity to record various contaminants at the same time, with the help of either multi-wavelength excitation or the use of advanced optical fibers.
- **Nanotechnology**: Sensitivity and selectivity of sensors can be enhanced by nanomaterial integration and can be used to detect minute levels of contaminants within complex water matrices.
- **Wireless and Remote Sensing**: By combining wireless technology, e.g., IoT sensors, with optical and fluorescent sensors, it will be possible to constantly monitor water quality remotely in real time, thereby simplifying the management of water resources and responding to contamination occurrences in the shortest possible time.
- **Artificial Intelligence**: The machine learning algorithms would be used to process sensor data, and more precise predictions of contamination occurrences would be made, as well as help to implement more efficient decision-making regarding the water treatment processes.

4.7 CONCLUSION

Fluorescent and optical sensors have emerged as powerful tools for real-time monitoring of waterborne contaminants, offering significant advantages over conventional analytical methods. Their exceptional sensitivity enables the detection of pollutants at trace levels, reaching parts per billion and even parts per trillion, which is critical for early warning, rapid intervention, and the prevention of waterborne disease outbreaks. The portability of these sensors allows on-site deployment across diverse settings, from municipal water treatment facilities to remote and field environments, eliminating reliance on costly and time-consuming laboratory analyses. Rapid data acquisition further enhances decision-making, particularly in pollution-prone aquatic systems and water-scarce regions where timely responses are essential. Despite challenges such as interference from coexisting substances, environmental variability, and limited detection ranges, ongoing technological advancements are addressing these constraints. Integration with nanomaterials, AI, and wireless platforms is markedly improving sensor sensitivity, selectivity, data interpretation, and remote monitoring capabilities. Collectively, these innovations position fluorescent and optical sensors as transformative technologies for sustainable water quality management and global public health protection.

REFERENCES

[1] S. Madhav, A. Ahamad, A.K. Singh, J. Kushawaha, J.S. Chauhan, S. Sharma, P. Singh, Water pollutants: sources and impact on the environment and human health. In D. Pooja, P. Kumar, P. Singh, & S. Patil (Eds.), *Sensors in Water Pollutants Monitoring: Role of Material* (pp. 43–62). Springer Nature Singapore. Advanced Functional Materials and Sensors Series. https://doi.org/10.1007/978-981-15-0671-0_4.

[2] Y. Fan, X. Wang, T. Funk, I. Rashid, B. Herman, N. Bompoti, M.S. Mahmud, M. Chrysochoou, M. Yang, T.M. Vadas, A critical review for real-time continuous soil monitoring: Advantages, challenges, and perspectives, *Environmental Science & Technology* 56(19) (2022) 13546–13564.

[3] N. Thirumalaivasan, S. Nangan, D. Verma, M. Shellaiah, S. Ali, S. Rajendran, K. Kanagaraj, R. Pothu, R. Boddula, A.B. Radwan, Exploring the diverse nanomaterials employed in dental prosthesis and implant techniques: An overview, *Nanotechnology Reviews* 14(1) (2025) 20250140.

[4] P.B. Stockwell, Abstracts of papers presented at the 2008 pittsburgh conference, *Journal of Automated Methods and Management in Chemistry* 2008 (2009) 502953.

[5] P. Nath, K.R. Mahtaba, A. Ray, Fluorescence-based portable assays for detection of biological and chemical analytes, *Sensors* 23(11) (2023) 5053.

[6] M. Kumar, K. Khamis, R. Stevens, D.M. Hannah, C. Bradley, In-situ optical water quality monitoring sensors—applications, challenges, and future opportunities, *Frontiers in Water* 6 (2024) 1380133.

[7] N. Thirumalaivasan, Collagen-composite scaffolds for alveolar bone and dental tissue regeneration: Advances in material development and clinical applications—A *narrative review, Dentistry Journal* 13(9) (2025) 396.

[8] M. Pradeep, A. Cecil, N. Suresh, R. Govindasamy, N. Thirumalaivasan, Exploring plant-derived carbon dots: Dual-functional nanomaterials for endodontic infections and enhanced bio-imaging, *Journal of Oral Biology and Craniofacial Research* 15(6) (2025) 1408–1414.

[9] N. Thirumalaivasan, Letter to the Editor "assessing the performance of virus-mimic nanoparticles against other nano-drug delivery systems in efficiency, biocompatibility, and patient results", *Oral Oncology Reports* 10 (2024) 100436.

[10] Y.-L. Oon, Y.-S. Oon, M. Ayaz, M. Deng, L. Li, K. Song, Waterborne pathogens detection technologies: advances, challenges, and future perspectives, *Frontiers in Microbiology* 14 (2023) 1286923.

[11] S.A. Popa, V. Herman, E. Tîrziu, A. Morar, A. Ban-Cucerzan, M. Imre, R.-T. Pătrînjan, K. Imre, Public health risk of Campylobacter spp. isolated from slaughterhouse and retail poultry meat: Prevalence and antimicrobial resistance profiles, *Pathogens* 14(4) (2025) 316.

[12] H.-Y. Lee, N. Thirumalaivasan, S.-P. Wu, H_2O_2-responsive boronic ester-modified mesoporous silica nanocarrier for TfR mediated tumor-specific drug delivery applications, *ACS Applied Bio Materials 8*(7) (2025) 6079–6087. https://doi.org/10.1021/acsabm.5c00645.

[13] P.A. Bower, C.O. Scopel, E.T. Jensen, M.M. Depas, S.L. McLellan, Detection of genetic markers of fecal indicator bacteria in Lake Michigan and determination of their relationship to *Escherichia coli* densities using standard microbiological methods, *Applied and Environmental Microbiology* 71(12) (2005) 8305–8313.

[14] A.I. Okoh, T. Sibanda, S.S. Gusha, Inadequately treated wastewater as a source of human enteric viruses in the environment, *International Journal of Environmental Research and Public Health* 7(6) (2010) 2620–2637.

[15] A. Hemphill, N. Müller, J. Müller, Comparative pathobiology of the intestinal protozoan parasites *Giardia lamblia*, *Entamoeba histolytica*, and *Cryptosporidium parvum*, *Pathogens* 8(3) (2019) 116.

[16] K. Kanagaraj, J. Rebek Jr., Y. Yu, Control of reactivity and selectivity in isomerization and rearrangement reactions inside confined spaces, *Chemical Communications* 59 (2023) 9056–9065.

[17] D. Verma, M. Okhawilai, N. Senthilkumar, N. Thirumalaivasan, A. Incharoensakdi, H. Uyama, Polyhydroxybutyrate/poly (ε-caprolactone)-based electrospun membranes loaded with amoxicillin-potassium clavulanate halloysite nanotubes for biomedical applications, *Polymer International* 74(1) (2025) 54–65.

[18] M. Balali-Mood, K. Naseri, Z. Tahergorabi, M.R. Khazdair, M. Sadeghi, Toxic mechanisms of five heavy metals: mercury, lead, chromium, cadmium, and arsenic, *Frontiers in Pharmacology* 12 (2021) 643972.

[19] S.M. Rad, A.K. Ray, S. Barghi, Water pollution and agriculture pesticide, *Clean Technologies* 4(4) (2022) 1088–1102.

[20] S.V. Kumar, C. Shobana, D. Rohini, J. Manjunathan, R. Ramasubburayan, N. Thirumalaivasan, S. Prakash, B. Usharani, Mukia maderaspatana (L.) M. Roem. phytoconstituents: Unveiling their anticancer potentials against hepatocellular carcinoma cells, *South African Journal of Botany* 169 (2024) 27–37.

[21] T. Matos, M. Martins, R. Henriques, L. Goncalves, A review of methods and instruments to monitor turbidity and suspended sediment concentration, *Journal of Water Process Engineering* 64 (2024) 105624.

[22] N. Thirumalaivasan, K. Kanagaraj, S. Nangan, R. Pothu, S.P. Rajendra, P. Karuppiah, R. Boddula, Bioactive Hydrogels (Bio-HyGs): Emerging Trends in Drug Delivery and Wound Healing Applications, *Polymers for Advanced Technologies* 36(4) (2025) e70132.

[23] R. Ramasubburayan, K. Kanagaraj, L. Gnanasekaran, N. Thirumalaivasan, N. Senthilkumar, Correction: Greener approach supported nitrogen-infused carbon dots for biocompatible cellular markers and dluorescent ink based spray-assisted fingerprint analysis, *Waste and Biomass Valorization* 15(12) (2024) 7139–7139.

[24] A.O. Olatunji, J.A. Olaboye, C.C. Maha, T.O. Kolawole, S. Abdul, Environmental microbiology and public health: Advanced strategies for mitigating waterborne and airborne pathogens to prevent disease, *International Medical Science Research Journal* 4(7) (2024) 756–770.

[25] K. Kanagaraj, R. Wang, M.-K. Zhao, P. Ballester, J. Rebek, Jr., Y. Yu, Selective binding and isomerization of oximes in a self-assembled capsule, *Journal of the American Chemical Society* 145(10) (2023) 5816–5823.

[26] F. Sun, Y. Mu, K.M. Leung, H. Su, F. Wu, H. Chang, China is establishing its water quality standards for enhancing protection of aquatic life in freshwater ecosystems, *Environmental Science & Policy* 124 (2021) 413–422.

[27] J. Ji, J.A. Schanzle, M.B. Tabacco, Real-time detection of bacterial contamination in dynamic aqueous environments using optical sensors, *Analytical chemistry* 76(5) (2004) 1411–1418.

[28] C. Xiao, W. Wu, W. Liang, D. Zhou, K. Kanagaraj, G. Cheng, D. Su, Z. Zhong, J.J. Chruma, C. Yang, Redox-triggered chirality switching and guest-capture/release with a pillar[6]arene-based molecular universal joint, *Angewandte Chemie International Edition* 59(21) (2020) 8094–8098.

[29] N. Turro, V. Ramamurthy, W. Cherry, W. Farneth, The effect of wavelength on organic photoreactions in solution. Reactions from upper excited states, *Chemical Reviews* 78(2) (1978) 125–145.

[30] A.P. De Silva, H.N. Gunaratne, T. Gunnlaugsson, A.J. Huxley, C.P. McCoy, J.T. Rademacher, T.E. Rice, Signaling recognition events with fluorescent sensors and switches, *Chemical Reviews* 97(5) (1997) 1515–1566.

[31] K. Kanagaraj, K. Bavanidevi, T.J. Chow, K. Pitchumani, Selective "turn-off" fluorescent sensing of mercury ion using aminocyclodextrin: 3-hydroxy-N-phenyl-2-naphthamide complex in aqueous solution, *RSC Advances* 4 (2014) 11714–11722.

[32] K. Kanagaraj, C. Xiao, M. Rao, C. Fan, V. Borovkov, G. Cheng, D. Zhou, Z. Zhong, D. Su, X. Yu, J. Yao, T. Hao, W. Wu, *J.J. Chruma, C. Yang,* A quinoline-appended cyclodextrin derivative as a highly selective receptor and colorimetric probe for nucleotides, *iScience* 23(3) (2020) 100927.

[33] M.-H. Nguyen, T.-N. Nguyen, D.-Q. Do, H.-H. Nguyen, Q.-M. Phung, N. Thirumalaivasan, S.-P. Wu, T.-H. Dinh, A highly selective fluorescent anthracene-based chemosensor for imaging Zn2+ in living cells and zebrafish, *Inorganic Chemistry Communications* 115 (2020) 107882.

[34] J. Manjunathan, M. Revathi, H. Sowmya, S. Meenambiga, R. Sudha, B. Prakash, S. Nangan, L. Gnanasekaran, N. Thirumalaivasan, Recent advancements in nanotechnological approaches for pollution monitoring and environmental sustainability, *Clean Technologies and Environmental Policy* 26(11) (2024) 3667–3683.

[35] G. Gauglitz, Direct optical sensors: principles and selected applications, *Analytical and Bioanalytical Chemistry* 381(1) (2005) 141–155.

[36] K. Kanagaraj, K. Pitchumani, Highly selective "turn-on" fluorescent and colorimetric sensing of fluoride ion using 2-(2-hydroxyphenyl)-2,3-dihydroquinolin-4(1H)-one based on excited-state proton transfer, *Chemistry – An Asian Journal* 9 (2014) 146–152.

[37] B.A. Samy, K. Raman, S. Velayutham, N. Senthilkumar, N. Thirumalaivasan, K. Kanagaraj, R. Pothu, R. Boddula, A.B. Radwan, N. Al-Qahtani, Natural product extract fractions as potential arthritis treatments: A detailed analysis using in-silico, in-vivo, and in-vitro methods, *International Immunopharmacology* 144 (2025) 113595.

[38] A. MariaJoseph, S. Nangan, D. Verma, L. Gnanasekaran, S. Rajendran, T. Natesan, P. Pattananuwat, M. Okhawilai, Rational coupling of selective electrochemical oxidation and reduction reactions for in-situ value-added chemical generation, *Fuel* 367 (2024) 131408.

[39] M. Shellaiah, K.W. Sun, N. Thirumalaivasan, M. Bhushan, A. Murugan, Sensing utilities of cesium lead halide perovskites and composites: A comprehensive review, *Sensors* 24(8) (2024) 2504.

[40] H. Gunter, C. Bradley, D.M. Hannah, S. Manaseki-Holland, R. Stevens, K. Khamis, Advances in quantifying microbial contamination in potable water: Potential of fluorescence-based sensor technology, *Wiley Interdisciplinary Reviews: Water* 10(1) (2023) e1622.

[41] C.I. Wen, C.N. Yu, N. Thirumalaivasan, H. Hiramatsu, 532-nm-excited hyper-Raman spectroscopy of amino acids, *Journal of Raman Spectroscopy* 52(3) (2021) 641–654.

[42] J. Zhang, M. Zhou, X. Li, Y. Fan, J. Li, K. Lu, H. Wen, J. Ren, Recent advances of fluorescent sensors for bacteria detection-A review, *Talanta* 254 (2023) 124133.

[43] P. Zhang, M. Yang, J. Lan, Y. Huang, J. Zhang, S. Huang, Y. Yang, J. Ru, Water quality degradation due to heavy metal contamination: Health impacts and eco-friendly approaches for heavy metal remediation, *Toxics* 11(10) (2023) 828.

[44] C. Peng, W. Liang, J. Ji, C. Fan, K. Kanagaraj, W. Wu, G. Cheng, D. Su, Z. Zhong, C. Yang, Pyrene-tiaraed pillar[5]arene: Strong intramolecular excimer emission applicable for photo-writing, *Chinese Chemical Letters* 32(1) (2021) 345–348.

[45] K. Kanagaraj, A. Affrose, S. Sivakolunthu, K. Pitchumani, Highly selective fluorescent sensing of fenitrothion using per-6-amino-β-cyclodextrin:Eu(III) complex, *Biosensors and Bioelectronics* 35 (2012) 452–455.

[46] F. Kordbacheh, G. Heidari, Water pollutants and approaches for their removal, *Materials Chemistry Horizons* 2(2) (2023) 139–153.

[47] B.P. Wong, B. Kerkez, Real-time environmental sensor data: An application to water quality using web services, *Environmental Modelling & Software* 84 (2016) 505–517.

[48] K.S. Aryamol, K. Kanagaraj, S. Nangan, J.T. Haponiuk, M. Okhawilai, S. Pandiaraj, M.B. Hanif, A. Alodhayb, S. Thomas, N. Thirumalaivasan, S. Gopi, Recent advances of carbon pathways for sustainable environment development, Environmental Research 250 (2024) 118513.

[49] M. Shamsudin, H. Sapingi, M. Aziz, Optical glass micro-fibre based transducers for dissolved oxygen sensing and monitoring: an overview, *Journal of Physics*: Conference Series, IOP Publishing, (2020) 012004.

[50] H.B. Glasgow, J.M. Burkholder, R.E. Reed, A.J. Lewitus, J.E. Kleinman, Real-time remote monitoring of water quality: a review of current applications, and advancements in sensor, telemetry, and computing technologies, *Journal of Experimental Marine Biology and Ecology* 300(1–2) (2004) 409–448.

[51] K. Kanagaraj, K. Lin, W. Wu, G. Gao, Z. Zhong, D. Su, C. Yang, Chiral buckybowl molecules, *Symmetry* 9(9) (2017) 174 (1–25).

[52] D.A. Abraham, A.-D. Li, A. Sanmugam, M.A. Wadaan, A. Baabbad, K. Kanagaraj, K. Karuppasamy, T. Maiyalagan, H.-S. Kim, D. Vikraman, Highly sensitive and selective detection of dopamine using atomic layer deposited HfO_2 ultra-thin films, *Electrochimica Acta* 497 (2024) 144574.

[53] S.N. Zainurin, W.Z. Wan Ismail, S.N.I. Mahamud, I. Ismail, J. Jamaludin, K.N.Z. Ariffin, W.M. Wan Ahmad Kamil, Advancements in monitoring water quality based on various sensing methods: a systematic review, *International Journal of Environmental Research and Public Health* 19(21) (2022) 14080.

[54] N. Thirumalaivasan, K. Kanagaraj, K. Logesh, S. Chandrasekaran, S. Kumar, R. Subramanian, N. Senthilkumar, A. Kumar, V.J. Angadi, A.A. Al-Kahtani, Exploring luminescent carbon dots derived from syrup bottle waste and curcumin for potential antimicrobial and bioimaging applications, *Chemosphere* 354 (2024) 141592.

[55] O.S. Wolfbeis, Materials for fluorescence-based optical chemical sensors, *Journal of Materials Chemistry* 15(27–28) (2005) 2657–2669.

[56] M.A. Butt, G.S. Voronkov, E.P. Grakhova, R.V. Kutluyarov, N.L. Kazanskiy, S.N. Khonina, Environmental monitoring: A comprehensive review on optical waveguide and fiber-based sensors, *Biosensors* 12(11) (2022) 1038.

Supramolecular Chemistry in Selective Capture and Purification of Water Pollutants

5

Shanmugam Paramasivam, Kumarasamy Chandraprakash, Arumugam Thangamani, and Kuppusamy Kanagaraj

5.1 INTRODUCTION: THE SPECIFICITY CHALLENGE IN WATER REMEDIATION

Water is life-sustaining and a constituent of human civilization, but the contamination of the water by industrial, agricultural, and domestic wastes has emerged as a significant challenge in the world today. The initial design of traditional water treatment systems was to eliminate pathogens, suspended particles, and organic substances that are easily degraded. Nevertheless, there is a growing trend of persistent micropollutants (pharmaceuticals, endocrine disruptors, and perfluoroalkyl substances (PFASs)) in the modern water sources at very low levels (ppb or ppt levels). Traditional adsorption media such as activated carbon in a granular form are typically not molecule-selective enough to be able to get these trace contaminants; this is where more selective and sophisticated methods of purification are necessary [1].

Selective adsorption is now an important demand in contemporary water remediation systems. Specifically designed materials may be used to target harmful micropollutants more efficiently, as they do not require interference with the large amounts of natural organic matter found in water. This selectivity is also able to regenerate adsorbents easily through reversible reactions, which give the ability to use the adsorbents again and again and limit operational expenses. Moreover, selective material may be used in extracting valuable metals or to safely dispose of dangerous contaminants [2,3]. There can also be possibilities of real-time monitoring of the pollution and better management of the environment using these properties.

DOI: 10.1201/9781003741626-5

The idea of supramolecular chemistry will become a viable solution to these problems, as it is aimed at developing molecular recognition when non-covalent interactions are involved. Based on biological systems, such as enzyme–substrate or antigen–antibody interactions, supramolecular methods employ hydrogen bonding, electrostatic forces, hydrophobic interactions, and van der Waals forces in order to selectively bind [4]. Rather, synthetic host molecules are programmed to contain preorganized cavities which are able to identify and trap the characteristic pollutant molecules in water and imitate the natural recognition processes.

This biomimetic approach is a novel water purification technology that is marked by the abandonment of conventional water purification methods that build on surface area to achieve adsorption but employs smart molecular capture technologies. Artificial receptors like cyclodextrins and calixarenes are able to selectively capture contaminants, be recycled to be used again, and possibly be built into sensors that will enable real-time monitoring of water quality [5–9]. All these developments demonstrate the promise of supramolecular chemistry as a starting point for the development of sustainable, efficient, and selective water remediation technologies.

Figure 5.1 is an abstract idea of a chapter on the use of supramolecular chemistry in water remediation. The principal idea is that synthetic receptors are designed, i.e., cyclodextrins and calixarenes, which are engineered to have particular binding sites to identify and pick up pollutant molecules in water. These receptors are non-covalently working, and this is what makes them selectively trap contaminants. One of the benefits mentioned is the possibility to regenerate those receptors, and this makes them regain their binding ability to be reused again, which will facilitate the treatment processes to be sustainable. The end product of this technology is that it will track the water quality in real time by identifying pollutants, resulting in the efficient purification of water as well as a safe or clean supply of water due to the methodology of molecular capture that is precise and specific.

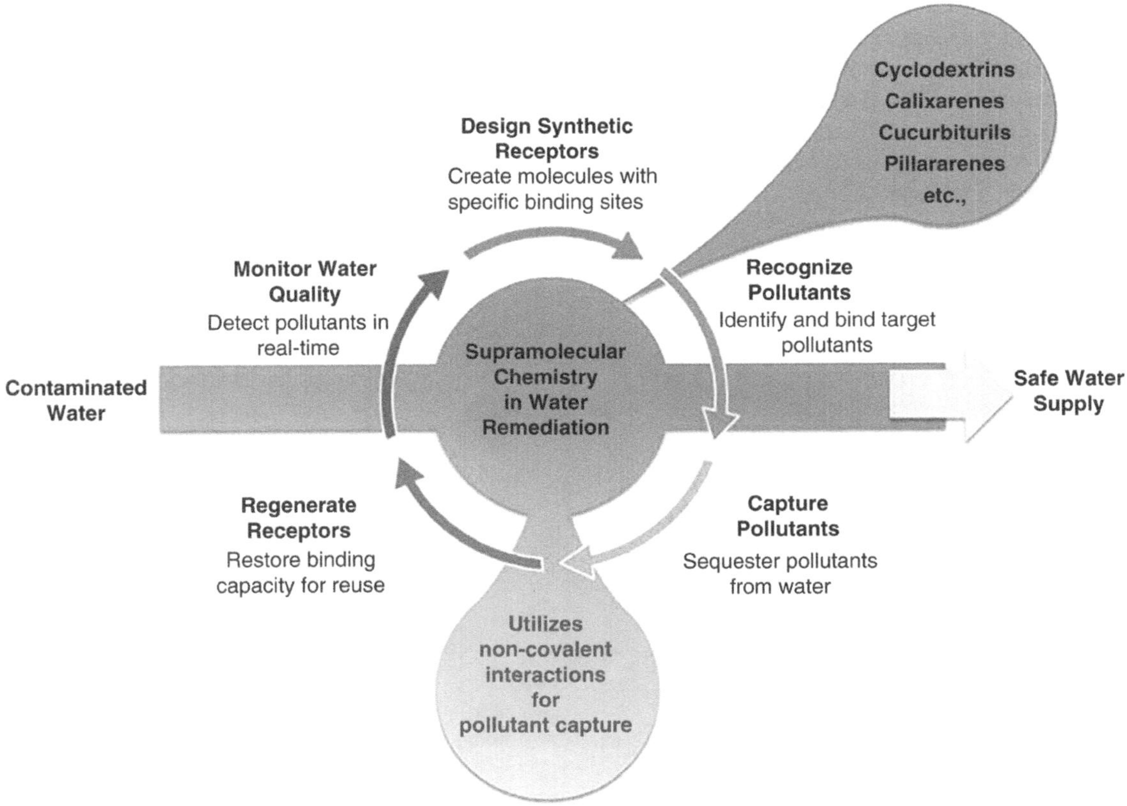

FIGURE 5.1 Summary of this chapter about the design of synthetic receptors to clean water, in which pollutant recognition and capture were demonstrated by supramolecular chemistry.

5.2 THE SUPRAMOLECULAR TOOLBOX: NON-COVALENT INTERACTIONS FOR MOLECULAR RECOGNITION

The shift from a traditional, non-selective adsorption process to precision capture in water remediation is based on a fundamental shift of design philosophy (Figure 5.2). Instead of the high surface area and physisorption, in the supramolecular approach, the sophisticated language of molecular recognition is utilized (Figure 5.3). The efficacy of any synthetic host, be it a macrocycle, a framework, or an imprinted polymer, is not determined by any one force controlling the system but by subtle, synergistic action of suite of non-covalent interactions. For the environmental engineer or chemist exploring this field, not only is a deep and intuitive understanding of these forces not purely academic, but it is also the essential key to the rational design of effective and well-functioning next-generation remediation agents. This section is designed to take apart some of the main guns in the supramolecular toolbox, encountering not textbook definitions but instead the impact this has in the complex world of water purification.

The arsenal of non-covalent interactions can be divided into those of origin, strength, and directionality, playing a different but often cooperative role. Hydrogen bonding is one of the most effective and directional tools available (Figure 5.3). It involves a donor (hydrogen atom, which is covalently bonded to an electronegative atom, e.g., O or N) and an acceptor (another electronegative atom with a lone pair). In a water environment the competitive hydrogen bonding provided by the enormous excess of water molecules is a major challenge and will often undermine this interaction. However, this very challenge can be put to good use in terms of selectivity [7, 10, 11]. A suitably designed host cavity can partially desolvate the target pollutant and self and form a microenvironment where a preorganized set of donor and acceptor groups can form several cooperative hydrogen bonds with a specific guest. This is of prime importance for the recognition of pharmaceuticals, which are often decorated with hydrogen-bonding pharmacophores. For example, a drug such as the beta-blocker atenolol, which has secondary amine and alcohol functionalities, is able to be selectively complexed with a host consisting of complementary crown ether-type oxygen atoms (acceptor) and aromatic hydroxyl groups (donor) in a good competition with the aqueous environment.

For capture of pervasive hydrophobic contaminants, the workhorses are the pi–pi interactions and the hydrophobic effect. pi–pi interactions occur as attractive forces between aromatic rings, and their strength

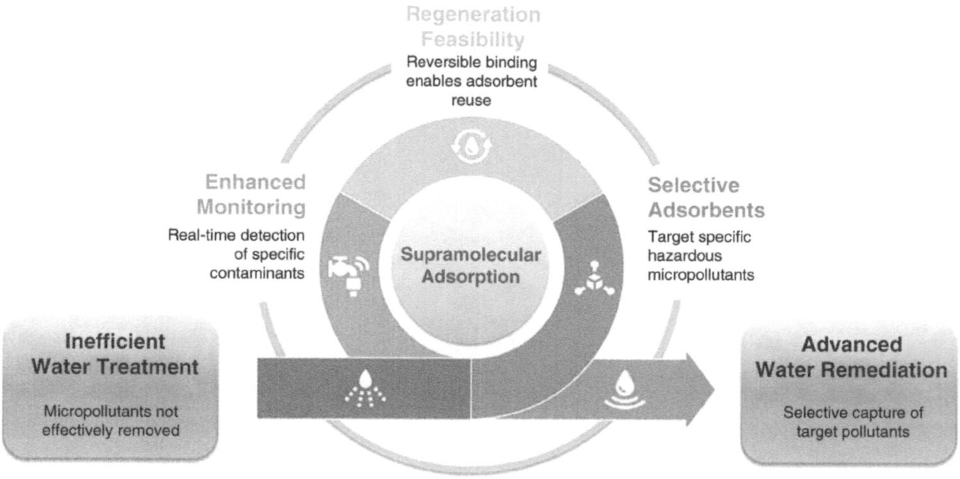

FIGURE 5.2 The diagram compares the inefficient removal of micropollutants with a supramolecular solution with selective adsorbents that allow regeneration and in situ detection of a contaminant.

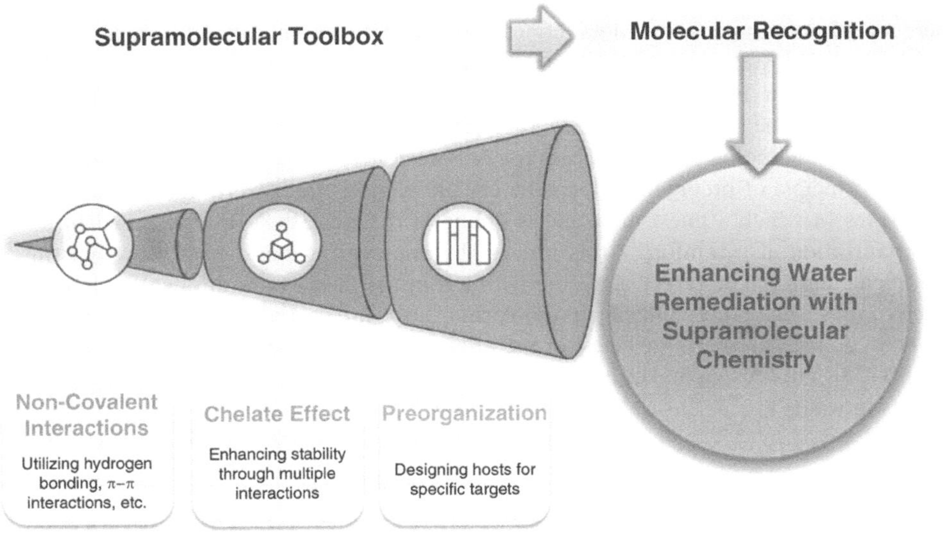

FIGURE 5.3 Illustration of the basic concepts of supramolecular chemistry, showing the importance of non-covalent interactions and the chelate effect in order to increase molecular recognition for water treatment.

is dependent on the electron density of the systems involved. Electron-rich (pi basic) hosts exhibit high affinity for electron-poor (pi acidic) guests and vice versa. This is of vital importance for the attack of the huge number of planar, aromatic pollutants, ranging from polycyclic aromatic hydrocarbons (PAHs) such as pyrene to industrial dyes and many endocrine-disrupting compounds (EDCs) such as bisphenol A. The geometrical fit is very important; if the ring system is contorted or off by even a bit, then the association will be much weaker. The hydrophobic effect, which is also a widely misunderstood "force," is an entropic driving force. It explains the tendency of non-polar molecules or molecular moieties to aggregate in aqueous solution in order to minimize the thermodynamically unfavorable disruption of the hydrogen-bonding network of water. This is the main mechanism responsible for non-polar pesticide (e.g., atrazine) or solvent molecule inclusion into hydrophobic cavities of hosts such as cyclodextrins or calixarenes [12]. The binding event is essentially a type of molecular dehydration, in which the release of the ordered water molecules out of the host cavity and guest surface into the bulk solvent is an appreciable entropic gain, driving the association [13–15].

When it comes to the ionic pollutants, a group of which contain some of the most toxic heavy metals and oxyanions, electrostatic and ion–dipole interactions are in the forefront. These are the Coulombic attractions of full or partial charges. For cationic heavy metals such as Pb^{2+}, Cd^{2+}, or Hg^{2+}, the host is functionalized with negatively charged groups, i.e., carboxylic groups, sulfonic groups, or polar groups with high dipole moments such as carbonyl groups in cucurbiturils. The main thing is to go beyond the simple non-specific ion exchange. A supramolecular design uses these charged groups as part of a preorganized design which can partially cage the ion and improve selectivity due to the chelate effect. For oxyanions such as chromate (CrO_4^{2-}) or arsenate (AsO_4^{3-}), however, the opposite is used; hosts have protonated amine groups or other positive features, which provide good electrostatic attraction [16,17]. This approach is also important for the capture of ionizable organic molecules such as herbicides (e.g., the dicationic paraquat) or antibiotics, the charge state of which is pH-dependent and thus tunable with respect to the capture and release.

Metal coordination is among the strongest types of interaction to achieve ultra-high affinity and selectivity in supramolecular pollutant capture. In this strategy, pollutants bind to the Lewis acidic metal centers directly incorporated into host frameworks to form strong coordination bonds, which act as precise docking sites. According to the hard–soft acid–base (HSAB) principle, soft heavy metal ions like Hg^{2+} ion or Pd^{2+} ion prefer to interact with soft donor atoms like sulfur or nitrogen. For example, zirconium-based metal–organic frameworks (MOFs) such as UiO-66 that have thiol (–SH) functional groups have exceptional mercury adsorption capacity. Metal coordination can also be used to absorb anionic pollutants at unsaturated metal sites, especially in lanthanide-based MOFs. In addition, weaker van der Waals interactions add cumulatively to host–guest stabilization in the confines of molecular cavities [18–20].

The efficiency of the supramolecular capture, however, stems from the sum of multiple interactions instead of a single binding mechanism. This type of cooperative phenomenon is explained by the chelate effect and the principle of preorganization. The chelate effect is the description of the increased stability of complexes in which a guest molecule is bound at different sites in a host structure at the same time. Preorganization also reinforces this process by engineering hosts with a pre-existing geometry that is complementary to the target molecule with a minimal loss of entropy while binding. Similar to enzyme active sites, such preorganized supramolecular hosts are spatially arranged binding sites, such as hydrogen-bond donors, aromatic surfaces, or metal-coordination centers, so that multi-point recognition is ensured [21–23]. This design makes it possible to selectively capture specific pollutants in very complex environments in water.

5.3 KEY SUPRAMOLECULAR PLATFORMS FOR POLLUTANT CAPTURE

The theoretical principles of molecular recognition, ruled by non-covalent interactions, find their physical expression in a variety of designed materials of very diverse and fast growth (Figure 5.4). For the environmental engineer or materials chemist, the right choice of supramolecular platform is a key decision to be made considering a combination of binding affinity, selectivity, scalability of synthesis, stability of materials, and ease of regeneration. These platforms can be separated into three broad subfamilies: discrete-type receptors of macrocyclic polymers, which provide individual but often isolated binding sites; polymeric networking systems, which add to the molecular recognition and transform into a bulk network material property; and crystalline self-assembled frameworks, which surpass the others by far in order and porosity. Each category has its own unique benefits as well as challenges that make them particularly well suited for certain applications in the complicated business of water purification [24].

5.3.1 Synthetic Macrocycles: Precision Tools for Molecular Recognition

Discrete macrocycles are basic parts of supramolecular chemistry in which their well-defined cyclic structures define internal cavities that are able to selectively bind guest molecules. The size, shape, and chemical functionality of these cavities regulate their interactions with the host and make them selective to different pollutants in the environment. Among them, cyclodextrins (CDs) are the most developed supramolecular hosts for water treatment in the market. These cyclic oligosaccharides, which are alpha, beta, and gamma forms with six, seven, and eight glucose molecules, respectively, have a hydrophobic cavity on the inside and a hydrophilic surface on the outside and thus have the ability to encapsulate organic contaminants such as pesticides, endocrine disruptors, and volatile organic compounds. Because of the high-water solubility of CDs, many studies have been conducted to immobilize them via cross-linking into insoluble polymers or grafted onto solid supports such as silica nanoparticles and polymer membranes in order to obtain high adsorption capacity and practical applicability in filtration and separation systems [25–27].

Other types of macrocycles like calix[n]arenes, cucurbit[n]urils (CB[n]s), and pillar[n]arenes offer other structural versatility for the capture of pollutants. Calixarenes have tunable cup-shaped cavities that can be functionalized at the upper and lower rims of the cavity, allowing the selective binding of heavy

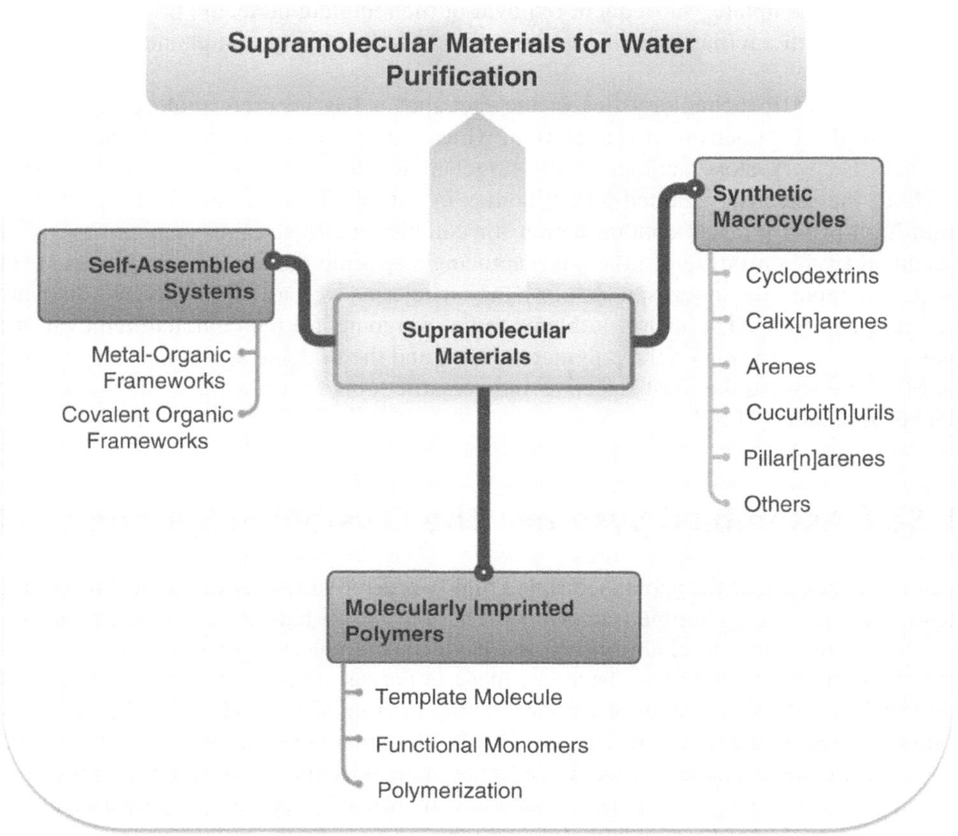

FIGURE 5.4 The diagram categorizes supramolecular water purification materials, showcasing self-assembled systems, various synthetic macrocycles, and the preparation process of molecularly imprinted polymers.

metal ions and aromatic pollutants [28]. Cucurbiturils have rigid hydrophobic cavities with carbonyl-lined portals that interact very strongly with cationic species, making them very good for removing compounds like paraquat and protonated pharmaceuticals. Pillar[n]arenes, which are a newer class of macrocycles that have symmetrical pillar-like structures, have rigid, electron-rich cavities that can be made into stimuli-responsive systems that permit pollutants to be captured and released. Collectively, these are examples of macrocycle hosts that demonstrate the potential of supramolecular design in the development of advanced and selective water purification technologies [29].

5.3.2 Molecularly Imprinted Polymers: Biomimetic Plastic Antibodies

Molecularly imprinted polymers (MIPs) are a radically different approach to supramolecular capture. Instead of synthesizing a predefined host and finding out what binds to it, the MIPs are designed around the pollutant of interest itself and produce a synthetic polymeric antibody. The process is triggered by the forming of a pre-polymerization complex: The target molecule (the pollutant to be removed, such as a particular antibiotic) is combined with functional monomers which have complementary groups which are designed to interact with the target through non-covalent interactions (e.g., methacrylic acid to hydrogen-bond with a ketone group). A high concentration of cross-linker is then added, and the mixture is made to polymerize, thereby "freezing" the functional monomers in place around the template. Subsequent removal of the template molecule produces the network of porous polymers with cavities, which are sterically and chemically complementary to the original target [30].

The power of MIP technology lies in the fact that it has accurate molecular memory. These cavities contain three-dimensional recognition, which can discriminate not only between the target molecule but also very close analogs to the target molecule (and in this respect the selectivity is comparable to the selectivity gained by antibodies in nature). This makes MIPs extremely valuable for certain high-priority target contamination for complex waste streams such as separation of one pharmaceutical from a mixture of others or capturing a specific metabolite of a pesticide. They may be fabricated as monoliths, microspheres, or nanoparticles as well as can be grafted onto surfaces or entrapped in membranes. The major problems to be overcome are the complete removal of the template, non-specific adsorption on the polymer surface, and the scalable synthesis. In spite of these difficulties, MIPs are among the most selective and versatile tools for ultra-trace analysis and extraction of particular pollutants [31,32].

5.3.3 Self-Assembled Systems: The Crystalline Sponge Strategy

This category makes use of the spontaneous and highly ordered assembly of molecular building blocks into extended and porous crystalline frameworks. These materials have the combined benefits of having supreme porosity, molecular-level uniformity, and having supramolecular binding sites incorporated into every unit of the structure. MOFs have become super-star porous materials during the past 20 years. They are made from ions of a metal atom or cluster (playing the role of the nodes) coordinated to multitopic organic linkers (playing the role of the struts). This is a modular construction which offers near-infinite tunability. The supramolecular design can be reflected in the selection of linker, because the linker with a large, p-conjugated system will favor the interaction with aromatic dyes; the linker with an amine group can be used as a Lewis base to interact with metal ions or CO_2; and the linker with a thiol group can be used to directly coordinate with soft heavy metals such as Hg^{2+}. The resulting materials often have staggering surface areas (in some cases more than 7,000 m^2/g), which means that an enormous amount of binding sites is available.

The historical Achilles' heel of MOFs has been hydrothermal stability, where many of them are prone to water-induced degradation. This challenge is being addressed head on in the form of robust frameworks using high-valent metal cations such as Zr^{4+} (e.g., the UiO-66 series), Fe^{3+} (e.g., MIL-100/101), and Al^{3+} (e.g., MIL-53) to form strong and hydrolysis-resistant metal–oxygen bonds [33]. These stable MOFs are now being seriously examined for actual use in water treatment, showing amazing capacities for pharmaceuticals, dyes, and heavy metals. Covalent organic frameworks (COFs) are similar to MOFs but are formed completely from light elements (C, H, O, N, B) linked together by strong irreversible covalent bonds (e.g., boroxine, imine, hydrazone linkages). Such organic composition often gives them a better chemical and thermal stability than many MOFs. Like MOFs, they are highly porous and crystalline, but they are completely organic and thus can be precisely and predictably functionalized in the pore walls. Amino groups, hydroxy groups, chelating groups, and catalytic sites can be directly incorporated into the organic struts de novo in the synthesis process to ensure the uniform distribution of functionality in the material [34–37]. This allows one to design COFs with pores that are not only empty channels but also lined with specific supramolecular recognition sites specific to a given target pollutant, such as a PFAS molecule or a specific metal ion. While synthesis and processability issues still need to be overcome, COFs are on the cutting edge of designed porous organic materials for selective sequestration in that they provide a compelling combination of stability, porosity, and chemical precision.

5.4 TARGETING SPECIFIC POLLUTANT CLASSES: CASE STUDIES IN SUPRAMOLECULAR DESIGN

The real performance of any water treatment technology is against the real-life contaminants. The supramolecular chemistry suffers only when its principles are applied to the practical solutions of the most urgent problems of pollution (Figure 5.5). These contaminants do not form a simple collection of contaminants with a single set of chemistries, toxicologies, and environmental behaviors; each class of contaminants requires a specific approach. This part provides more in-depth insight into using supramolecular platforms on three severe types of water pollutants and how the molecular toolbox is used to solve a particular recognition and capture problem. It is with these case studies that we shift beyond the esoteric ideas as they are applied into real design solutions that have become the keystone of environmental remediation [38].

5.4.1 Heavy Metals: Leveraging Coordination Chemistry and Preorganization

The impact of heavy metal contamination by industrial processes and mining, as well as electronic waste, is also an environmental threat because of the toxicity, persistence, and bioaccumulation of ions, including Pb^{2+}, Hg^{2+}, Cd^{2+}, and As species. Traditional types of remediation such as precipitation and ion-exchange resins are usually not selective and produce secondary wastes. A more effective alternative approach is supramolecular methods, which exploit preorganized ligand systems which preferentially bind target metal ions by coordination chemistry. Calixarenes, functionalized with either carboxylate, amide, or thiol, are a type of macrocyclic host that forms well-defined binding cavities, attracting soft metal ions, such as Pb^{2+} and Hg^{2+}, selectively. This principle is extended to MOFs, where Hg^{2+} is selectively trapped in pores by thiol moieties, e.g., thiol-functionalized UiO-66-$(SH)_2$, in which zirconium nodes are sufficiently

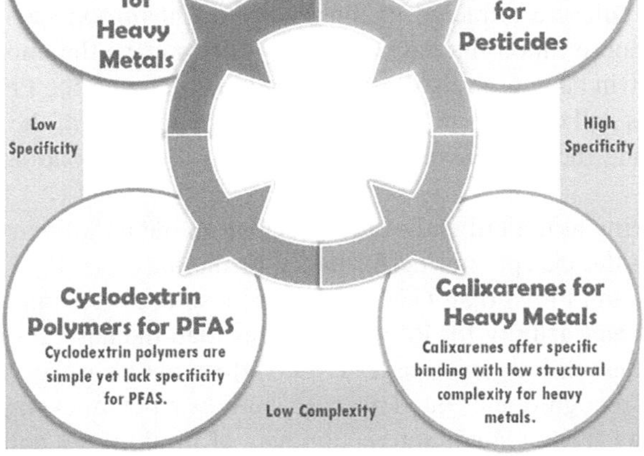

FIGURE 5.5 The diagram evaluates different supramolecular materials, MIPs, and cyclodextrin polymers to trap the heavy metals, pesticides, and PFASs, emphasizing their specificity and complexity of structure.

robust to offer structural stability and thiol groups isolate Hg^{2+}. The ability to maintain high adsorption capacity, rapid kinetics, and the capability to have the mercury concentrations reduced to traces of mercury are evidence of the high level of efficacy of supramolecular design in the evolution of the water purification process [39].

5.4.2 Organic Micropollutants (OMPs): The Challenge of Complexity and Trace Concentration

The organic micropollutants (OMPs), including pharmaceuticals, personal care products, endocrine disruptors, and pesticides, are found in water at extremely low levels yet have high ecological impact and are hard to dispose of through standard methods of treating water. One of the effective approaches is provided in supramolecular chemistry in which host molecules capture the contaminants selectively via molecular recognition. As an example, cucurbiturils are remarkably effective in the binding of β-blocker drugs such as propanol due to hydrophobic inclusion and ion–dipole interactions, and cyclodextrin-based polymers are remarkably efficient in the binding of anionic drugs such as diclofenac since the polymers have hydrophobic inclusion and electrostatic interactions [40–44]. Endocrine disruptors such as bisphenol A are well accommodated in 2-cyclodextrin cavities through the influence of hydrophobic interactions, although pillararenes give superior binding with additional hydrogen bonding. In structurally diverse pesticides, MIPs form highly selective cavities complementing the shape and functional groups of target molecules, allowing the efficient and selective elimination of pollutants.

5.4.3 Per- and Polyfluoroalkyl Substances (PFASs): Confronting the "Forever Chemicals"

PFASs represent perhaps the most formidable challenge in water treatment today. Their extreme persistence is due to the incredibly strong C–F bonds, and their amphiphilic nature – a hydrophobic, lipophobic fluorinated carbon tail coupled with a hydrophilic head group (e.g., carboxylate, sulfonate) – makes them highly mobile and bioavailable. Traditional sorbents like activated carbon show limited capacity and are easily outcompeted by background organic matter.

Supramolecular chemistry has taken up this challenge with a brilliantly biomimetic approach – dual recognition. The masterful work in the development of the beta-cyclodextrin polymers cross-linked with decafluorobiphenyl is a master class in rational design. This material is not based on one interaction. First, the fluorinated cross-linker makes the environment of the polymer matrix "fluorous." The fluorocarbon chains of the PFAS molecule are attracted to this environment through a special and strong type of hydrophobic interaction: fluoro–fluoro interactions. This is countering the unique chemistry of the PFAS tail. Second, the cyclodextrin cavities are still available to interact with the PFAS head group. While the nature of this interaction is still debated, it is probably through inclusion of the head group or the junction between the head group and the tail, possibly stabilized by hydrogen bonding with the CD's secondary hydroxyl rim [45].

A dual approach looking at the tail end of the PFAS molecule as well as the head end is something revolutionary. It describes the exceptional performance of the material: high binding capacities (orders of magnitude higher for some PFAS than GAC), fast kinetics and, most importantly, excellent selectivity. The material has a greater affinity for PFAS molecules than the natural organic matter that plagues activated carbon, because natural organic matter (NOM) does not have the specific fluorinated tail that is able to provide the first recognition event. This case study is a potent example of how supramolecular chemistry, by moving away from a single interaction way of thinking by holistic and multi-point host–guest design, could offer a solution to a problem that has stymied conventional technologies. It seems like a very promising approach for addressing the whole class of PFAS contaminants, from the long chain to the emerging short chain, and is a beacon of light for the future of selective water remediation [46,47].

5.5 FROM BENCHTOP TO REAL-WORLD IMPLEMENTATION: CHALLENGES AND FRONTIERS

The transition from a high-performing material in a controlled laboratory experiment to an in situ technology in a municipal water treatment plant or industry effluent line is a large and often unmerciful chasm (Figure 5.6). While the scientific literature is full of supramolecular adsorbents with unbelievably high binding constants, staggering capacities, and impressive selectivity in purified water spiked with one single pollutant, these are only the starting point. The real measure of these sophisticated materials can be found in their performance in the complex, chaotic, and chemically hostile environment of water matrices in the real world. Translating the promise of supramolecular into reality means a fundamental change in outlook from chemistry to holistic engineering and systems thinking [48]. This transition is marked by major interconnected challenges that characterize the state-of-the-art frontiers of research and development in this area.

5.5.1 Fouling, Stability, and Longevity in Complex Matrices

The major problem with the high-performance adsorbent materials is how they behave in real-water systems, where complex mixtures of NOM, inorganic ions, microorganisms, and particulates exist, and not in simplified laboratory environments. NOM can foul, whether by non-specific adsorption on exterior surfaces or in larger pores, and block selective adsorption sites, and greatly decrease capacity with repeated cycles. To solve this, there can be materials of hydrophilic surfaces to repel NOM or size-selective pores to permit target pollutant but inhibit larger organic molecules. Also, the long-term chemical and mechanical

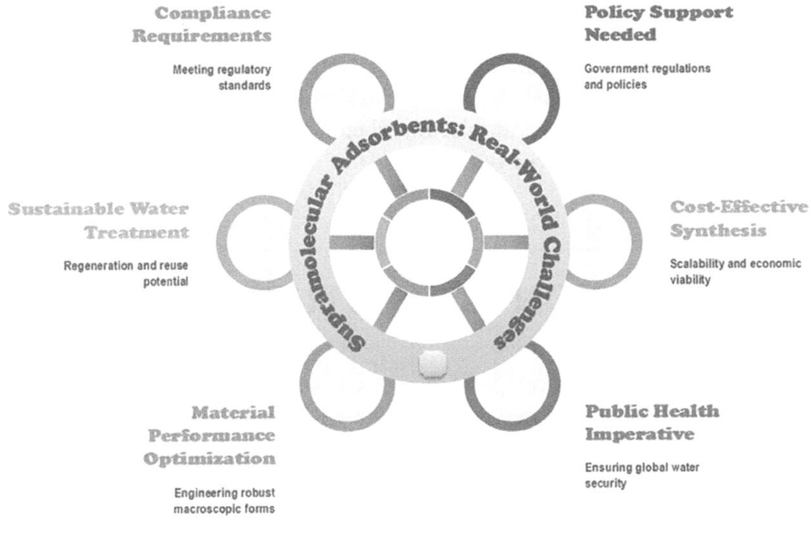

FIGURE 5.6 Illustration of the path to sustainable water treatment, connecting technical factors like material synthesis with regulatory needs and the overarching public health imperative.

stability of such materials in continuous flow is unknown as structures could hydrolyze, oxidize or collapse in the course of numerous adsorption–desorption cycles. Even though certain materials including zirconium-based MOFs demonstrate promising stability, their stability with respect to persevering through real environmental conditions over prolonged periods has yet to be assessed fully [48–51].

5.5.2 The Regeneration Imperative and Closing the Loop

Supramolecular systems provide a significant advantage for regeneration and reuse that is essential for economic and environmental sustainability by providing the possibility of the multiple removal of pollutants without degradation of the host material. Efficient regeneration is focused on breaking the host/guest complex by stimuli that result in the reduction of non-covalent interactions. Some typical methods include pH swing, i.e., protonation of ionizable groups to liberate trapped metals; solvent washing, where polarity change is achieved to alter hydrophobic or electrostatic interactions; electrochemical stimulation, i.e., stimulation of applied potential for changes in oxidation state of the host/guest species to disrupt binding; and temperature swing, i.e., application of mild heat to favor desorption. Beyond recovering adsorbent functionality, regeneration also concentrates pollutants located within dilute streams into smaller quantities that can be more easily overcome using downstream methods (advanced oxidation, electrochemical degradation, and incineration) and therefore also supports the development of a more sustainable closed-loop remediation process [52–55].

5.5.3 The Multi-Pollutant Reality and Future Frontiers

Water contamination usually does not include a single pollutant, so multi-pollutant capture systems are necessary for actual water treatment. One strategy is the development of "omnivorous" host materials with broad affinity for a class of contaminants, such as the polymers composed of beta-cyclodextrins that are capable of capturing diverse PFAS compounds. Another possible way is the treatment train concept, where different selective materials are stacked one after another, such as a thiol-based MOF for Hg^{2+} removal, a cucurbituril-based material for cationic pharmaceuticals, and an MIP for particular pesticides. This is a modular design that enables treatment systems to be optimized to a contaminant profile of a water source. Looking into the future, it can be expected that there will be some developments in stimuli-responsive smart materials that are able to selectively capture pollutants on demand with sensing and remediation integration, and materials will be able to signal when they need to regenerate. Conquering these challenges will decide whether supramolecular chemistry is developed from an academic idea into a useful technology for global water security [56,57].

5.6 CONCLUSION AND FUTURE PERSPECTIVES

Supramolecular chemistry has been developed from a basic scientific idea to an effective tool in environmental engineering, especially to advanced and selective water purification. By means of precise molecular recognition mechanisms, it allows the targeted removal of very persistent pollutants that are not removed by other means of treatment. New developments have resulted in the development of innovative host molecules with excellent selectivity and efficiency to several environmental contaminants. The future direction of this field is interdisciplinary collaboration between chemistry, materials science, and environmental engineering, and some directions of development are as follows: biomimetic design of hosts, stimuli-responsive "smart" adsorbents for controlled pollutant capture and release, hybrid design combining supramolecular hosts with membrane technologies, and computational methods like machine learning and molecular modeling to enhance host–guest design. Collectively, these advances tip the supramolecular chemistry field as a revolutionizing technology that is able to better define efficiency, selectivity, and sustainability in modern water treatment systems.

REFERENCES

[1] K.G. Steinhäuser, A. Von Gleich, M. Große Ophoff, W. Körner, The necessity of a global binding framework for sustainable management of chemicals and materials—Interactions with climate and biodiversity, *Sustainable Chemistry* 3(2) (2022) 205–237.

[2] N. Thirumalaivasan, P. Venkatesan, S.-P. Wu, Highly selective turn-on probe for H 2 S with imaging applications in vitro and in vivo, *New Journal of Chemistry* 41(22) (2017) 13510–13515.

[3] K. Kanagaraj, A. Affrose, S. Sivakolunthu, K. Pitchumani, Highly selective fluorescent sensing of fenitrothion using per-6-amino-β-cyclodextrin:Eu(III) complex, *Biosensors and Bioelectronics* 35 (2012) 452–455.

[4] F. Liu, J.J. Pignatello, R. Sun, X. Guan, F. Xiao, A comprehensive review of novel adsorbents for per-and polyfluoroalkyl substances in water, *ACS ES&T Water* 4(4) (2024) 1191–1205.

[5] M. Harun-Ur-Rashid, I. Jahan, T. Foyez, A.B. Imran, Bio-inspired nanomaterials for micro/nanodevices: a new era in biomedical applications, *Micromachines* 14(9) (2023) 1786.

[6] S. Noor, A. Fatima, I. Fatima, A. Javed, *Antigen-antibody interaction in cell signaling, Cell Signaling*, CRC Press 2025, pp. 92–125.

[7] K. Kanagaraj, K. Pitchumani, Highly selective "turn-on" fluorescent and colorimetric sensing of fluoride ion using 2-(2-hydroxyphenyl)-2,3-dihydroquinolin-4(1H)-one based on excited-state proton transfer, *Chemistry – An Asian Journal* 9 (2014) 146–152.

[8] J.M. Lehn, Supramolecular chemistry—Scope and perspectives molecules, supermolecules, and molecular devices (Nobel Lecture), *Angewandte Chemie International Edition in English* 27(1) (1988) 89–112.

[9] L. Escobar, P. Ballester, Molecular recognition in water using macrocyclic synthetic receptors, *Chemical Reviews* 121(4) (2021) 2445–2514.

[10] K. Kanagaraj, J. Rebek Jr., Y. Yu, Control of reactivity and selectivity in isomerization and rearrangement reactions inside confined spaces, *Chemical Communications* 59 (2023) 9056–9065.

[11] P. Peluso, B. Chankvetadze, Recognition in the domain of molecular chirality: From noncovalent interactions to separation of enantiomers, *Chemical Reviews* 122(16) (2022) 13235–13400.

[12] K. Kanagaraj, R. Wang, M.-K. Zhao, P. Ballester, J. Rebek, Jr., Y. Yu, Selective binding and isomerization of oximes in a self-assembled capsule, *Journal of the American Chemical Society* 145(10) (2023) 5816–5823.

[13] H.-Y. Lee, N. Thirumalaivasan, S.-P. Wu, H_2O_2-responsive boronic ester-modified mesoporous silica nanocarrier for TfR mediated tumor-specific drug delivery applications, *ACS Applied Bio Materials* 8(7) (2025), 6079–6087. https://doi.org/10.1021/acsabm.5c00645.

[14] K. Kanagaraj, J. Rebek, Jr., Y. Yu, Self-assembled hexameric capsule: A highly β-selective O-glycosylation reaction enabled via proton wire mechanism, *Green Synthesis and Catalysis (GSC)* 4(1) (2023) 7–9.

[15] F.M. Haque, J.S. Ishibashi, C.A. Lidston, H. Shao, F.S. Bates, A.B. Chang, G.W. Coates, C.J. Cramer, P.J. Dauenhauer, W.R. Dichtel, Defining the macromolecules of tomorrow through synergistic sustainable polymer research, *Chemical Reviews* 122(6) (2022) 6322–6373.

[16] M.-M. Tang, K. Kanagaraj, J. Rebek, Jr., Y. Yu, Role of rim functions in recognition and selectivity of small molecule guests in water-soluble cavitand hosts, *Chemistry – An Asian Journal* 17(15) (2022) e202200466.

[17] J. Chen, Z. Gu, J.M. Perez-Aguilar, Y. Luo, K. Tian, Y. Luo, Molecular dynamics simulations reveal efficient heavy metal ion removal by two-dimensional Cu-THQ metal-organic framework membrane, *Scientific Reports* 15(1) (2025) 199.

[18] J. Ji, X. Wei, W. Wu, C. Fan, D. Zhou, K. Kanagaraj, G. Cheng, K. Luo, X. Meng, C. Yang, The more the slower: self-inhibition in supramolecular chirality induction, memory, erasure, and reversion, *Journal of the American Chemical Society* 144(3) (2022) 1455–1463.

[19] B. Farasati Far, *Antioxidant activity of metal-organic frameworks, Logic for Metal–Organic Framework Selection: MOFs for Biomedical Applications*, ACS Publications 2024, pp. 107–146.

[20] C. Peng, W. Liang, J. Ji, C. Fan, K. Kanagaraj, W. Wu, G. Cheng, D. Su, Z. Zhong, C. Yang, Pyrene-tiaraed pillar[5]arene: Strong intramolecular excimer emission applicable for photo-writing, *Chinese Chemical Letters* 32(1) (2021) 345–348.

[21] K.A. Schug, W. Lindner, Noncovalent binding between guanidinium and anionic groups: Focus on biological- and synthetic-based arginine/guanidinium interactions with phosph [on] ate and sulf [on] ate residues, *Chemical Reviews* 105(1) (2005) 67–114.

[22] C. Xiao, W. Wu, W. Liang, D. Zhou, K. Kanagaraj, G. Cheng, D. Su, Z. Zhong, J.J. Chruma, C. Yang, Redox-triggered chirality switching and guest-capture/release with a pillar[6]arene-based molecular universal joint, *Angewandte Chemie International Edition* 59(21) (2020) 8094–8098.

[23] X. Niu, R. Zhao, S. Yan, Z. Pang, H. Li, X. Yang, K. Wang, Chiral materials: Progress, applications, and prospects, *Small* 19(38) (2023) 2303059.

[24] W.-Y. Yang, K. Anusuyadevi, P.-H. Lu, N. Thirumalaivasan, W.-H. Lin, S. Velmathi, S.-P. Wu, A two photon fluorescent probe for highly selective detection and endogenous imaging of hydrogen sulfide, *Spectrochimica Acta Part A: Molecular and Biomolecular Spectroscopy* 273 (2022) 121043.

[25] K. Kanagaraj, C. Xiao, M. Rao, C. Fan, V. Borovkov, G. Cheng, D. Zhou, Z. Zhong, D. Su, X. Yu, J. Yao, T. Hao, W. Wu, *J.J. Chruma, C. Yang,* A quinoline-appended cyclodextrin derivative as a highly selective receptor and colorimetric probe for nucleotides, *iScience* 23(3) (2020) 100927.

[26] S. Nangan, K. Kanagaraj, G. Kaarthikeyan, A. Kumar, M. Ubaidullah, B. Pandit, R. Govindasamy, T. Natesan, Sustainable preparation of luminescent carbon dots from syringe waste and hyaluronic acid for cellular imaging and antimicrobial applications, *Environmental Research* 237 (2023) 116990.

[27] A. Nair, K. Kuppusamy, S. Nangan, T. Natesan, J.T. Haponiuk, S. Thomas, R. Ramasubburayan, L. Gnanasekaran, M. Selvaraj, S. Gopi, Multifunctional natural derived carbon quantum dots from Withania somnifera (L.)–Antiviral activities against SARS-CoV-2 pseudoviron, *Environmental Research* 239 (2023) 117366.

[28] N. Thirumalaivasan, S. Mahapatra, G. Ramanathan, A. Kumar, T. Raja, M. Muthuramamoorthy, B. Pandit, S. Pandiaraj, S. Prakash, Exploring antimicrobial and biocompatible applications of eco-friendly fluorescent carbon dots derived from fast-food packaging waste transformation, *Environmental Research* 244 (2024) 117888.

[29] R.K. Gupta, Z. Wang, B. Mohan, C.H. Tung, D. Sun, Advancements in atomically precise nanocluster protected by thiacalix [4] arene, *Advanced Materials* 36(45) (2024) 2410054.

[30] M.V. Balli, F. Biedermann, L. Prodi, P. Picchetti, Supramolecular chemistry for optical detection and delivery applications in living plants, *Chemical Society Reviews* 54(17) (2025), 7769–7869. https://doi.org/10.1039/D4CS00500G

[31] N. Thirumalaivasan, K. Kanagaraj, K. Logesh, S. Chandrasekaran, S. Kumar, R. Subramanian, N. Senthilkumar, A. Kumar, V.J. Angadi, A.A. Al-Kahtani, Exploring luminescent carbon dots derived from syrup bottle waste and curcumin for potential antimicrobial and bioimaging applications, *Chemosphere* 354 (2024) 141592.

[32] K. Kanagaraj, K. Lin, W. Wu, G. Gao, Z. Zhong, D. Su, C. Yang, Chiral buckybowl molecules, *Symmetry* 9(9) (2017) 174 (1–25).

[33] Z. Hassan, Y. Matt, S. Begum, M. Tsotsalas, S. Bräse, Assembly of molecular building blocks into integrated complex functional molecular systems: structuring matter made to order, *Advanced Functional Materials* 30(26) (2020) 1907625.

[34] K. Aryamol, K. Kanagaraj, S. Nangan, J.T. Haponiuk, M. Okhawilai, S. Pandiaraj, M.B. Hanif, A.N. Alodhayb, S. Thomas, N. Thirumalaivasan, Recent advances of carbon pathways for sustainable environment development, *Environmental Research* 250 (2024) 118513.

[35] K. Kanagaraj, K. Bavanidevi, T.J. Chow, K. Pitchumani, Selective "turn-off" fluorescent sensing of mercury ion using aminocyclodextrin: 3-hydroxy-N-phenyl-2-naphthamide complex in aqueous solution, *RSC Advances* 4 (2014) 11714–11722.

[36] D.A. Abraham, A.-D. Li, A. Sanmugam, M.A. Wadaan, A. Baabbad, K. Kanagaraj, K. Karuppasamy, T. Maiyalagan, H.-S. Kim, D. Vikraman, Highly sensitive and selective detection of dopamine using atomic layer deposited HfO2 ultra-thin films, *Electrochimica Acta* 497 (2024) 144574.

[37] P.J. Waller, F. Gándara, O.M. Yaghi, Chemistry of covalent organic frameworks, *Accounts of Chemical Research* 48(12) (2015) 3053–3063.

[38] Q. Lin, X. Ding, Y. Hou, W. Ali, Z. Li, X. Han, Z. Meng, Y. Sun, Y. Liu, Adsorption and separation technologies based on supramolecular macrocycles for water treatment, *Eco-Environment & Health* 3(3) (2024) 381–391.

[39] G.I. Edo, P.O. Samuel, G.O. Oloni, G.O. Ezekiel, V.O. Ikpekoro, P. Obasohan, J. Ongulu, C.F. Otunuya, A.R. Opiti, R.S. Ajakaye, Environmental persistence, bioaccumulation, and ecotoxicology of heavy metals, *Chemistry and Ecology* 40(3) (2024) 322–349.

[40] H.Q. Anh, I. Watanabe, L.H. Tuyen, N.M. Tue, N.T. Nghia, P.H. Viet, T.B. Minh, S. Takahashi, Comprehensive monitoring of more than 1000 organic micro-pollutants in drainage water: Case study in a rural village with end-of-life vehicle processing activities in northern Vietnam, *Water, Air, & Soil Pollution* 232(1) (2021) 14.

[41] N. Thirumalaivasan, L. Gnanasekaran, S. Kumar, R. Durvasulu, T. Sundaram, S. Rajendran, S. Nangan, K. Kanagaraj, Utilization of fungal and bacterial bioremediation techniques for the treatment of toxic waste and biowaste, *Frontiers in Materials* 11 (2024) 1416445.

[42] N. Thirumalaivasan, S. Nangan, K. Kanagaraj, S. Rajendran, Assessment of sustainability and environmental impacts of renewable energies: Focusing on biogas and biohydrogen (Biofuels) production, *Process Safety and Environmental Protection* 189 (2024) 467–485.

[43] N. Thirumalaivasan, S. Gopi, K. Karthik, S. Nangan, K. Kanagaraj, S. Rajendran, Nano-PCM materials: Bridging the gap in energy storage under fluctuating environmental conditions, *Process Safety and Environmental Protection* 189 (2024) 1003–1021.

[44] M.S. Kang, J.-H. Lee, K.S. Kim, Small toxic molecule detection and elimination using molecularly imprinted polymers (MIPs), *Biosensors* 15(6) (2025) 393.

[45] M.G. Evich, M.J. Davis, J.P. McCord, B. Acrey, J.A. Awkerman, D.R. Knappe, A.B. Lindstrom, T.F. Speth, C. Tebes-Stevens, M.J. Strynar, Per-and polyfluoroalkyl substances in the environment, *Science* 375(6580) (2022) eabg9065.

[46] J. Manjunathan, M. Revathi, H. Sowmya, S. Meenambiga, R. Sudha, B. Prakash, S. Nangan, L. Gnanasekaran, N. Thirumalaivasan, Recent advancements in nanotechnological approaches for pollution monitoring and environmental sustainability, *Clean Technologies and Environmental Policy* 26(11) (2024) 3667–3683.

[47] R. Ramasubburayan, K. Kanagaraj, L. Gnanasekaran, N. Thirumalaivasan, N. Senthilkumar, Correction: Greener approach supported nitrogen-infused carbon dots for biocompatible cellular markers and fluorescent ink based spray-assisted fingerprint analysis, *Waste and Biomass Valorization* 15(12) (2024) 7139–7139.

[48] Q. Wu, X. Yao, X. Duan, X. Chen, J. Zhang, Nanomedicine reimagined: Translational strategies for precision tumor theranostics, *Advanced Materials* 38(8) (2025) e202510293. https://doi.org/10.1002/adma.202510293.

[49] B.A. Samy, K. Raman, S. Velayutham, N. Senthilkumar, N. Thirumalaivasan, K. Kanagaraj, R. Pothu, R. Boddula, A.B. Radwan, N. Al-Qahtani, Natural product extract fractions as potential arthritis treatments: A detailed analysis using in-silico, in-vivo, and in-vitro methods, *International Immunopharmacology* 144 (2025) 113595.

[50] J.R. Sonawane, R. Jundale, A.A. Kulkarni, Continuous flow synthesis of metal nanowires: Protocols, engineering aspects of scale-up and applications, *Materials Horizons* 12(2) (2025) 364–400.

[51] S.A. Snyder, P. Westerhoff, Y. Yoon, D.L. Sedlak, Pharmaceuticals, personal care products, and endocrine disruptors in water: implications for the water industry, *Environmental Engineering Science* 20(5) (2003) 449–469.

[52] S. Mukherjee, S. Ghosh, S. Biswas, Amine-rich porous MOF nanocrystals for the selective capture of carcinogenic anions and organo-pollutants from the waste water environment at neutral pH, *ACS Applied Nano Materials* 6(23) (2023) 22231–22240.

[53] E. Ploetz, H. Engelke, U. Lächelt, S. Wuttke, The chemistry of reticular framework nanoparticles: MOF, ZIF, and COF materials, *Advanced Functional Materials* 30(41) (2020) 1909062.

[54] Y.-Q. Chen, H.-W. Guan, K. Kanagaraj, J. Rebek, Jr., Y. Yu, Metal coordination to a deep cavitand promotes binding selectivities in water, *Chinese Chemical Letters* 33(11) (2022) 4908–4911.

[55] N. Thirumalaivasan, K. Kanagaraj, S. Nangan, R. Pothu, S.P. Rajendra, P. Karuppiah, R. Boddula, Bioactive hydrogels (Bio-HyGs): Emerging trends in drug delivery and wound healing applications, *Polymers for Advanced Technologies* 36(4) (2025) e70132.

[56] D. Lacalamita, C. Mongioví, G. Crini, N. Morin-Crini, Cyclodextrins for the removal of per-and polyfluoroalkyl substances: a review, *Environmental Chemistry Letters* 23(6) (2025) 1713–1743. https://doi.org/10.1007/s10311-025-01868-x.

[57] N. Thirumalaivasan, S. Nangan, D. Verma, M. Shellaiah, S. Ali, S. Rajendran, K. Kanagaraj, R. Pothu, R. Boddula, A.B. Radwan, Exploring the diverse nanomaterials employed in dental prosthesis and implant techniques: An overview, *Nanotechnology Reviews* 14(1) (2025) 20250140.

6 Detection Strategies and Removal Technologies for Chiral Contaminants in Aquatic Environments

Kandavalli Babu, A.V. Roger Krupa Bhushan, Jesu Dakshin Thippiah K., Bairi Levi Rakshith, Lazarus Godson Asirvatham, G. Lawrance, and S. Vincent Sam Jebadurai

NOMENCLATURE

Symbol/Abbreviation	Full Form/Meaning
AOP	Advanced oxidation process
AI	Artificial intelligence
CD	Cyclodextrin
CE	Capillary electrophoresis
CE–MS	Capillary electrophoresis–mass spectrometry
CE–MS/MS	Capillary electrophoresis–tandem mass spectrometry
CE–UV	Capillary electrophoresis–ultraviolet detection
COF	Covalent organic framework
DLLME	Dispersive liquid–liquid microextraction
EF	Enantiomeric fraction
EE	Enantiomeric excess
EDC	Endocrine-disrupting compound
GC	Gas chromatography

DOI: 10.1201/9781003741626-6

HPβCD	Hydroxypropyl-β-cyclodextrin
LC	Liquid chromatography
LC–MS	Liquid chromatography–mass spectrometry
LC–MS/MS	Liquid chromatography–tandem mass spectrometry
MBBR	Moving bed biofilm reactor
MIP	Molecularly imprinted polymer
MBC@MIP	Magnetic biochar–molecularly imprinted polymer composite
MOF	Metal–organic framework
NF	Nanofiltration
NSAID	Non-steroidal anti-inflammatory drug
QA/QC	Quality assurance/quality control
Qmax	Maximum adsorption capacity
R2	Coefficient of determination
RO	Reverse osmosis
SERS	Surface-enhanced Raman spectroscopy
SPE	Solid-phase extraction
SPE–LC–MS/MS	Solid-phase extraction–liquid chromatography–tandem mass spectrometry
UHPLC	Ultra-high-performance liquid chromatography
UV/O3	Ultraviolet/ozone oxidation
WWTP	Wastewater treatment plant

6.1 INTRODUCTION

Chirality is a stereochemical feature whereby a molecule can be seen to be composed of two non-superimposable mirror images referred to as enantiomers. Though enantiomers have the same molecular formula and they are similar in most physicochemical properties, they can have significant differences regarding biological activity, metabolic pathways, environmental persistence, and toxicity (Zhang et al., 2024). Many pharmaceuticals, pesticides, and personal care products that are discharged into water are chiral compounds and are commonly produced and sold as racemic mixtures. Nonetheless, the aquatic environments are stereoselective in nature. The enantiomers are often converted at a different rate by microbial metabolism, enzymatic degradation, photochemical reactions, and sorption reactions, leading to non-racemic distribution of the environment. As a result, the environmental behavior of a chiral contaminant cannot be comprehensively known based on bulk concentration analysis only.

The studies carried out in the emerging fields indicate that enantiomeric structures of drugs and pesticides drift relative to rivers and coastal areas affected by wastewater, which are stereoselectively attenuated during transit and treatment (Wang et al., 2024; Wu et al., 2024). Such changes are especially significant since the ecological toxicity can also be enantiomer-specific. Through experimental analysis, it is demonstrated that some stereoisomers of typical identified pharmaceuticals, including ibuprofen, have a higher ecotoxicological impact than corresponding mirror isomers, indicating that assumptions of racemic toxicity may not be sufficiently aware of environmental risk (Chen et al., 2024). The same has been stated in regard to chiral pesticides, where enantiomers vary in their bioaccumulation capacity as well as their biochemical interactions with aquatic organisms (Zhang et al., 2024).

Although this continues to provide mounting evidence, the regular environmental monitoring initiatives and regulatory risk assessment highly depend on achiral analysis and racemic concentration reporting. This racemic supposition creates an ambiguity of hazard characterization and can eliminate the existence of the more inactive or biologically active enantiomer. This gap is further augmented by the absence of enantiomer-specific monitoring systems and the inability to employ stereochemistry in environmental policy. With the improved detection methodologies, removal technologies, and regulatory approaches that ought to keep pace with the stereochemistry of aquatic compounds, there is a pressing need to align the methods of detection, removal, and regulation with the stereochemistry of aquatic contaminants.

The review thus provides a synthesis of existing information on the sources of chiral contaminants, environmental behavior, detection, and elimination by analysis of the contaminant in aquatic systems. The discussion will encompass stereoselective fate mechanisms of water-quality management over and above racemic reporting by applying the concept to treatment performance and risk implications to enhance water-quality management in support of stereochemistry-based environmental protection.

6.2 DETECTION TECHNIQUES FOR CHIRAL CONTAMINANTS IN AQUATIC ENVIRONMENTS

The development of enantioselective detection of chiral contaminants in water systems has transformed the lab-level development of chiral tracers into fieldwork-level detectors. Recent reports show that the detection accuracy is a factor not only of chromatographic resolution but also of extraction strategy, compatibility of the matrix, and interpretation of the enantiomeric fraction (EF) as well.

A good example of the early use of automated monitoring at large scale, where an online solid-phase extraction was combined with chiral liquid chromatography–tandem mass spectrometry (SPE–LC–MS/MS) of multi-class chiral pharmaceuticals in wastewater. Their work revealed that automation can play a huge role in the decreased risk of contamination and enhancing reproducibility in effluent monitoring programs. Notably, their effort provided the operational feasibility of routine enantiomer-resolved wastewater monitoring, which is one of the greatest obstacles of environmental stereochemistry. In accordance with this, Wu et al. (2024) used chiral LC–MS/MS in the analysis of antidepressants and antihistamines in coastal waters and showed how enantiomeric profiling advanced probabilistic ecological risk profiling. Instead of reporting total concentrations only, they demonstrated that EF shifts changed the results of hazard characterization. This represented a shift to an analytical detection stage to an interpretation of the environment that is decision relevant. Linear-scale stereoselective transformation in the field level was also expressed by new research topics by Mac Loughlin et al. (2025), which explored triazole fungicides within the horticulture streams. Under natural environmental conditions, they were able to observe sharp spatial EF shifts using SPE–chiral LC–MS/MS after procedures and discover that stereoselective degradation processes indeed take place under natural environmental conditions. Their conclusions confirmed the suggestion that enantioselective monitoring is not a farfetched concept but can be observed in the environment.

The extraction strategies have been the change agents in an enhancement in detection limits. Emulsive liquid-liquid microextraction with UHPLCMS/MS is one such approach. Song et al. (2025) proposed emulsive liquid microextraction and UHPLCMS/MS/MS of the pydiflumetofen enantiomers in natural waters. These provided their work with evidence that the design of extraction can serve a number of functions to increase sensitivity and minimize solvent consumption, which contributes to better sustainability of analytical processes. Also, Lucci et al. (2025) used ferrofluid-based microextraction followed by chiral LC–MS/MS to preconcentrate agrochemicals, which the authors performed before using for complex matrices, like river water contaminated by agricultural runoff. Capillary electrophoresis (CE) has been coming in as an alternative that is more flexible, especially to ionic contaminants. A methodological framework was presented by García-Cansino et al. (2024), which was based on the analysis of chiral pesticides with selectors, cyclodextrins in CE, in which the order of migration and the consistency of EF reports were given importance. Clivillé-Cabré et al. (2026) demonstrated the use of CE–MS/MS on amphetamine-type stimulants in surface waters, and the use of the so-called selector tuning was shown to be useful in optimization of the enantiomers. In spite of the fact that generally CE is less sensitive compared to LC–MS/MS, low solvent consumption and the adjustability of the method make it useful in targeted applications.

Volatile analytes are still of interest to gas chromatography (GC). Critto et al. (2024) optimistically assessed cyclodextrin-based GC stationary phases regarding the agrochemical enantioseparation strategy, emphasizing the backbone composition of a polymer on the resolution. Juvancz et al. (2024) also made interactions between selectors and analytes of ibuprofen derivatives, which contributes to basic information regarding chiral recognition mechanisms. Although GC is not commonly used to monitor aqueous pharmaceuticals, it is potent where the analyses runnable are volatile. In addition to chromatographic enantioselective sensors, future directions in the sphere of enantioselective sensing are determined by material-based and optical methods of detection. Montes-García et al. (2025) identified optical isomers using surface-enhanced Raman spectroscopy (SERS) and machine learning. These methods do not require chromatographic separation, yet they now need validation to LCMS/MS/MS standards.

The recent integration of optics with microfluidics was more recently developed by Liang et al. (2025), who created a chiral photonic platform with multiple specters that aided in screening chiral molecules in aqueous conditions with a high throughput. Even though the concept has only been shown to be applied to model systems, there is a way forward to portable enantiomeric monitoring devices. In general, the analyzed studies attest to a definite methodological order. The best method used in environmental surveillance is still automated SPE–LC–MS/MS (Wu et al., 2024). Microextraction techniques have greater sensitivity and capability for sustainability (Song et al., 2025; Lucci et al., 2025). CE offers flexible ionic separation of ionic species through use of selectors (García-Cansino et al., 2024). GC is also useful with volatile contaminants (Critto et al., 2024). New optical and nanomaterial solutions provide a way to achieve miniaturization of detection in the future, but at present, they do not replace chromatographic ones but complement them (Chen et al., 2024; Montes-García et al., 2025). Irrespective of these improvements, EF alculation, inter-laboratory validation, and harmonized reporting structures are yet to be developed. The majority of the studies are method-sensitive, and few studies have long-term monitor data required to incorporate into the regulations. Figure 6.1 provides the summary of chiral contaminant removal methods, and Table 6.1 provides the summary of recent research in the elimination methods. Thus, although the analytical ability has been developed to a high level, it is not popular on a large scale in ecological policy and the application of daily water-quality check-ups.

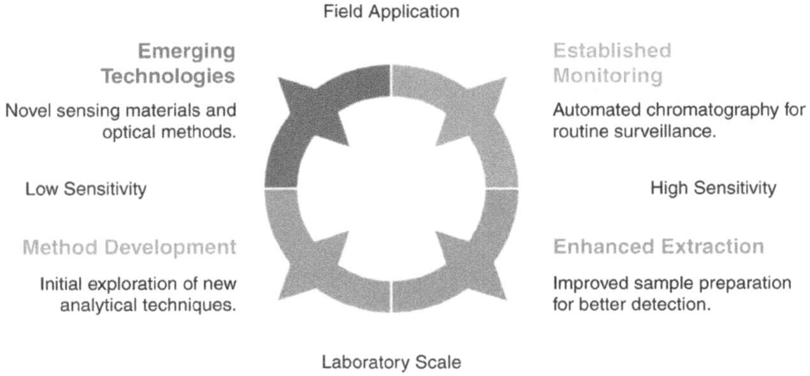

FIGURE 6.1 Overview of chiral contaminant detection techniques.

TABLE 6.1 Summary of detection techniques for chiral contaminants in aquatic environments

REFERENCE	*TARGET ANALYTE CLASS*	*WATER MATRIX*	*SAMPLE PREPARATION/ ENRICHMENT*	*CHIRAL SEPARATION PRINCIPLE*	*DETECTOR/ READOUT*	*PRACTICAL NOTE (WHY IT MATTERS)*
García-Cansino et al. (2024)	Pesticides + emerging contaminants (method focus)	Environmental waters	SPE (common)	CE with chiral selectors (e.g., CDs)	CE–UV/CE–MS (varies)	Practical framework for CE selector choice and EF reporting
Clivillé-Cabré et al. (2026)	Amphetamine-type stimulants	Surface/urban waters	SPE	CE (CD selector)	CE–MS/MS	Strong for cationic drugs; fast selector tuning
Song et al. (2025)	Chiral agrochemical (fungicide)	Natural waters	Microextraction (DLLME/ variants)	Chiral LC	UHPLC–MS/MS	Extraction-driven sensitivity improvements with lower solvent
Critto et al. (2024)	Agrochemicals (review/critical)	Environmental samples	Varies	Chiral GC strategies	GC-based	When analytes are volatile, GC can be very powerful
Juvancz et al. (2024)	NSAID model (ibuprofen derivatives)	Method paper	—	Chiral GC stationary-phase design	GC	Useful for understanding selector–analyte interactions
Chen et al. (2024)	Chiral small molecules (sensor/material)	Aqueous test systems	—	Chiral MOF/COF interaction	Electrochemical/ optical (study-specific)	Shows direction for compact enantioselective sensing
Zhang et al. (2024)	Chiral contaminant model	Aqueous samples	—	Chiral polymer (HPβCD) material separation	chromatographic/ material	Stationary-phase innovation relevant to green separations
Lucci et al. (2025)	Chiral agrochemicals	Natural waters	Ferrofluid-based microextraction	Chiral LC	LC–MS/MS	High preconcentration; good for complex matrices
Abdel-Gawad (2024)	Pharmaceutical models	Water (lab)	SPE/ microextraction	Chiral LC	LC–UV/LC–MS	Method adaptations where MS is limited
Can Doğan et al. (2025)	Broad EDC/pharma monitoring	Water systems	SPE	LC methods	LC–MS/MS	Shows integration into water-quality workflows
Ishaq et al. (2024)	Water contaminants (method-focused)	Water	SPE	LC methods	LC–MS	Useful supporting reference for aquatic analytical QA/QC
Mac Loughlin et al. (2025)	Triazole fungicides (enantioselective)	Streams/ horticulture waters	SPE	Chiral LC	LC–MS/MS	Illustrates real EF shifts in impacted catchments

Saawarn et al. (2024)	Pharmaceuticals	Wastewater/river	SPE	Chiral LC	LC–MS/MS	Shows monitoring logic + stereoselective signals
Wu et al. (2024)	Antidepressants/ antihistamines	Coastal/metro waters	SPE	Chiral LC	LC–MS/MS	Links EF with probabilistic risk framing
Wang et al. (2025)	p-Phenylenediamines/ quinones	WWTP context	SPE	LC (incl. chiral signatures where relevant)	LC–MS	Demonstrates "signature chemistry" in treatment trains
Montes-García et al. (2025)	Isomer/enantiomer discrimination	Aqueous test systems	—	Plasmonic chiral response	SERS + ML	Screening-style approach; needs LC–MS validation
Liang et al. (2025)	Chiral optical detection	Microfluidic/ aqueous	—	Chiral photonic resonance	Optical	Strong for miniaturized sensing concepts
Zhao et al. (2024)	Chiral polymer/ membrane recognition	Water	—	Chiral polymer discrimination	(study-specific)	Material route to "separate + sense" concepts
Song et al. (2025)	Environmental chiral fungicide	Water	Microextraction	Chiral LC	UHPLC–MS/MS	Shows how extraction choice changes method robustness

6.3 REMOVAL TECHNOLOGIES FOR CHIRAL CONTAMINANTS IN AQUATIC ENVIRONMENTS

Chiral contaminants of aquatic environments are a far more difficult challenge to address than detection. Whereas detection is concerned with solving enantiomers analytically, removal technologies have to take into account how treatment processes are degrading, transforming, concentrating, or selective to individual enantiomers. There is growing evidence showing that most traditional treatment systems eliminate contaminants in a non-enantioselective fashion that may change the EFs without necessarily eliminating ecological risk. The barrier to chiral pharmaceuticals and pesticides is the conventional biological wastewater treatment in the first place. Activated sludge systems are mostly based on biodegradation and sorption to biomass flocs (heterotrophic). Nevertheless, it has been demonstrated that these systems often have low or low enantioselectivity. As an example, the studies of biofilm and moving bed biofilm reactors (MBBRs) showed that the degradation kinetics and EF transitions are highly sensitive to the working conditions, including feast–famine cycles (Liang et al., 2024). Stereoselective transformation was only observed to proceed in secondary metabolic steps in other instances, indicating that the relationship between the dynamics of microbial communities and competition between enzymes affects the fate of enantiomers. These results are pointing to the fact that the chirality patterns can be guided by biological systems, yet they are lowly predictable.

A more advanced biological alternative has been introduced by electroactive biofilters. Zhang et al. (2024) showed an increase in chiral emerging contaminants removal with electroactive biofilters, whereby they reported around 80% pharmaceutical removal and a higher enantiomeric alteration with the system compared to conventional systems. The integration of microbial metabolism with electrochemical redox reactions seems to facilitate a stereoselective metabolic reaction, which decreases overall ecotoxicity. This is a shift toward passive degradation to regulated bioelectrochemical remediation. Another pathway of removal is the advanced oxidation process (AOPs), whereby a non-selective radical attack takes place. Their enantioselectivity is, however, not fixed. Accordingly, ozonation and UV/O_3 treatment resulted in less than 29% depletion of the chiral cathinone butylene, with little selectivity of the enantiomers. Electro-Fenton systems, on the contrary, showed good degradability of pharmaceutical substances using hydroxyl radical reaction routes, producing predetermined intermediates. Even though AOPs have the potential of significantly decreasing the overall contaminant loads, they can also produce chiral transformation products with different toxicity profiles. Consequently, oxidative treatment can playact the ecological risk instead of eradicating it unless it is monitored so that the enantiomers shift. Removal methods based on adsorption have continued to be used extensively because they are easy to operate and scalable. Carbon materials that are activated still exhibit high pharmaceutical removal capacities. Arslan et al. (2024) observed a highest adsorption capacity of 252.7 g^{-1} of naproxen on micro-mesoporous carbonaceous material with a pore-filling mechanism. But the majority of adsorption experiments measure the total concentration removal and not the uptake of an enantiomer. The lack of stereoselective adsorption data is a serious gap in research.

Sustainable alternatives have been suggested, such as biochar and modified carbon material. Modekwe et al. (2024) showed the high absorption of the emerging contaminants by the graphene oxide composites with a strong focus on π–π interactions and electrostatic interactions. Equally, Jiao et al. (2024) produced molecularly imprinted magnetic biochar (MBC@MIPs) with the capability of selective removal of oxytetracycline with a higher binding capacity than non-imprinted materials. Even though not explicitly enantioselective, the molecular imprinting strategies provide structure specificity, which can be modified into stereoselective removal of pollutants. Carbon nanotubes and graphene oxide are additional nanomaterials that can be used to improve adsorption by virtue of the high surface area and configurable functional groups. In optimal conditions, Ishaq et al. (2024) found that 92% of the moxifloxacin was removed with the help of graphene–polypyrrole composites. Although the systems exhibit good removal efficiencies, enantiomer-resolved adsorption data are still scarce, and it is therefore important to have a stereospecific analysis. A different removal mechanism offered by membrane technologies is based on size exclusion and solution diffusion. Nanofiltration (NF) has demonstrated compound-specific rejection behavior. According to Fang et al. (2024), rejection using high-recovery NF with azithromycin was 99.7% with electrochemical degradation of the concentrate. Nevertheless, the degree of rejection diffused substantially across pharmaceuticals, which means removal is highly affected by membrane–solute interactions (Can Doğan et al., 2025). Reverse osmosis (RO) tends to reach a rejection of organic micropollutants and is highly energy-demanding with a reject stream that is concentrated. Notably, the vast majority of the studies on membranes do not record any differences in enantiomer rejection, which leaves the issue of stereoselective membrane behavior pretty much uncharted.

Another more specific type includes membranes and chiral recognition materials, which are enantioselective. Zheng et al. (2025) created chiral covalent organic framework nanochannel membranes with the capability to achieve the values of enantiomeric excess up to 86% of selected chiral drugs. Equally, Zhao et al. (2024) have shown cellulose-based membranes that have been measured to perform stereoselective separation of phenylglycine enantiomers. The mechanisms of works of these materials are chiral binding sites and host–guest interactions upon fabricated pore architectures. Even though they are currently on a laboratory level, these technologies are the first real enantioselective removal platforms, as opposed to non-specific contaminant rejection. Another promising method is enzyme degradation. Zhang et al. (2024) revealed that *Rhodococcus* sp. D32 was a great biodegrade of dexamethasone utilizing a new metabolic route to minimize its ecotoxicity. These results point to the possibility of enzyme-mediated and microbial stereoselective biotransformation. Nevertheless, quantitative reporting of enantiomeric excess also is not consistent across studies. Optimized remediation systems based on artificial intelligence (AI) are starting to drive the process optimization and the discovery of materials faster. Although such studies are not always directly concerned with the idea of chirality, combining AI-based parameter optimization with enantioselective treatment material has the potential to play a crucial and dominant role in future remediation efforts.

In general, there are four functional classes of removal technologies for chiral contaminants, namely: (i) non-selective biological degradation, (ii) radical-driven oxidation, (iii) adsorption-based capture, and (iv) membrane-based separation. Among them, chiral membrane materials and some bioelectrochemical systems are found to exhibit steady stereoselective potential. The vast majority of standard strategies eliminate pollutants without a marked enantiomer selectivity, which may alter the EF values and shift the toxicity profiles. The general idea of chiral contaminant removal methods can be seen in Figure 6.2, and Table 6.2 provides the summary of the recent research in the removal methods. In such a way, the contemporary task is not only how to make the removal more efficient but also how to make treatment processes less ecologically hazardous with specific enantiomers and more reasonable in terms of their distribution in aquatic environments.

6.4 FUTURE OPPORTUNITIES AND CHALLENGES

This area of research on chiral contaminants in aquatic environments has evolved significantly in regard to the precision of the analysis and the removal efficiency. Stereoselective behavior within wastewater systems has been improved because of high-resolution enantiomeric methods of detection and enhanced treatment approaches. However, there exist serious scientific, technological, and regulatory gaps. Overcoming these drawbacks presents great potential to interdisciplinary innovation cutting across the fields of analytical chemistry, environmental engineering, toxicology, materials science, and policy formulation.

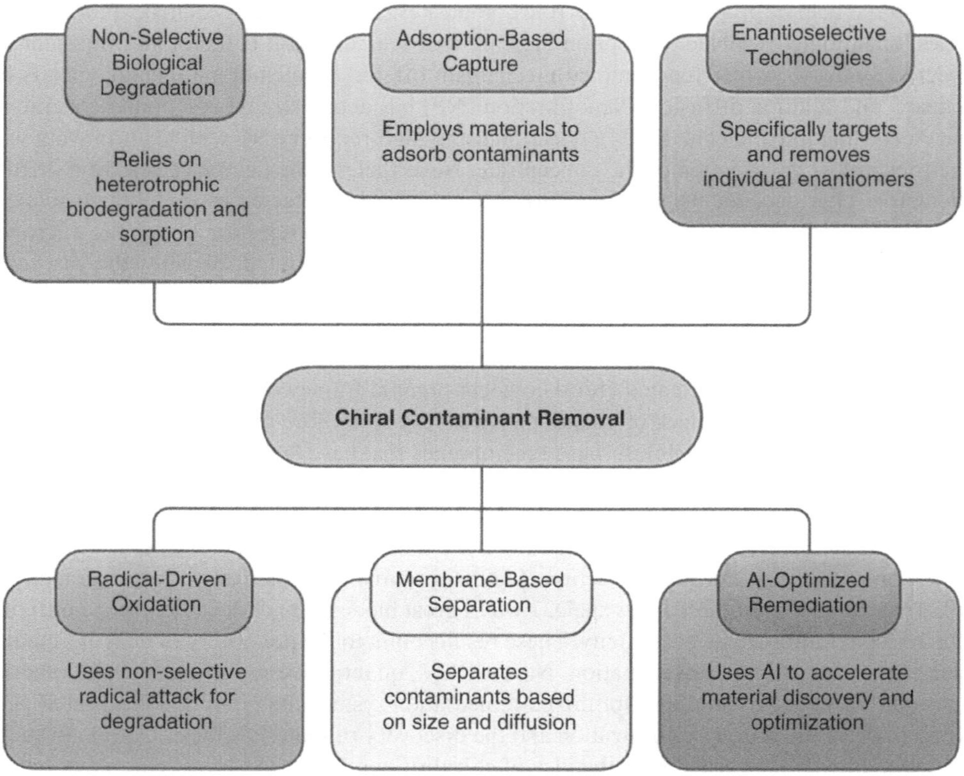

FIGURE 6.2 Overview of chiral contaminant removal techniques.

The lack of standardized procedures in reporting EFs in environmental monitoring is one of the most urgent problems in this field. Numerous studies are still method-specific and short-term, and frequently inter-laboratory validation and harmonized analytical models are missing. Although growing evidence is suggesting that individual enantiomers can be very different in terms of toxicity, environmental persistence, and bioaccumulation potential, regulatory systems are still heavily dependent on data on racemic concentrations. Future studies should focus on the application of harmonized methods to calculate EFs, the setting of inter-laboratory calibration standards, the creation of certified reference materials of chiral pollutants, and the generation of long-term enantiomer-resolved monitoring data. Enantioselective surveillance would help in the adoption of stereochemical information in the evaluation of environmental risks and water-quality regulations.

The present-day wastewater treatment methods such as activated sludge, AOPs, adsorption, and membrane filtration are not designed to exhibit stereoselectivity. In spite of the fact that changes in EF are sometimes seen during treatment, they are seldom optimized or regulated. The next research challenge is to come up with truly enantioselective removal technologies. The development of chiral membranes with designed nanochannels, molecularly imprinted polymers that can be used in stereospecific adsorption, enzyme systems that can be used to selectively degrade, and bioelectrochemical reactors that are designed to follow stereoselective routes of transformation are also promising directions. Chiral covalent organic framework membranes and cellulose-based selective membranes have been demonstrated successfully on the laboratory scale, but issues of scalability, stability, durability of operations, and economic viability must first be overcome before large-scale operations can be considered feasible.

Extensive oxidation and biodegradation reactions frequently produce transformation products whose stereochemistry and the toxicological characteristics are not well understood. The present ecotoxicological assessments often do not consider enantiomer-specific ecotoxicology responses to

TABLE 6.2 Summary on removal technologies for chiral contaminants in aquatic environments

REFERENCE	*TARGET ANALYTE CLASS*	*WATER MATRIX*	*REMOVAL MECHANISM*	*TECHNOLOGY TYPE*	*REMOVAL EFFICIENCY*	*PRACTICAL NOTE (WHY IT MATTERS)*
Liang et al. (2024)	Chiral pharmaceuticals	Wastewater	Biodegradation (feast–famine cycles)	Moving bed biofilm reactor (MBBR)	Moderate; EF shifts observed	Demonstrates operationally dependent stereoselective degradation
Zhang et al. (2024)	Emerging chiral pharmaceuticals	Wastewater	Bioelectrochemical oxidation + biodegradation	Electroactive biofilter	~80% removal	Enhanced stereoselective transformation compared to conventional sludge
Arslan et al. (2024)	Naproxen (NSAID)	Industrial wastewater	Adsorption (pore filling)	Micro-mesoporous carbon	$Q_{max} \approx 252.7$ mg g^{-1}	High adsorption capacity; enantioselectivity not evaluated
Modekwe et al. (2024)	Emerging contaminants	Water	π–π interaction/ electrostatic adsorption	Graphene oxide composite	High (>85%)	Sustainable adsorption platform; stereospecificity unclear
Jiao et al. (2024)	Oxytetracycline	Aqueous systems	Molecular recognition	Magnetic biochar–MIP composite	Enhanced selective binding	Demonstrates template-driven specificity; potential for enantioselective design
Ishaq et al. (2024)	Moxifloxacin	Water	Adsorption (electrostatic + π interaction)	Graphene–polypyrrole composite	~92%	High removal; scalable nanocomposite application
Fang et al. (2024)	Azithromycin	Wastewater	Size exclusion + diffusion	Nanofiltration (NF)	99.7% rejection	High rejection; concentrate requires secondary treatment
Can Doğan et al. (2025)	Multi-class pharmaceuticals	Water systems	Pressure-driven separation	Nanofiltration/RO	Compound-dependent	Membrane selectivity varies; enantiomer-specific rejection underexplored
Zheng et al. (2025)	Chiral drugs	Aqueous systems	Chiral nanochannel recognition	COF membrane	Enantiomeric excess up to 86%	True enantioselective separation platform
Zhao et al. (2024)	Phenylglycine enantiomers	Water	Chiral membrane binding	Cellulose-based membrane	Significant enantiomer discrimination	Demonstrates membrane-level stereoselectivity
Zhang et al. (2024)	Dexamethasone	Wastewater	Microbial biodegradation	*Rhodococcus* sp. D32	Significant degradation	Identified metabolic pathway; ecotoxicity reduction confirmed

an organism, thus underrating ecotoxicological risk. Future studies should thus aim at determining chiral transformation products in the course of treatment, carrying out enantiomer-resolved toxicity assessment of mixture toxicity under ecologically realistic conditions, and conducting bioaccumulation studies expressly taking into account the stereochemical aspect. In the absence of such careful evaluation, the treatment systems can only help to redistribute ecological risk instead of remove it successfully.

Machine learning and AI have the potential to change the state of chiral contaminant management. Predictive modeling may be useful in determining stereoselective degradation pathways, optimization of the adsorption material, chiral membrane design, and prediction of EF changes in wastewater treatment plants. Recent research of AI-based material screening and process optimization has shown a high level of predictive accuracy, which implies high applicability in this area. The next-generation systems need to incorporate mechanistic models, kinetic frameworks of stereoselective degradation, multi-omics datasets of microbes, or real-time sensor data. This kind of integration may allow dynamic and smart water treatment systems that are responsive to changes in contaminant loads and stereochemical changes.

Currently, enantiomeric analysis mostly relies on liquid–liquid chromatography in the laboratory that uses tandem mass spectrometry systems. These techniques are very sensitive and selective, but they are resource intensive and cannot be deployed in the field in real time. New optical spectroscopic methods, nanoplatforms of SERS, and nanosensor technologies indicate future possibilities of portable enantiomeric detection. The next generation of chiral-sensing miniaturized systems that can monitor EF in real time at the treatment plants and can be incorporated into Internet-of-Things-based networks monitoring water quality should be the topic of future investigations. There is, however, a major technological challenge of providing adequate sensitivity, selectivity, reproducibility, and stability in the calibration at field conditions.

The shortcoming of converting analytical and technological innovations into regulatory systems perhaps remains the biggest challenge in the long-term perspective. The existing water-quality standards fail to differentiate the individual enantiomers, notwithstanding scientific data suggesting that regulatory levels grounded entirely on racemic mixtures might not be correlative to ecological and human health concerns. Regulatory evolution must, however, take into account the setting of enantiomer-specific risk limits, the incorporation of the EFs in the environment monitoring standards, and the risk prioritization of chiral pharmaceuticals and pesticides. Such a development will be necessitated by concerted efforts by chemists, toxicologists, engineers, and policymakers to make sure that stereochemical risk is properly addressed in the governance of the environment. Figure 6.3 represents the summary of the problems and the opportunities in the chiral contaminant study.

The next step in the development of chiral contaminant management will be the intersection of analytical chemistry and materials science, biotechnology, AI, and environmental policy. Detection technologies are very mature, taking into consideration that scalable and economical enantioselective remediation is still at its infantile stage. To address this gap, a more mechanistic insight into stereoselective environmental processes, globalized analytical schemes, scalable chiral separating materials, AI-based design and monitoring platforms, and regulatory response to stereochemical risk implications will be necessary. In case of these issues being properly resolved, the ecological protection policies can no longer remain concentration-based evaluation tools but be developed with chirality listed among the key factors of ecological and general health hazards in water.

6.5 CONCLUSION

Chiral contaminants are one of the most vital aspects of aquatic pollution that are poorly addressed. Even though enantiomers have the same physicochemical characteristics, their biological activity,

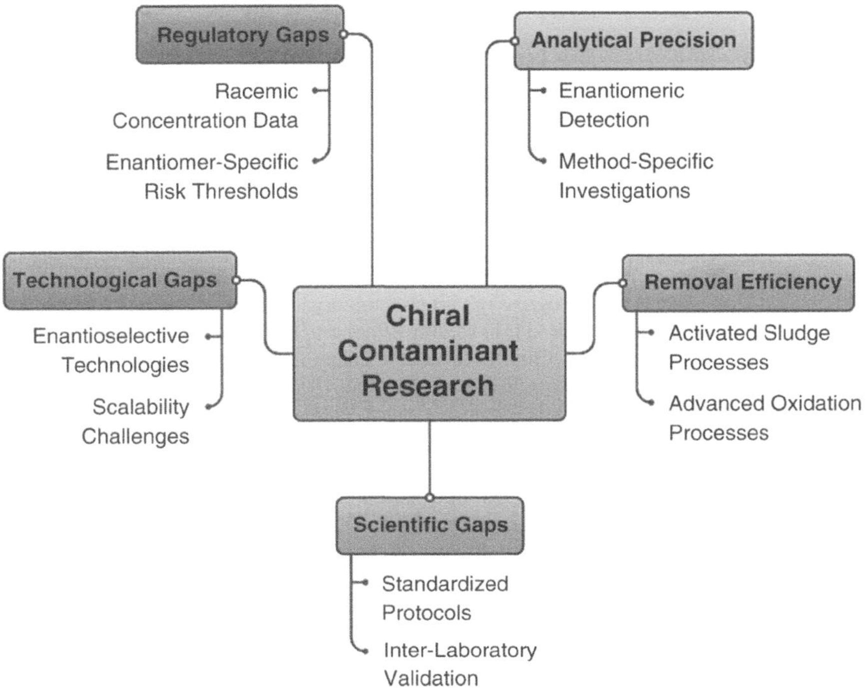

FIGURE 6.3 Chiral pollutant studies: opportunities and challenges.

environmental persistence, degradation pathway, and toxicity may differ significantly. Much evidence has been gathered in the field to indicate that aquatic systems are stereoselective in nature and so exhibit quantifiably varying EFs shifts throughout the treatment of wastewater, transportation via a river, and dispersal along the coast. These results are a clear indication that the racemic concentration reporting is not enough to describe the environmental fate and environmental risk. Enantiomer-resolved detection technologies have made a great breakthrough. Automated solid-phase extraction combined with liquid chromatography–tandem mass spectrometry is the standard of the art in environmental monitoring, complemented by microextraction methods, CE, and GC, which become more analytically flexible. Future advancements in field-deployable, miniature monitoring candidates. The future of miniaturized and field-deployable monitoring is signaled by emerging material-based and optical sensing platforms. However, there are limited standardized protocols of EF calculations, inter-laboratory validations, and long-term surveillance databases, which limit the integration of regulations.

The stereoselective removal technologies exhibit inconsistent stereo activity. Biological systems that are traditional and sophisticated oxidation processes tend to alter distributions of enantiomers uncontrollably, potentially creating transformation products whose toxicity is unknown. Systems using adsorption and membranes can be successfully used to minimize the total contaminant loads, but the evaluation of the enantiomers is often not provided. At the scale of such enantioselective removal, chiral membranes, molecularly imprinted materials, bioelectrochemical systems, and selective enzymatic degradation pathways have shown true enantioselective removal, although these technologies are mostly not yet beyond laboratory scale. The next generations of chiral contaminant management must be based on the combination of mechanistic stereochemical knowledge with the design of the treatment on a scale. AI provides potent prospects of predictive stereoselective degradation, material performance optimization, and adaptive treatment systems. At the same time, the regulatory framework should be adjusted to the concept that goes beyond the racemic assumption, integrating an enantiomer-specific threshold of risk and surveillance principles.

REFERENCES

Abdel-Gawad, S. A. (2024). Determination of some non-steroidal anti-inflammatory drug residues in industrial wastewater effluents in Al-Kharj City using capillary-zone electrophoresis technique. *Asian Journal of Chemistry*, *36*(6), 1366–1370.

Arslan, Y., Tomul, F., Kınaytürk, N. K., Dong, N. T., Trak, D., Kabak, B., & Tran, H. N. (2024). Important role of pore-filling mechanism in separating naproxen from water by micro-mesoporous carbonaceous material. *Water Environment Research*, *96*(1), e10966. https://pubmed.ncbi.nlm.nih.gov/38226502/

Chen, H., Xia, L., & Li, G. (2024). Recent progress of chiral metal–organic frameworks in enantioselective separation and detection. *Microchimica Acta*, *191*(11), 1–18. https://doi.org/10.1007/s00604-024-06729-y

Clivillé-Cabré, P., Borrull, F., Aguilar, C., & Calull, M. (2026). Enantioselective determination of amphetamine-type stimulants in environmental waters using SPE and CE–MS/MS. *Electrophoresis*, *47*(3), 250–260. https://doi.org/10.1002/elps.70074

Critto, E. F., Giovannoni, S., Lancioni, C., & Castells, C. B. (2024). Enantioseparation of agrochemicals by gas chromatography. Exploring columns based on cyclodextrin derivatives dissolved into polysiloxanes. *Journal of Separation Science*, *47*(1), 2300804. https://pubmed.ncbi.nlm.nih.gov/38234022/

Chen, H., Xia, L., & Li, G. (2024). Recent progress of chiral metal–organic frameworks in enantioselective separation and detection. *Microchimica Acta*, *191*(11), 640. https://doi.org/10.1007/s00604-024-06729-y

Doğan, E. C., Yaşar, A., Narcı, A. O., Mert, B. K., & Yeşildağlı, B. U. (2025). Optimisation of nanofiltration process for pharmaceutically active compounds: Removal of paracetamol, diclofenac and metoprolol. *Water and Environment Journal*, *39*(1), 5–19. https://doi.org/10.1111/wej.12954

Fang, C., Garcia-Rodriguez, O., Yang, L., Zhou, Y., Imbrogno, J., Swenson, T. M., Lefebvre, O., & Zhang, S. (2024). Sequential high-recovery nanofiltration and electrochemical degradation for the treatment of pharmaceutical wastewater. *Water Research*, *259*, 121832. https://doi.org/https://doi.org/10.1016/j.watres.2024.121832

García-Cansino, L., Marina, M. L., & García, M. Á. (2024). Chiral analysis of pesticides and emerging contaminants by capillary electrophoresis—Application to toxicity evaluation. *Toxics*, *12*(3), 185. https://doi.org/10.3390/toxics12030185

Ishaq, S., Nadim, A. H., Amer, S. M., & Elbalkiny, H. T. (2024). Optimization of graphene polypyrrole for enhanced adsorption of moxifloxacin antibiotic: An experimental design approach and isotherm investigation. *BMC Chemistry*, *18*(1), 113. https://doi.org/10.1186/s13065-024-01208-0

Jiao, Y., Yi, Y., Fang, Z., & Tsang, P. E. (2024). Selective removal of oxytetracycline by molecularly imprinted magnetic biochar. *Bioresource Technology*, *395*, 130394. https://doi.org/10.1016/j.biortech.2024.130394

Juvancz, Z., Bodane-Kendrovics, R., Agoston, C., Czegledi, B., Kaleta, Z., Jicsinszky, L., & Riszter, G. (2024). Chiral selectivities of permethylated α-, β-, and γ-cyclodextrins containing gas chromatographic stationary phases towards ibuprofen and its derivatives. *International Journal of Molecular Sciences*, *25*(14), 7802. https://doi.org/10.3390/ijms25147802

Liang, C., Svendsen, S. B., de Jonge, N., Carvalho, P. N., Nielsen, J. L., & Bester, K. (2024). Eat seldom is better than eat frequently: Pharmaceuticals degradation kinetics, enantiomeric profiling and microorganisms in moving bed biofilm reactors are affected by feast famine cycle times. *Journal of Hazardous Materials*, *468*, 133739. https://doi.org/10.1016/j.jhazmat.2024.133739

Liang, X., Zhao, Z., Tang, X., Yang, H., Liang, L., Zhao, M., … & Ding, X. (2025). Multispectral chiral quasi-bound states in the continuum enabled microfluidics for high-throughput molecular screening and quantification. *Advanced Science*, *12*(47), e15443. https://doi.org/10.1002/advs.202515443

Lucci, E., Falcinelli, G., Antonelli, L., Dal Bosco, C., Felli, N., De Cesaris, M. G., & Gentili, A. (2025). Hydrophobic deep eutectic solvent-ferrofluid microextraction followed by liquid chromatography-mass spectrometry for the enantioselective determination of chiral agrochemicals in natural waters. *Analytical and Bioanalytical Chemistry*, *417*(7), 1341–1357. https://doi.org/10.1007/s00216-024-05619-9

Mac Loughlin, T. M., Navarro, M., Rosero Garces, R. A., Ramos, M., Salimbeni, A., & Peluso, M. L. (2025). From degradation to detection: Assessing enantioselective behavior of chiral triazole fungicides in horticultural stream waters. *Chemosphere*, *381*, 144486. https://doi.org/https://doi.org/10.1016/j.chemosphere.2025.144486

Modekwe, H. U., Ayeleru, O. O., & Olubambi, P. A. (2024). Adsorptive removal of emerging pollutants from water using graphene-oxide and graphene oxide-based composites and its adsorption mechanisms. In *Sustainable Remediation Technologies for Emerging Pollutants in Aqueous Environment* (pp. 137–162). Elsevier. https://doi.org/10.1016/B978-0-443-18618-9.00017-6

Montes-García, V., Martín, V. F., Obelleiro-Liz, M., Pérez-Juste, I., Ciesielski, A., & Samorì, P. (2025). Ultrasensitive isomer discrimination: A joint Surface-Enhanced Raman Scattering (SERS) spectroscopy and machine learning strategy. *Advanced Sensor Research*, *4*(3), 2400160. https://doi.org/10.1002/adsr.202400160

Saawarn, B., Mahanty, B., & Hait, S. (2024). Adsorptive removal of perfluorooctanoic acid from aqueous matrices using peanut husk-derived magnetic biochar: Statistical and artificial intelligence approaches, kinetics, isotherm, and thermodynamics. *Chemosphere*, *360*, 142397. https://doi.org/10.1016/j.chemosphere.2024.142397

Song, R., Fu, J., Zhu, G., Wang, M., Wang, Y., Jing, X., & Li, L. (2025). Emulsive liquid-liquid microextraction coupled with UHPLC-MS/MS for chiral analysis of pydiflumetofen in environmental water samples. *Journal of Chromatography A*, *1761*, 466412. https://doi.org/https://doi.org/10.1016/j.chroma.2025.466412

Wang, J., Li, Y., Nie, C., Liu, J., Zeng, J., Tian, M., Chen, Z., Huang, M., Lu, Z., & Sun, Y. (2025). Occurrence, fate and chiral signatures of p-phenylenediamines and their quinones in wastewater treatment plants, China. *Water Research*, *276*, 123272. https://doi.org/https://doi.org/10.1016/j.watres.2025.123272

Wang, W., Xie, T., Ma, N., Jiang, X., Zhang, H., Sun, T., & Cui, B. (2024). In-stream attenuation and enantioselective fractionation of psychiatric pharmaceuticals in a wastewater effluent-dominated river basin. *Science of the Total Environment*, *951*, 175521. https://doi.org/10.1016/j.scitotenv.2024.175521

Wu, R., Sin, E. Y. Y., Zhang, K., Xu, S., Ruan, Y., Mak, Y. L., … & Lam, P. K. (2024). Medicating the coast in a metropolitan city: Enantiomeric profiles and joint probabilistic risk assessment of antidepressants and antihistamines. *Environment International*, *184*, 108434. https://doi.org/10.1016/j.envint.2024.108434

Zhang, J. N., Gong, X. F., Ying, G. G., Wang, A. G., Dong, M. J., Liu, Y. S., … & Xu, M. Y. (2024). Innovative bioremediation of dexamethasone in aquatic ecosystems using Rhodococcus sp. D32: Pathway discovery and reduction of ecotoxicity. *Journal of Cleaner Production*, *444*, 141319. https://doi.org/10.1016/j.jclepro.2024.141319

Zhao, H. W., Cui, X. F., & Yuan, L. M. (2024). Chiral separation of D, L-phenylglycine by cellulose triacetate membranes. *Journal of Applied Polymer Science*, *141*(26), e55571. https://api.semanticscholar.org/CorpusID:269064224. https://doi.org/10.1002/app.55571

Zheng, C. Y., Qian, H. L., Yang, C., & Yan, X. P. (2025). Design of self-standing chiral covalent-organic framework nanochannel membrane for enantioselective sensing. *Small Methods*, *9*(3), 2401120. https://doi.org/10.1002/smtd.202401120

7 Biodegradability and Resilience Assessment Framework for Decentralized Wastewater Treatment Systems in Rural India

Roshini Praveen Kumar, N. Anand, Cyril Samuel J.S., and A. Diana Andrushia

State	Abbreviations
AOP	Advanced oxidation process
BI	Biodegradability index
BOD	Biochemical oxygen demand
BOD5	5-day biochemical oxygen demand
COD	Chemical oxygen demand
CPCB	Central pollution control board
DWTS	Decentralized wastewater treatment systems
EC	Electrical conductivity
EU	European Union
GAC	Granular activated carbon
IoT	Internet of Things
LC–MS/MS	Liquid chromatography–tandem mass spectrometry
MLD	Million liters per day
O&G	Oil and grease
STP	Sewage treatment plant

DOI: 10.1201/9781003741626-7

STP-1 to STP-5	Individual sewage treatment plants (nodes 1–5)
TDS	Total dissolved solids
TSS	Total suspended solids
USEPA/US EPA	United States Environmental Protection Agency
WET	Whole effluent toxicity
WWTP	Wastewater treatment plant
ZLD	Zero liquid discharge

7.1 INTRODUCTION

7.1.1 The Challenge of Decentralized Water Quality

Decentralized wastewater treatment systems (DWTSs) recently have become a common circular water management system, especially in rural and peri-urban regions (Capodaglio, 2017). These systems are economical and flexible; they are often facing operational realities that differ from urban plants. The decentralized systems are vulnerable to varying high inflows and localized shock loads. Unlike large facilities that benefit from high dilution. To detect the chemically complex influents, mainly the emerging contaminants like pharmaceuticals, personal care products, and surfactants, is a complex challenge. These substances are found in low concentrations and freely bypass the conventional monitoring. However, when treated water is reused, it gradually contaminates the soil and water. Conventional indicators, such as biochemical oxygen demand (BOD) and chemical oxygen demand (COD), are not enough to effectively track general biodegradable waste that is insensitive to recalcitrance. Even though a system may meet the legal limits, it can still release a chemical shadow, which later builds up over time (U.S. EPA, 2002). To address this, advanced regulatory frameworks have shifted toward metrics that can identify these missing loads. For instance, the US EPA's Whole Effluent Toxicity (WET) testing mandates the monitoring of the aggregated biological effects of effluents, which were missed by the chemical listing of the BOD. The European Union (EU)'s Urban Wastewater Treatment Directive now requires quaternary treatments to physically remove micropollutants (European Union, 2024). In India, this shift is evident in the Central Pollution Control Board's Zero Liquid Discharge (ZLD) guidelines; this explicitly recognizes that standard treatments are inadequate for toxicity or high COD streams (Central Pollution Control Board, 2025). Furthermore, national polices now push industries beyond simple discharge standards toward "fitness for reuse." These frameworks aim to remove advanced recalcitrant organics to protect irrigation and industrial recycling (Ministry of Jal Shakti, 2022). Despite these advancements, many regions are still struggling to enforce and obtain long-term sustainability (Lourenço & Nunes, 2020). This chapter introduces a two-part approach to interpret water-quality data to forensically identify chemical loads and test the systems' resilience under hydraulic stress.

Instead of relying on expensive, high-end sensors, this framework uses low-cost, existing data as a low-cost early warning system. For example, by detecting the ratio and pollutant interdependency of COD/BOD (chemical stress). If the threshold of 3.0 is exceeded for COD to BOD, it is flagged as 'chemical stress.' This would be because metabolic inhibition prevents standard degradation (Abdalla & Gina, 2014). This metric helps the operators to fully diagnose complex compound relationships without high-cost equipment. However, the indicator cannot characterize complex chemical stresses on their own; it requires advanced chemical analysis. Hence, a robust diagnosis is required to pair the physicochemical parameters and targeted pollutant analyses. This approach offers a practical and scalable method for identifying water-quality risks in resource-constrained settings.

7.1.2 Novel Contribution of the Framework

To understand the framework proposed in this study, a clear understanding of decentralized wastewater management should be made familiar. In general, the methods that are used to identify the characteristics of wastewater (COD, BOD, and effluent thresholds) often provide insights but fail to provide results for system-level resilience. Recent studies state the need for integrated socio-technical perspectives that link treatment performance with operational stability, governance capacity, and adaptive monitoring (Starkl et al., 2022; Matesun et al., 2024; Sha et al., 2024).

The forensic-resilience framework proposed in this study is integrated with routine water-quality diagnostics with longitudinal hydraulic performance and STP-specific interpretation. While traditional biodegradability assessments incorporate COD/BOD ratio as an isolated indicator of organic treatability. However, the present framework enhances its value through three analytical dimensions.

- Diagnostic decoupling: This employs statistical interdependencies among physicochemical parameters that are used to differentiate benign biodegradable fluctuations from traces of recalcitrant or chemically complex organic loads. These are majorly overlooked by conventional monitoring.
- Integrated hydraulic–quality resilience: This combines effluent quality indicators with IoT (Internet of Things) incorporated hydraulic stability metrics (flow continuity, loss minimization), providing parallel assessment of treatment performance and operational resilience.
- STP-specific governance diagnostics: This approach aims to identify the localized biodegradability stress within individual STPs, targeting tertiary interventions instead of network-wide upgrades. This approach can be more cost-effective in decentralized settings.

This three-dimensional framework reframes routine monitoring data toward predictive resilience diagnostics in resource-constrained decentralized wastewater systems. This directly addresses the gaps highlighted in recent works (Sravan et al., 2024; Etafo et al., 2025).

7.2 STUDY AREA AND METHODOLOGY

7.2.1 Study Setting: Site Characteristics and Infrastructure

The study is considered for a large residential education campus in rural Southern India. This acts as a functional model for small-town decentralized sanitation. A stable population of roughly 8,500 residents lives in and around the campus. The diverse wastewater stream from various campus facilities (hostels, campus buildings, mess halls, etc.) highlights the institute's usage and circularity. This distinct mix mirrors the loading profiles of typical wastewater treatments in peri-urban regions. The network consists of five decentralized sewage treatment plants (STP 1 through STP 5) with an overall capacity of approximately 2.45 MLD. High-density digital meters are installed on the campus, with 119 enabled level controllers. These networks are distributed across sumps, overhead tanks, and treatment outlets, allowing accurate tracking of flow events. By reusing the treated water for irrigation, toilet flushing, and agriculture, the campus can measure effectively the efficiency and circularity goals.

7.2.2 Analytical Framework

This study is a hybrid monitoring strategy that is incorporated to obtain reliable and forensic depth for a 36-month period (2021–2024). The approach analyses volumetric flow, reuse efficiency, and overflow events using data from campus digital flowmeters. Targeted chemical audits during peak loading conditions, a 90-day operational window (August–October), are opted for to capture the maximum stress scenario. Influent and effluent samples were analyzed for pH, BOD_5, COD, total suspended solids (TSS), oil and grease (O&G), total dissolved solids (TDS), and electrical conductivity (EC).

7.2.2.1 Statistical Forensics

A Kruskal–Wallis test was applied to assess consistent performance across the five treatment units. This approach was selected based on the standard practice of comparing three or more independent groups when water quality is not normally distributed or has unequal variance. This test is a standard practice in water-quality analysis, especially when normality assumptions are violated. Pearson correlation matrices were also generated to identify pollutant interdependencies, particularly to decouple organic load behaviors (BOD/COD) from inorganic salinity markers. Prior to correlation analysis, normality was assessed using the Shapiro–Wilk test, which confirmed non-normal distribution ($p < 0.05$) for several parameters, justifying the subsequent use of Spearman's rank correlation for robustness.

7.2.2.2 Forensic Indices

The biodegradability index (BI), expressed as the COD/BOD_5 ratio, was used to characterize the organic nature of wastewater and to identify potential biodegradability stress within the decentralized network. The COD/BOD_5 ratio is a widely accepted screening indicator of organic treatability in wastewater systems. Established literature categorizes BI ranges from highly biodegradable to chemically stressed conditions. The thresholds adopted in this study are summarized in Table 7.1 (Lacalamita et al., 2024; Płuciennik-Koropczuk & Myszograj, 2019).

The biodegradability of the influent and effluent was assessed using the BI, calculated as shown in Equation 7.1:

$$Biodegradability\ Index = \frac{Chemical\ Oxygen\ Demand\ (COD)}{Biochemical\ Oxygen\ Demand\ (BOD)_5} \qquad (1)$$

TABLE 7.1 Biodegradability index (BI) thresholds and interpretation

BI RANGE (COD/BOD5)	*BIODEGRADABILITY CATEGORY*	*OPERATIONAL SIGNIFICANCE*	*REFERENCE(S)*
<2.0	Highly biodegradable	Optimal for biological treatment	Lacalamita et al. (2024)
2.0–3.0	Moderately biodegradable	Typical for domestic wastewater	Diwyanjalee et al. (2024)
3.0–4.0	Low biodegradability	Presence of recalcitrant organics	Maddah (2022)
>4.0	Chemical stress	Inhibitory or non-biodegradable load	Płuciennik-Koropczuk & Myszograj (2019)

7.3 BASELINE PERFORMANCE AND REMOVAL EFFICIENCY

7.3.1 High-Efficiency Removal of Macropollutants

The system's resilience was evaluated against emerging contaminants, as it is essential to validate the functional integrity of the biological treatment units. If biological oxidation fails, the forensic ratio becomes unreliable. As illustrated in the performance distribution analysis in Figure 7.1, the decentralized network highlights a satisfactory baseline. The systems achieved near-complete removal of suspended solids, reducing average inlet TSS from 114 mg/L to 2 mg/L (98.2% reduction). This result confirmed that the solids were stable and the treatment process functioned correctly to support biological compliance. Similarly, the systems had effectively oxidized biodegradable waste fractions. The inlet BOD (mean: 74.5 mg/L) was reduced to 9.31 mg/L (87.5% removal), which consistently followed

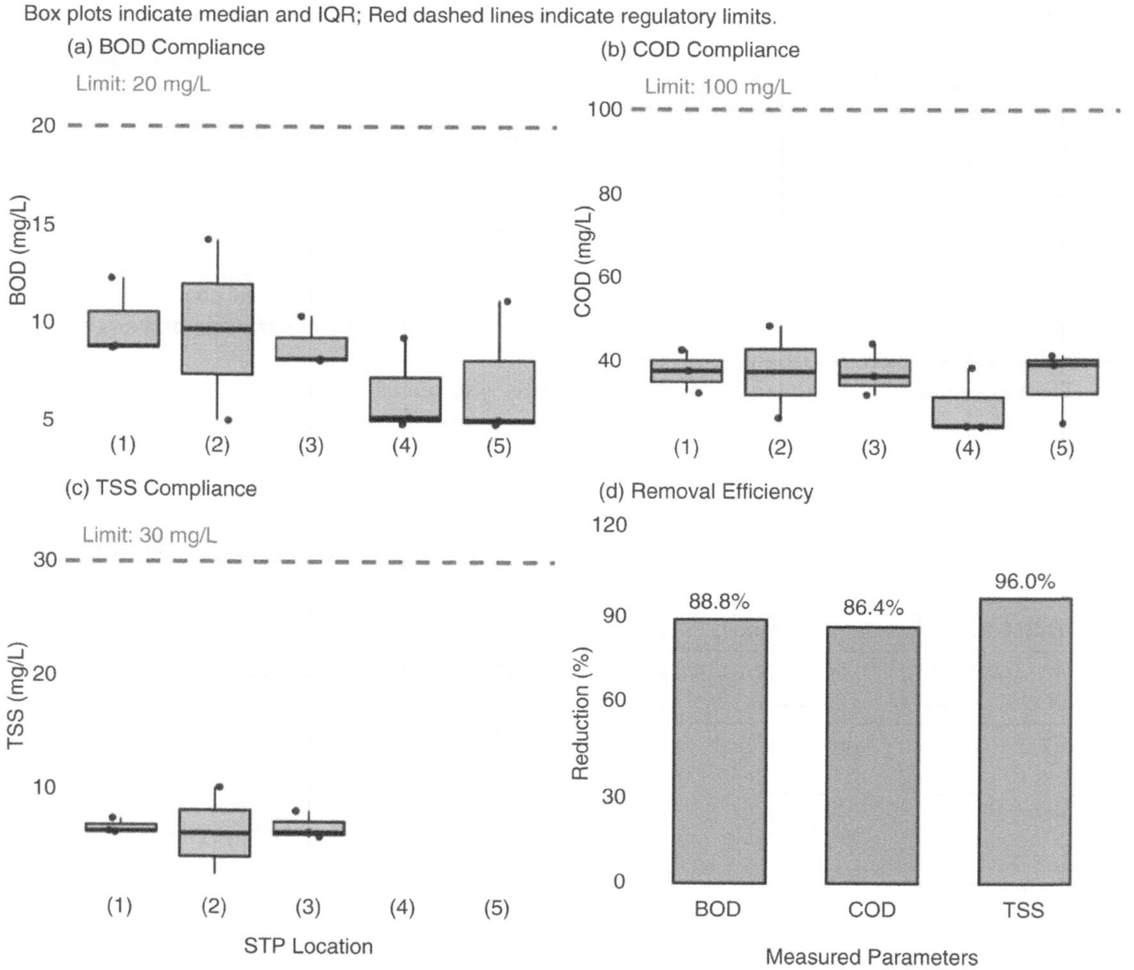

FIGURE 7.1 Box-plot distribution of influent versus effluent concentrations for key physicochemical parameters across the five STPs. The median TSS effluent value was 2 mg/L (IQR: 1.5–3.0 mg/L), representing a 98.2% reduction from the influent median of 114 mg/L (IQR: 98–132 mg/L).

TABLE 7.2 Kruskal–Wallis test results for effluent quality comparison

PARAMETER	X^2	*DF*	*P-VALUE*	*EFFECT SIZE* (E^2)	*INTERPRETATION*
BOD (mg/L)	2.560	4	0.63	0.029	Not significant
COD (mg/L)	3.112	4	0.54	0.035	Not significant
pH	1.442	4	0.83	0.016	Not significant
TDS (mg/L)	4.021	4	0.40	0.045	Not significant
TSS (mg/L)	8.523	4	0.071	0.096	Borderline significant

the regulatory discharge limit of 20 mg/L that is required for unrestricted irrigation; this verifies compliance.

7.3.2 Statistical Verifications of Process Uniformity

In a decentralized wastewater management system, treatment performance of STPs differs based on the locations. To identify this, the Kruskal–Wallis test is conducted on the discharged effluent data to five spatially distinct STPs. As shown in Table 7.2, the analysis revealed no statistically significant differences between the plants across any key parameters. The statistical audit was based on 90 daily composite samples that were collected during the peak operational window and evaluated by a 95% confidence interval ($p < 0.05$). A preliminary normality check was conducted by applying the Shapiro–Wilk test, which confirmed non-normal distribution ($p < 0.05$) for BOD, COD, and TSS (hence, justifying the choice of method). The effect sizes were calculated further by supporting these findings. All the variables remained below 0.05, representing negligible practical variance between the facilities. Among the variables, TSS showed a borderline p-value of 0.071, but it showed only a modest effect size ($\varepsilon^2 = 0.096$). These findings overall demonstrate the campus treatment infrastructure's uniform performance. This consistency is a critical baseline for further analysis, as the STPs operate identically; any anomalies, if detected in later chemical audits, could be attributed to the influent compositions rather than mechanical or operational failures.

7.4 POLLUTANT INTERDEPENDENCIES AND CLUSTERING ANALYSIS

7.4.1 Decoupling Salinity from Organics

The treatable organic loads from the recalcitrant chemical pressures are analyzed using a Spearman correlation matrix (ρ) generated to identify the parameter association (Figure 7.2). The results of the correlation provided three diverse understandings of the STP effluent behavior.

The first behavior identified is that effective remediation requires understanding pollutant transport. The organic cluster (COD, BOD, and TSS) is strongly correlated with their ρ values ranging from 0.64 to 0.86. The strong relationship between COD and TSS ($\rho > 0.6$) confirms that the biological sedimentation process consistently removes the dominant organic fraction in the influent. Here, parameters rise and fall in unison, confirming that particulate organic matter is the dominant fraction, which is effectively managed by the biological sedimentation process.

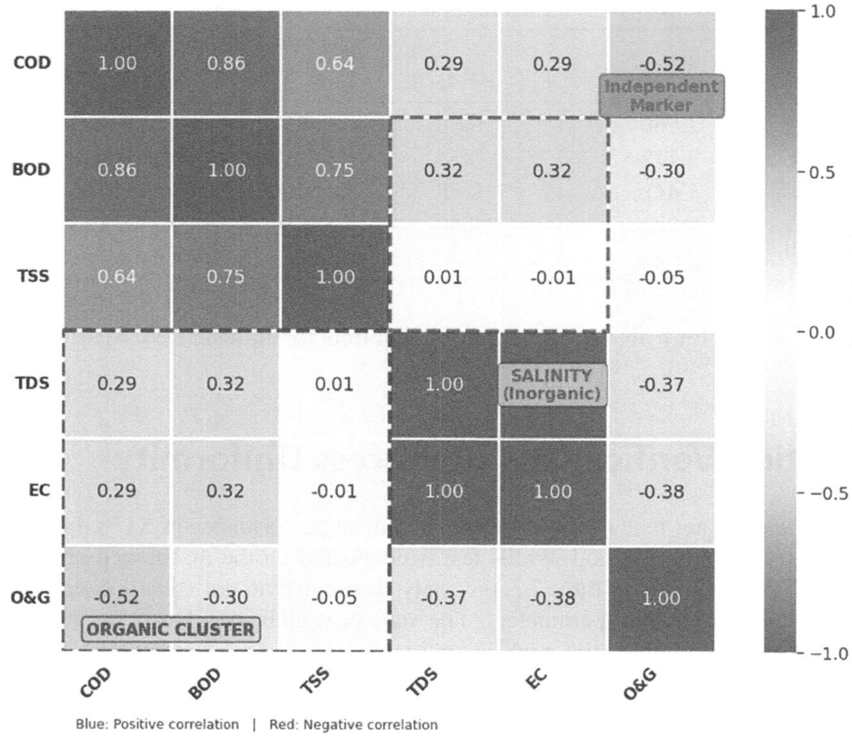

FIGURE 7.2 Spearman correlation heatmap (ρ) illustrating distinct clustering of organic load vectors (COD–BOD–TSS) versus inorganic salinity markers (EC–TDS), revealing the independence of dissolved solids from biological treatment mechanisms. Correlation strengths are categorized as follows: weak ($|\rho| < 0.3$), moderate ($0.3 \leq |\rho| < 0.7$), and strong ($|\rho| \geq 0.7$).

The second behavior is observed due to salinity decoupling; in contrast, the EC and TDS shape an independent cluster ($\rho = 1.0$ between themselves). These parameters showed negligible correlations with organic markers (e.g., EC vs. TSS $\rho \sim 0.01$). Dissolved inorganic solids are not tracked based on organic load. This implies that whenever a biological system handles food (organic), it tends to offer no resistance to the dissolution of inorganic vectors. These are also known as pharmaceutical salts or heavy metals, which pass untreated in this system.

The third behavior showed a significant discrepancy in the oil and grease behavior. This in general is considering kitchen-derived interferences. This is the most critical finding of the study, as it identifies a negative correlation, in contrary to the relation, high COD influenced by kitchen lipids. O&G displayed a moderately negative correlation with COD ($\rho = -0.52$) and BOD ($\rho = -0.30$).

These decoupled sources indicate that O&G are present due to the episodic and distinct events, such as canteen wash-down cycles, even though these outlets are not connected to the main chemical load.

In addition, COD peaks were independent of grease, which means that high COD load cannot be directly attributed to lipids. This statement strengthens the hypothesis that dissolved non-lipid anthropogenic chemicals (surfactants or emerging contaminants) are the primary drivers of chemical oxygen demand. This identification fits the principles of stoichiometry of oxidation, phase partitioning, biological selectivity, and source tracking. The organic and inorganic transport vectors investigated provide deeper understanding for forensic auditing. The objective of the further phase is to distinguish between benign organic loadings (e.g., food waste) and recalcitrant chemical stress using ratio-based diagnostics.

7.5 FORENSIC DIAGNOSTICS OF BIODEGRADABILITY STRESS

This section identifies signals that are consistent with chemically complex and slowly degradable influent conditions in the WWTP.

7.5.1 Evaluating the Role of Oil and Grease

In an institutional environment, kitchen-derived lipids can artificially increase COD values and interfere with the treatment performances. There would be failure in physical separations rather than chemical treatment limitations if COD values were lipid-driven. Therefore, to isolate the emerging contaminants due to the higher COD/BOD ratios to emerging contaminants, lipid interference was ruled out.

Forensic source analysis (Figure 7.3a) showed that O&G concentrations across all systems remained consistently low during routine sampling, with an overall mean of 2.8 mg/L (IQR: 2.1–3.4 mg/L) and a maximum of 5.0 mg/L. This low lipid baseline, combined with the negative correlation established in Section 7.4 ($\rho = -0.52$ with COD), suggests that lipids are unlikely to be the dominant contributor to elevated COD values during normal operating conditions.

An important caveat must be noted. Episodic lipid discharge, such as a periodic canteen deep-cleaning event, may not align temporally with our routine sampling windows. The consistently low O&G measurements indicate that if such events occur, they are not captured in this monitoring dataset. What we can say with confidence is that during periods of representative operation, O&G does not explain the elevated COD/BOD ratios at STP-4.

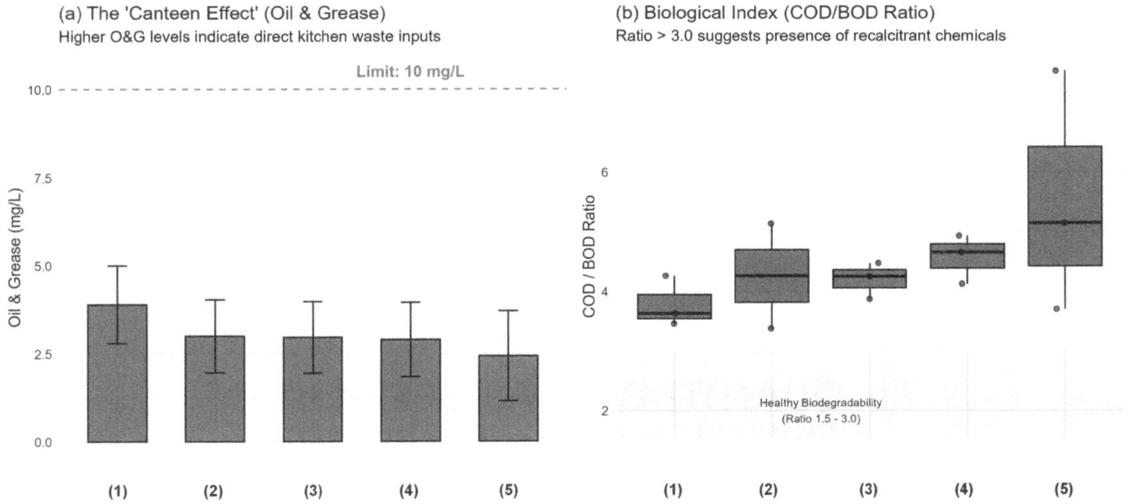

FIGURE 7.3 Forensic source tracking diagnostics: (a) Box-plot distribution of oil and grease concentrations across all STPs. O&G levels remained consistently low across all plants, with an overall mean of 2.8 mg/L (IQR: 2.1–3.4 mg/L) and a maximum of 5.0 mg/L. (b) Longitudinal tracking of COD/BOD ratios (2021–2024) comparing STP-4 against the network baseline. STP-4 exhibited a mean ratio of 5.2 (IQR: 4.8–5.7) with occasional peaks exceeding 6.0, consistently above the chemical stress threshold of 4.0. In contrast, the other four STPs maintained mean ratios of 2.3–2.8 (IQR: 2.1–3.1), within the moderately biodegradable range. Shaded bands represent ±1 standard deviation.

This group matters for two reasons. First, as it supports the interpretation, the observed COD/BOD divergence during routine monitoring dissolves chemically complex constituents rather than particulate fat. Second, it highlights a key limitation of conventional grab sampling: episodic pollution events may go undetected. This reinforces the value of the forensic ratio approach as a screening tool that flags the need for more targeted investigation.

The chemical stress signal at STP-4, therefore, points toward other recalcitrant organics—a hypothesis that warrants targeted investigation in future studies using higher-frequency sampling or event-triggered monitoring.

7.5.2 Screening-Level Indicator

The COD/BOD ratio was used as the primary BI, providing a baseline check for screening. A typical domestic wastewater ratio ranges between 1.5 and 2.5 (sometimes up to 3.0, Rudaru et al., 2022). Figure 7.3b displays the four STPs are operated within an acceptable range, while one of the plants (STP-4) exhibited a consistently higher ratio (5.0), exceeding the occasional outlier (6.0). Longitudinal monitoring (2021–2024) confirmed that the 'chemical stress' signal remained persistent across STP-4 during the seasonal cycles. Even though being stable hydraulic flows, the STP behaves chronically in nature rather than being an episodic source of recalcitrant matter. The results from this approach should be considered as screening hypotheses rather than definitive conclusions. The COD/BOD ratio is an indirect indicator, suggesting the chemically complex organic matter. These indicators cannot identify specific compounds. In residential and institutional settings, the higher ratios of the complex organic matters can be formed due to quaternary ammonium compounds (disinfectants), refractory organics, laundry surfactants, or non-domestic contributors such as workshops, laboratories, or hospital discharges (Płuciennik-Koropczuk & Myszograj, 2019; AbdulRazzak, 2023). This study has not incorporated a direct chemical validation via liquid chromatography–tandem mass spectrometry (LC–MS/MS). Without a molecular confirmation, the presence of emerging contaminants—such as pharmaceuticals or microplastics—remains an inference, not a directive conclusion. These forensic signals identified should be viewed as early warning indicators that can justify targeted follow-up investigations, not as definitive proof.

For example, STP-4 as a potential chemical stressor was cross-referenced with O&G data. The O&G levels at this STP remained consistently low (<3.0 mg/L), suggesting that the "chemical shadow" is not driven by other recalcitrant organics rather than kitchen-related fats or oils. This indicator can serve as a practical early warning signal. From a management perspective for strategic implications, the identified specific STPs (such as STP-4) indicate the need for advanced monitoring or tertiary treatment enhancements (e.g., activated carbon), rather than opting for an expensive upgrade for the entire network.

7.6 QUANTITATIVE ASSESSMENT OF RESILIENCE AND CIRCULARITY

7.6.1 System Resilience and Water Mass Balance

Even though the biodegradability stress signals found above represent that the decentralized network states a strong functional resilience, Figure 7.4 shows that the water mass-balance analysis found approximately 385,204 m^3 of wastewater recorded, treated, and productively reused annually.

The system's circularity was quantified using a mass-balance approach derived from longitudinal flow monitoring. Reuse efficiency was calculated by Equation 7.2:

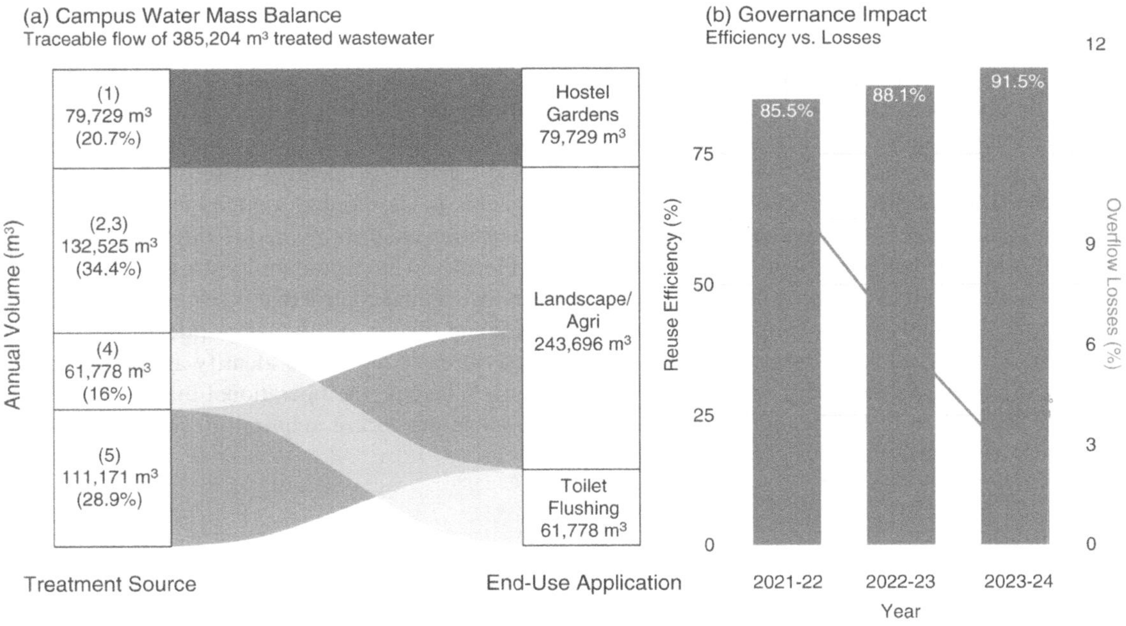

FIGURE 7.4 Annual water mass balance (Sankey diagram) demonstrating physical separation of wastewater streams and quantifying system-wide reuse efficiency using Equation 7.2. Total wastewater treated: 420,988 m³/year.

$$Reuse\ efficiency(\%) = \frac{V_{reuse}}{(V_{total})} \times 100 \qquad (2)$$

where V_{reuse} represents the metered annual volume reused for irrigation and flushing (≈ 385,204 m^3 year^{-1}) and V_{total} denotes the total treated effluent volume. The resulting reuse efficiency of 91.5% reflects the high level of hydraulic recovery achieved through the integrated distribution and storage network, as illustrated in the Sankey mass-flow diagram (Figure 7.4). Overflow and minor conveyance losses accounted for the remaining ~8.5% of the treated effluent.

Nearly 91.5% of the treated effluent is reused for non-potable applications, which are mainly contributed to landscape irrigation (~243,696 m^3) and toilet flushing (~61,778 m^3), along with agricultural and horticultural reuse within the campus. Based on location-based source separation, Figure 7.4 shows high-strength canteen wastewater is physically isolated and directed to STP-2, separate from the residential sewage lines. This confirmation supports the chemical stress observed in STP-4. This clearly indicates that the treatment plant is not influenced by the lipids but rather by localized point-source inputs.

The campus treatment system highlights the adaptive nature of circularity that is beyond absolute volume. Multi-year operational data also indicate an improvement in reuse efficiency from 85.5% in earlier phases to 88.1% and are currently efficient at 91.5% during the monitoring period. This trajectory is influenced by learning-by-operation and infrastructure optimization and incorporates the routing of reuse flows.

7.6.2 Governance, Digital Monitoring, and Hydraulic Resilience

These improvements strengthen hydraulic resilience where decentralized systems have the ability to reduce physical losses while maintaining operational control. The campus maintains multi-year operational logs that record 8.5% overflow losses have been reduced to 5.0% because of the monitoring sensor-based systems. The results are an established framework obtained by socio-technical systems theory. These compare engineering resilience (the ability to return to equilibrium after distribution) and adaptive resilience (the capacity to reorganize and adapt under stress) (Holling, 1973; Folke et al., 2010). In decentralized wastewater infrastructure, hydraulic resilience is nothing but an ability to maintain service continuity despite variable inflows; this represents an engineering resilience perspective.

This study is extended toward adaptive resilience by identifying latent chemical stresses before they trigger system failure. This aligns with Johannessen and Wamsler's (2017) observation that urban water systems require both structural reliability and diagnostic capacity to identify disturbances. Similarly, Butterworth et al. (2018) argued that decentralized systems in developing systems must evolve from reactive maintenance toward proactive monitoring—a transition enabled by the IoT infrastructure documented here. The campus network has an ability to detect localized biodegradability stress at STP-4 while maintaining overall hydraulic stability. These results provided a dual control that is seen when the system is both mechanically robust (minimizing physical losses) and diagnostically aware (identifying chemical stress hotspots). This bifurcation is important because forensic indicators identify areas of potential chemical and biodegradability stress (quality dimension). Whereas, the operational evidence confirms that the network is hydraulically mature and stable. A stable network is required to support advanced treatment and handle complex chemical loads.

7.7 CONCLUSION

This chapter processes and highlights a biodegradability-resilience forensic framework for studying the DWTSs used in a large residential campus in rural Southern India. The study area functioned as a microcosm that found critical distinction between regulatory compliance and safety. The hidden chemical stress during standard compliance monitoring was the robust baseline performance (with removal efficiencies exceeding 87% for BOD and 98% for TSS). The forensic application of the COD/BOD ratio highlighted the latent vulnerabilities. The STP-4 exhibited a ratio approaching 5.0, indicating a presence of recalcitrant and non-biodegradable organic fractions in the compost. Source differentiation identified that oil and grease (<3.0 mg/L) with negative correlation were ruled out as dietary lipids. Since they cause operational stresses. The approach further confirmed that the observed divergences were driven by dissolved anthropogenic chemicals, not by the kitchen waste. This caused localized chemical stresses on the treatment units, and the campus demonstrated exceptional hydraulic integrity. The integration of digital governance (119 IoT sensors) reduced physical overflow losses from 8.5% to 2.0%. This study validates that decentralized systems can be hydraulically stable but also chemically stressed at the same time. The finding of this study presents that low-cost monitoring data, when interpreted through a forensic lens, can serve as an effective early warning screening tool. This tool identifies potential hotspots of emerging contaminants before they lead to systemic failure. While screening can be used as a preliminary tool, validation with molecular confirmation techniques such as LC–MS/MS is still required to confirm the specific pollutants and right treatment plan.

7.8 FUTURE RESEARCH AND STRATEGIC OUTLOOK

In order to advance in the sustainability domain for decentralized sanitation systems that undergo the pressure of emerging contaminants, future research and strategies should focus on targeted tertiary treatment units, integrating IoT sensing technology and advanced molecular-level analysis.

The implementation of expensive setups can be replaced by investments in diagnostic areas like focusing on the stressed STPs and incorporating advanced oxidation processes (AOP), activated carbon beds, and filters. The source points can be targeted and managed to avoid the operational and maintenance load on the systems. The next level of governance is integrating digital grids for effective management of flows in the units. Low-cost quality sensors to monitor real-time EC and TDS can provide instant alerts for inorganic shocks. Future studies should aim to correlate the high COD/BOD ratios using molecular markers, which can include predictive analysis via machine learning, and calibrate the forensic index for refining diagnostic precision in the frameworks.

REFERENCES

Abdul Razzak, A. M. (2023). Performance evaluation of Al-Khadraa' wastewater treatment plant, Mosul-Iraq. *Journal of Engineering,* 29(4), 15–28. https://doi.org/10.31026/j.eng.2023.04.02

Abdalla, K. Z., & Gina, H. (2014). Correlation between biochemical oxygen demand and chemical oxygen demand for various wastewater treatment plants in Egypt to obtain the biodegradability index. *International Journal of Sciences: Basic and Applied Research*, 13(1), 42–48.

Butterworth, J., Suteethorn, K., & Leal, M. C. (2018). Resilience and decentralized wastewater management in peri-urban areas: A framework for analysis. *Water Policy*, 20(3), 487–503.

Capodaglio, A. G. (2017). Integrated, decentralized wastewater management: Trends and challenges. *Journal of Reliability and Safety,* 6(2), Article 22. https://doi.org/10.3390/resources6020022.

Central Pollution Control Board. (2025). *Draft guidelines for the pharmaceutical industry in India* (Draft for public consultation). Ministry of Environment, Forest and Climate Change, Government of India. https://lexplosion.in/wp-content/uploads/2025/12/Draft-Guidelines-for-Pharmaceutical-Industry-in-India-1.pdf

Diwyanjalee, G. R., Bellanthudawa, B. K. A., De Silva, D. K. N. S., & Gunawardena, A. R. (2024). Biodegradability index (BDI) as an indicator for effluents quality measurement: A case study based on different industry sectors in Matara District, Sri Lanka. *Water Practice & Technology*, 19(8), 3092–3108. https://doi.org/10.2166/wpt.2024.183

Etafo, N. O., Adekanmi, D. G., Awobifa, O. S., Torres, J. R. P., Herrera, L. A. I., & Awobifa, O. A. (2025). Clean and green: The multifaceted solution of the electrocoagulation technology in emerging contaminants in wastewater. *Discover Civil Engineering*, 2(1), Article 103. https://doi.org/10.1007/s44290-025-00261-5

European Parliament and Council of the European Union. (2024, November 27). Directive (EU) 2024/3019 of the European Parliament and of the Council of 27 November 2024 concerning urban wastewater treatment (recast). *Official Journal of the European Union, L*, 2024/3019. https://eur-lex.europa.eu/eli/dir/2024/3019/oj

Folke, C., Carpenter, S. R., Walker, B., Scheffer, M., Chapin, T., & Rockström, J. (2010). Resilience thinking: Integrating resilience, adaptability and transformability. *Ecology and Society*, 15(4), *Article* 20. https://doi.org/10.5751/ES-03610-150420

Holling, C. S. (1973). Resilience and stability of ecological systems. *Annual Review of Ecology and Systematics*, 4(1), 1–23. https://doi.org/10.1146/annurev.es.04.110173.000245

Johannessen, Å., & Wamsler, C. (2017). What does resilience mean for urban water services? *Ecology and Society*, 22(1), Article 1.

Lacalamita, D., Mongioví, C., & Crini, G. (2024). Chemical oxygen demand and biochemical oxygen demand analysis of discharge waters from laundry industry: Monitoring, temporal variability, and biodegradability. *Frontiers in Environmental Science*, 12, 1387041. https://doi.org/10.3389/fenvs.2024.1387041

Lourenço, N., & Nunes, L. (2020). Review of dry and wet decentralized sanitation technologies for rural areas: Applicability, challenges and opportunities. *Environmental Management*, 65, 642–664. https://doi.org/10.1007/s00267-020-01268-7.

Maddah, H. A. (2022). Predicting optimum dilution factors for BOD sampling and desired dissolved oxygen for controlling organic contamination in various wastewaters. *International Journal of Chemical Engineering*, 2022, 1–14. https://doi.org/10.1155/2022/8637064

Matesun, J., Petrik, L., Musvoto, E., Ayinde, W., & Ikumi, D. (2024). Limitations of wastewater treatment plants in removing trace anthropogenic biomarkers and future directions: A review. *Ecotoxicology and Environmental Safety*, 281, 116610. https://doi.org/10.1016/j.ecoenv.2024.116610

Ministry of Jal Shakti. (2022). *National framework on safe reuse of treated water.* Department of Water Resources, River Development and Ganga Rejuvenation, Government of India. https://nmcg.nic.in/

Płuciennik-Koropczuk, E., & Myszograj, S. (2019). New approach in COD fractionation methods. *Water*, 11(7), 1484. https://doi.org/10.3390/w11071484

Rudaru, D., Lucaciu, I., & Fulgheci, A. (2022). Correlation between BOD_5 and COD – Biodegradability indicator of wastewater. *Romanian Journal of Ecology & Environmental Chemistry*, 4(2), 80–86. https://doi.org/10.21698/rjeec.2022.207

Sha, C., Shen, S., Zhang, J., Zhou, C., Lu, X., & Zhang, H. (2024). A review of strategies and technologies for sustainable decentralized wastewater treatment. *Water*, 16(20), 3003. https://doi.org/10.3390/w16203003

Sravan, J. S., Matsakas, L., & Sarkar, O. (2024). Advances in biological wastewater treatment processes: Focus on low-carbon energy and resource recovery in biorefinery context. *Bioengineering*, 11(3), 281. https://doi.org/10.3390/bioengineering11030281

Starkl, M., Brunner, N., Das, S., & Singh, A. (2022). Sustainability assessment for wastewater treatment systems in developing countries. *Water*, 14(2), 241. https://doi.org/10.3390/w14020241

Chemical Tools for Tracking and Targeting Emerging Water Contaminants

8

Soham Kundu, Jyotiraditya Mallik, and Shampa Sen

8.1 INTRODUCTION

In the 21st century, with the advent of technology in sectors like petrochemicals, pharmaceuticals, cosmetics, and electronics, new types of pollutants have emerged. These are termed as "emerging" contaminants (ECs). An EC is a pollutant that has recently been identified or has gone unnoticed in the environment for some time. It has not been a reason for interest, because of the small number of identified adverse health and environmental implications. With rising concentrations in environmental niches, these contaminants are proving detrimental to the organisms living there. ECs generally include pollutants such as pharmaceuticals, nanoparticles, and personal care products (Krishnan et al., 2023). The treated water from wastewater treatment plants (WWTPs) still contains ECs, as they are equipped with analytical techniques that can detect and remediate only conventional pollutants (Figure 8.1). This leads to unaccounted discharge of ECs into the environment in large amounts.

Acetamiprid, a cyano-substituted neonicotinoid, has been used in Japan since 1995 and was later approved in other countries. This is now considered an emerging pollutant (EP) owing to its rising adverse impacts. It has been found to interfere with the reproductive success of honeybees and solitary bees and thus significantly disrupts the foraging behavior, which in turn affects agriculture (Shi et al., 2020; Mayr et al., 2025). Studies with respect to human health implications are showing accumulation in the breast, which promotes cancer and also in the cerebrospinal fluid of children's brains, which leads to developmental neurotoxicity (DNT) (Li et al., 2022a; Altman and Bayer, 1985). This happened earlier with another pesticide, DDT (Dichlorodiphenyltrichloroethane). DDT was introduced in 1945 by the U.S. government as a "miracle drug" capable of wiping out pests which cause diseases like malaria, dengue, and typhus. It was not until 1962 that its detrimental effects were determined by an American marine biologist, Rachel Carson. Later, the Environmental Protection Agency (EPA) completely banned the use of DDT in 1972 when it was conceived as an EP (Wurster, 2015; Conis, 2016).

DOI: 10.1201/9781003741626-8

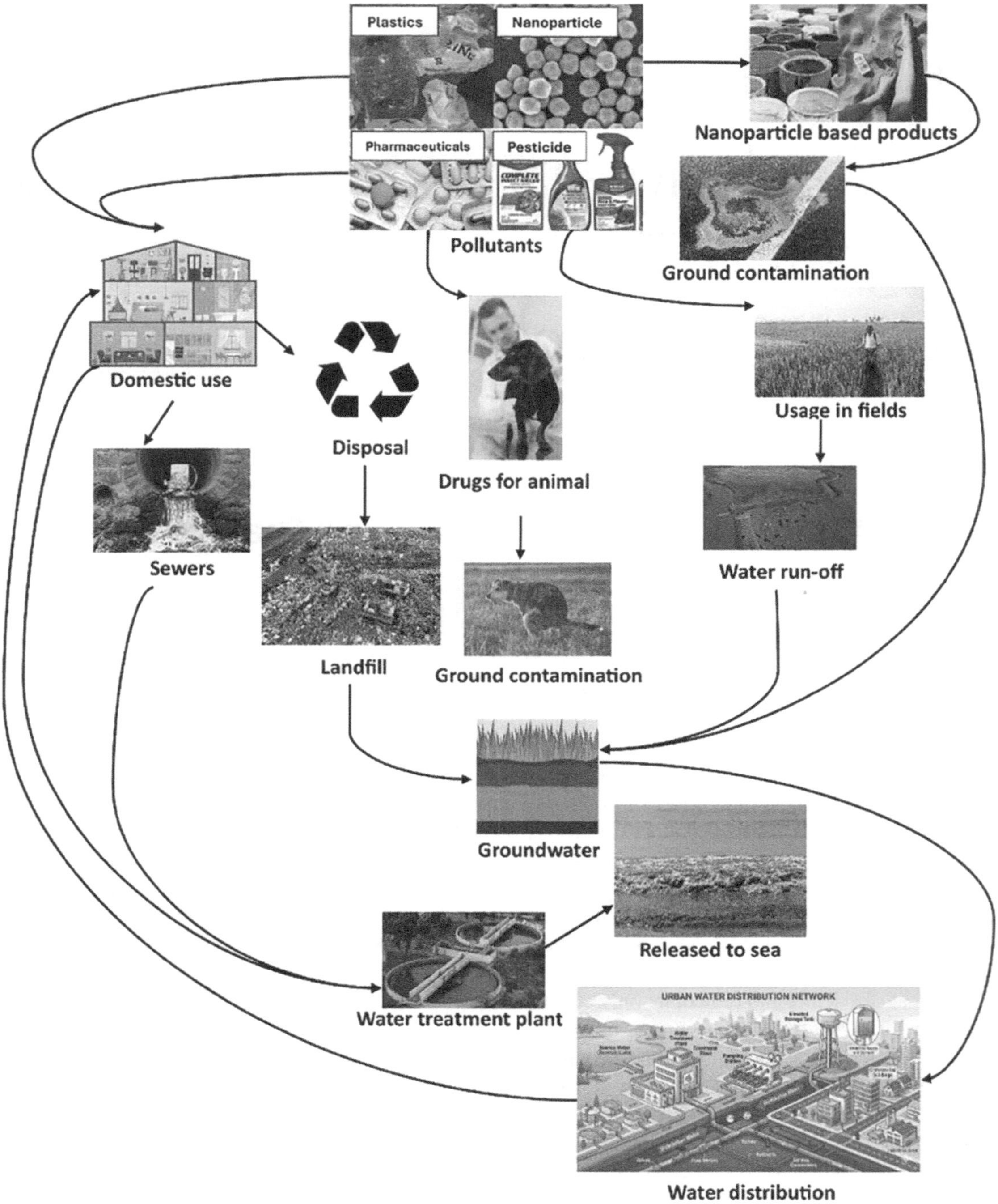

FIGURE 8.1 Dispersion of ECs into the environment.

Pollutants such as microplastics (MPs), pharmaceuticals, and nanoparticles came to light in the late 90s and early 2000s. The term microplastics was coined by Thompson et al. (2004) to refer to microscopic particles of various polymers while working on water samples collected from various sources. The improper disposal of commercial items such as clothing and packaging materials, which typically contain plastic, is responsible for this type of water pollution. Personal protective equipment (PPE) and face masks are also responsible for causing MP pollution in water sources (Li et al., 2022b; Ray et al., 2022).

Improper disposal of pharmaceuticals results in adverse environmental conditions due to their persistence. It gained a brief spotlight in the late 90s, when vulture populations in the genus *Gyps* across the Indian subcontinent experienced a catastrophic decline (Prakash et al., 2003). On investigation, it was revealed that the cause was diclofenac, a non-steroidal anti-inflammatory drug used as a veterinary drug at livestock farms (Oaks et al., 2004; Swan et al., 2006). During the COVID-19 pandemic, pharmaceutical pollution surged. Large amounts of medical waste were generated worldwide and, when improperly disposed of, found their way into water sources. Drugs like antivirals, antimicrobials, and psychotropics were detected in the waters of Greece, Vietnam, China, the USA, Austria, Portugal, etc. (Xia et al., 2024). In countries such as Australia, the Netherlands, the USA, and India, SARS-CoV-2 RNA was detected in wastewater (Wu et al., 2020; Ahmed et al., 2020; Kumar et al., 2020; Medema et al., 2020).

Nanoparticles are not yet known to have adverse effects on the environment or human health. However, their numerous uses across industries such as paints, sunscreens, and the coal industry and improper disposal envisage their prospects of becoming EC in the near future. Nanoparticle pollution was brought to attention in 2017, when Michael Hochella and his team demonstrated how industrial coal burning causes TiO_2 nanoparticles in their Magnéli phases ($Ti_x O_{2x-1}$ with $4 \leq x \leq 9$) to be released into the environment, which eventually finds its way into living organisms (Yang et al., 2017).

With greater understanding, scientists have begun to highlight these pollutants to raise public awareness of the potentially disastrous effects of ECs on the environment and human health (Table 8.1). On June 28, 2017, the U.S. EPA acknowledged the imminent threat ECs pose to the world at large. The United Nations Environment Assembly (UNEA) passed a resolution on March 15, 2019, which placed heavy emphasis on global cooperation to better manage chemicals and all wastes throughout their life cycle, while maintaining the agreed international frameworks. It also vouched for their reduction by releasing them into the air, water, and soil to minimize their detrimental effects on human health and the ecosystem (UNEP, 2019). Respectively, organizations such as the U.S. EPA, the European Union, and the UNEA began dedicating resources to research on novel contaminants and laid out new guidelines for regulation, detection, and possible elimination of ECs from the environment.

There are two challenges with ECs: (a) to remove and (b) to detect (primary focus of this chapter). New techniques are to be implemented to tackle the ineffectiveness and inefficiencies of current techniques. The scientific community is seeking new analytical detection techniques to expand research into these new pollutant groups.

8.2 CONVENTIONAL EC DETECTION TECHNIQUES

Conventional detection techniques for ECs in wastewater include Raman spectroscopy, Fourier transform infrared (FTIR) spectroscopy, GC–MS, LC–MS, and LC–MS.[2] These techniques are highly sophisticated and sensitive. Use of these techniques for this purpose dates back to the 1990s.

8.2.1 Raman Spectroscopy

Raman spectroscopy has been used to detect pollutants in water since the late 1960s. Bradley and Frenzel (1970) preferred RS over infrared spectroscopy for contaminant detection in water, as Raman spectra do not get absorbed by molecules of water, unlike infrared spectra. Raman spectroscopy is used to detect spectra resulting from the scattering of monochromatic radiation by molecules that exhibit a dipole moment not arising from prior absorption or emission of electromagnetic radiation (Figure 8.2). Analysis of spectra shows that, apart from inelastic scattering (scattering without a change in wavenumber of the incident light), these also contain radiation of altered wavenumbers. The altered wavenumbers may be greater than the unchanged wavenumber (anti-Stokes lines) or less than it (Stokes lines). RS is non-invasive and

TABLE 8.1 Types of EC and their methods for detection

TYPES OF CONTAMINANTS	*EXAMPLES*	*SOURCES*	*HARMFUL EFFECTS*	*DETECTION TECHNIQUES*	*REFERENCES*
Microplastics	Polyamide, low-density polyethylene, polyethylene terephthalate, polypropylene, polyurethane, polystyrene, polycarbonate	Urban runoff, discharge from industries like petrochemical industries and the degradation of larger plastic objects like plastic bottles, plastic bags, and fishing gear	Causes alterations in gastrointestinal tract physiology, immune system depression, oxidative stress, cytotoxicity, differential gene expression, and growth inhibition in aquatic animals	Metal oxide sensors, FTIR-Raman spectroscopy, gas chromatography–mass spectrometry (GC–MS), liquid chromatography–mass spectrometry (LC–MS)	Kim et al. (2026), Osman et al. (2023), Muthulakshmi et al. (2023), Biswas (2025)
Pharmaceuticals	NSAIDs and analgesics (Ibuprofen, Diclofenac, Ketoprofen), antibiotics (Tetracycline, Azithromycin, Ofloxacin), antiretrovirals (Lamvudine, Zidovudine, Nevirapine)	Urban wastewater, runoff from agricultural and livestock farms, and discharge from pharmaceutical industries	Disrupts naturally occurring microbiota of aquatic ecosystems, spreads antibiotic resistance genes to potentially pathogenic microorganisms and causes Alzheimer's disease, cancer, and thyroid problems in humans	Electrochemical sensor based on silver–palladium bimetal nanoparticles and reduced graphene oxide, electrochemical sensor based on carbon screen-printed electrodes (C-SPEs), FTIR, liquid chromatography–mass spectrometry (LC–MS), LC with tandem MS, layered double hydroxide-based electrochemical and optical (bio)sensors	Ortúzar et al. (2022), Fdez-Sanromán et al. (2025), Dou et al. (2019), Fatta et al. (2007), Sohrabi et al. (2022)
Per- and polyfluoroalkyl substances (PFAS)	Trifluoroacetic acid, perfluorooctanoic acid (PFOA), and perfluorooctanesulfonic acid	Discharge from chemical, paper and textile manufacturing industries, urban wastewater, landfills, consumer products like food wrappers, food containers, and non-stick cookware	Immune system suppression, development delays, and endocrine disruption in infants	Field-effect transistor sensors, solid-phase extraction, and liquid chromatography–tandem mass spectrometry, eosin Y-based "turn-on" fluorescent sensor	Gautam et al. (2025), Wang et al. (2025), Rehman et al. (2023), Zhang et al. (2021), Liang et al. (2015), Menger et al. (2021)
Nanoparticles	Ag, AgO, and TiO_2, as well as ZnO, SiO_2, and CuO	Waste water treatment plants, electronic waste, and construction materials	Induce necrotic plasma membrane rupture or apoptosis, disrupt cell signaling, and accumulate in the respiratory system, leading to lung inflammation, fibrosis, and even endocrine disruption by attaching to hormone receptors	Single-particle inductively coupled plasma mass spectrometry, fluorescent labeling	Doskocz et al. (2025), Singh et al. (2024), Long et al. (2016), Gómez-Pertusa et al. (2025), Chen et al. (2023)

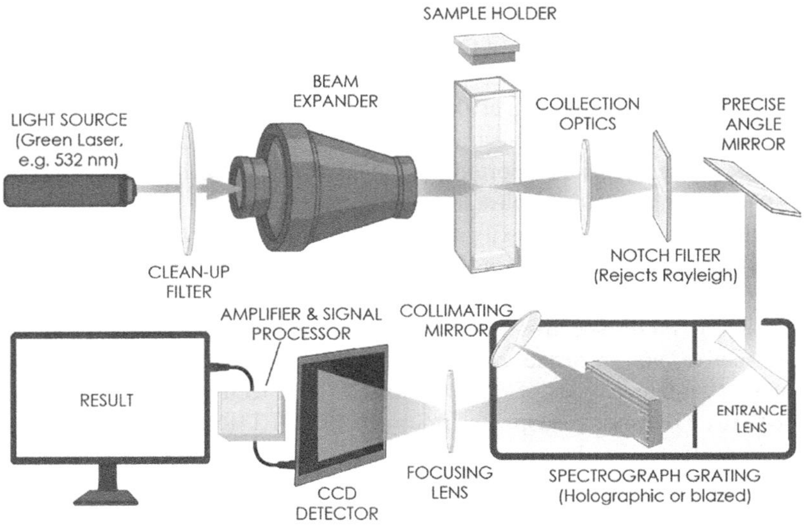

FIGURE 8.2 Raman spectroscopy (drawn by BioRender©).

non-destructive, and samples generally do not require pre-processing, but it has its limitations. Alterations in the Raman spectrometer can disrupt the reproducibility of spectroscopic signals (Guo et al., 2020). Fluorescence from samples, if present, often interferes with weaker RS signals, leading to inaccurate readings (Wei et al., 2015). Variations of the RS technique are much more robust than the basic RS used in the late 20th to the early 21st century. Variations that have been found useful for EC detection include micro-Raman spectroscopy and surface-enhanced Raman scattering (SERS) spectroscopy.

SERS spectroscopy is a widely used variation of Raman scattering. SERS spectroscopy utilizes metallic nanoparticles or nanostructures that adsorb contaminant particles (Xu et al., 2013). Photons, upon interaction with metal nanostructures, give rise to localized surface plasmon resonance, which in turn leads to field enhancement (Lal et al., 2007). Raman spectroscopy combined with a sol system of silver nanoparticles (Ag nanoparticles) and sodium iodide (NaI) was used to detect trace amounts of plastic particles as small as 20 nm, qualitatively and semi-quantitatively (Ruan et al., 2024). Huang et al. (2022) used 40 nm silver nanoparticles (AgNPs) drop-casted on aluminum substrates to detect trace amounts (20 femtograms per liter) of PFAS in water. However, SERS has its drawbacks, as silver nanoparticles rapidly oxidize in ambient air, degrading their plasmonic properties (Hosomi et al., 2019).

Micro-RS (μRS) is a variation in which an optical microscope is integrated into a surface-enhanced Raman spectrometer, which measures Raman spectra and generates chemical images of samples (Chakraborty et al., 2023). In Germany, Schymanski et al. (2018) used micro-RS to detect MPs (5μ) in packaged drinking water. RS, in combination with a convolutional neural network (CNN), was used to quantify MP contamination in water from various sources (Luo et al., 2024). μRS had been shown to characterize MPs. Danial Adib et al. (2021) compared various spectra obtained from samples of drinking water treatment plants. The data were compared with the Infrared and Raman Users Group library, which showed ten different polymers (polyethylene, polypropylene, polyethylene terephthalate, polystyrene, polytetrafluoroethylene, polyurethane, polyamide, polybutylene terephthalate, polyvinylidene fluoride, polyvinyl chloride, and polycarbonate) to be present.

8.2.2 Fourier Transform Infrared (FTIR) Spectroscopy

FTIR spectroscopy, like Raman spectroscopy, is based on the vibration and rotation of atoms. Also, molecules with a dipole moment can be detected by this technique. When a sample is irradiated with infrared light of a certain frequency, electrons absorb energy, resulting in a differential absorbance spectrum that depends on the functional groups present in the molecule. FTIR spectroscopy has been applied to detect a variety of ECs, including MPs such as polyamide, polyethylene terephthalate, acrylonitrile butadiene styrene, acrylic, and many more (Sefiloglu et al., 2024). Detection of pharmaceutical compounds, namely, paracetamol, desloratadine, ibuprofen, β-estradiol, ethynylestradiol, carbamazepine, sulfamethoxazole, and herbicides like atrazine, etc., was reported (Quintelas et al., 2020). The major advantage of FTIR is the rapid non-destructive analysis of samples. But FTIR offers functional group identification, not detailed structural information. Structural complexity of samples and interference from water and other interfering substances may hinder the analysis using FTIR. To overcome these limitations, various upgrades to FTIR spectroscopy are being studied for the detection of EC in wastewater.

Various improvised FTIR spectroscopy methods eventually emerged as fully digitalized, more sensitive, non-destructive, non-intrusive, cost-effective, and rapid, capable of simultaneously detecting hundreds of solutes. Two major types of these advanced FTIR techniques are, namely, attenuated total reflection FTIR (ATR-FTIR) spectroscopy and micro-FTIR (μ-FTIR) spectroscopy.

In ATR-FTIR spectroscopy, the sample is placed in contact with a crystal having a higher refractive index. An infrared beam is directed onto the crystal, which strikes the crystal–sample interface at an angle greater than the angle of incidence. Then, the beam undergoes total internal reflection, which creates an evanescent wave at the crystal–sample interface. This wave probes into the sample to a depth of a few micrometers to reveal the molecular structure. Obinaju and Martin (2016), detected various types of MPs and aromatic hydrocarbons (PAHs) using ATR-FTIR spectroscopy. ATR-FTIR spectroscopy can thus analyze water-based samples and is therefore more promising for screening ECs in wastewater (Kaur et al., 2021). ATR-FTIR spectroscopy, coupled with partial least squares regression (PLSR) and support vector machine (SVR) models, was used for in situ monitoring and quantitative prediction of nitrate content in industrial wastewater (Gan et al., 2022). One major drawback is that particles should be at least 200–500 μm in size for proper identification and to avoid damaging the ATR instrumentation (Meyers et al., 2024). Integration of an optical brightfield microscope with high resolution and the FTIR spectroscopy (μ-FTIR) device enables focusing on specific areas of a sample to classify chemical alterations. Thus, analysis of samples across 2D regions or 3D cubes became possible (Talukdar et al., 2024; Bhuiyan and Ngadi, 2023). As this variation follows the same principles of classic FTIR spectroscopy, it still suffers from the same drawbacks of FTIR. These upgraded FTIR techniques are often used in combination with other detection technologies to address the drawbacks of FTIR spectroscopy. Liu et al. used large area attenuated total reflectance-Fourier transform infrared spectroscopy (LAATR-FTIR) by combining μ-FTIR and ATR-FTIR spectroscopy for the detection of 1.3 μm MPs. Zhang et al. (2024) developed an eco-friendly, cost-effective, and precise detection technology for screening of MPs in the environment by combining μ-FTIR and μ-Raman spectroscopy. Both techniques are complementary.

8.2.3 GC–MS and LC–MS

Gas chromatography–mass spectroscopy (GC–MS) combines two techniques to identify each component in an unknown mixture. In gas chromatography, the sample is first vaporized and carried by an inert gas, generally helium. The sample gas is carried through a column packed or coated with a solid stationary phase. One or more components are preferentially attached to the solid phase due to non-covalent interactions, while other components of the gas mixture pass through the column without interruption. The resolved components are then ionized by electron ionization and sent to the quadrupole mass analyzer, which ensures that only fragments with a specific mass-to-charge (m/z) ratio travel toward the detector at a given time.

The liquid chromatography–mass spectrometry (LC–MS) technique is similar to GC–MS, except that GC is replaced by high-performance liquid chromatography (HPLC) and the way the sample is introduced into the detector in the mass spectrometer (Figure 8.3). In HPLC, the liquid sample is resolved

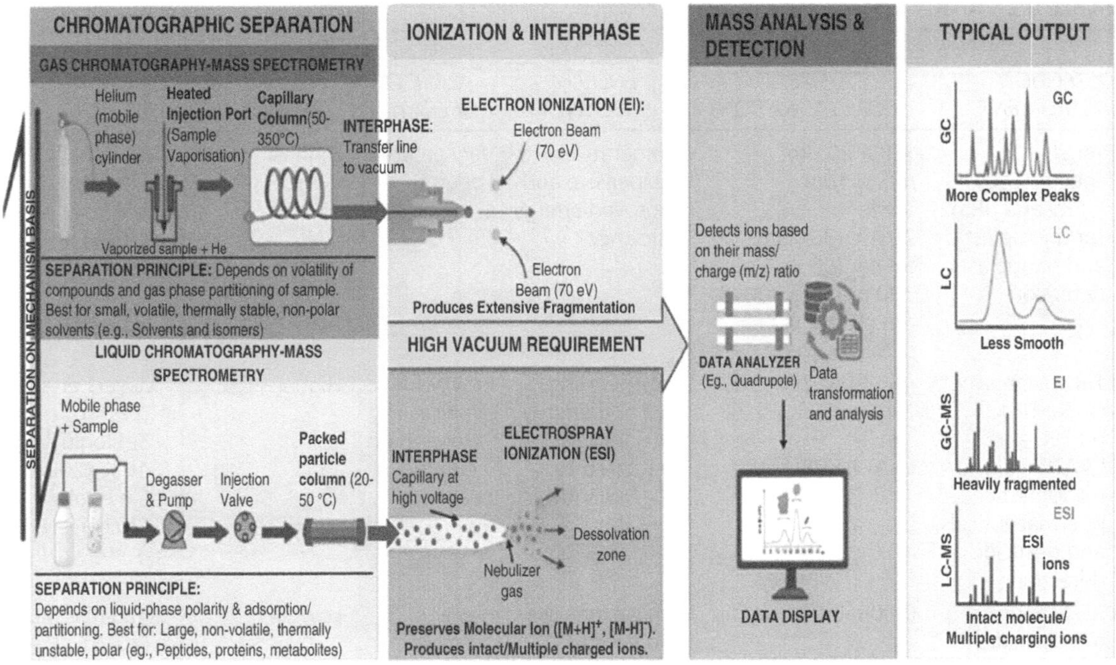

FIGURE 8.3 Differential analysis of contaminants based on their characteristics (drawn by BioRender©).

based on the preferential affinity of one or more components toward the stationary phase. The mobile phase carries other components to the MS. However, MS can only identify compounds in the gas phase. This phase change is done by heating the analyte. The ionization method used in LC–MS is different from that in GC–MS, owing to the different nature of the analyte (identifies more volatile compound mixtures). Electrospray ionization (ESI) and atmospheric pressure chemical ionization (APCI) are the most commonly used ionization methods for this technique. But the method for detecting the ions is the same as in GC–MS.

8.2.4 LC with Tandem MS (LC–MS^2)

Quadrupole instruments have become widely used analytical tools for the analysis of pollutants such as pesticides. However, several limitations must be overcome (Picó et al., 2004). The most important drawback is the inability to analyze the multi-component spectrum generated by the co-elution of weak fragment ions or single-molecule adducts (Hogenboom et al., 2000; Sandra et al., 2002). Advanced technologies such as LC–MS^2 are becoming more popular because of their high sensitivity and ability to confirm compounds even when compounds having the same molecular mass but different product ions co-elute. For this technique, there is no requirement for complete separation of target analytes, as the presence of a compound is confirmed by monitoring at least two transition states between precursor and product ions (Gros et al., 2006). This technique is useful for detecting pharmacologically active substances and pesticides (organochlorine compounds, including PCBs; organophosphorus compounds).

Among the conventional detection techniques discussed above, the choice of technique for a particular wastewater stream depends on resource availability and funding. A comparative analysis of these techniques is represented in Table 8.2 based on limit of detection (LOD), cost, energy efficiency, ease of operation, and environmental friendliness.

TABLE 8.2 Analysis of the conventional techniques

DETECTION INSTRUMENT	*LOD (PARTICLE SIZE/ CONCENTRATION)*	*COST AND ENERGY EFFICIENCY*	*EASE OF OPERATION*	*ENVIRONMENT FRIENDLINESS*	*REFERENCES*
Raman spectroscopy (SERS and μRS) for microplastic and nanoplastic detection	SERS: 20 nm μRS ≥ 1μm SERS: 20 nm (30 ng/L) 50 nm (50 ng/L) 100 nm and 300 nm (100 ng/L) μRS: 6.5 μg/mL	Comparatively inexpensive and improved energy efficiency	Very easy to operate	Highest	Wang et al. (2024), Mikac et al. (2023)
FTIR spectroscopy (ATR-FTIR, μ-FTIR, and LAATR-FTIR) for microplastic, pharmaceuticals, and pesticide detection	ATR-FTIR: 200–500 μm. μ-FTIR: 25–50 μm. LAATR-FTIR: 1.3–2.1 μm	Low operating cost and energy efficiency as compared to GC–MS and LC–MS	Easy, with minimal preparation required	High	Meyers et al. (2024), Sefiloglu et al. (2024), Liu et al. (2024)
GC–MS for microplastics and pharmaceuticals detection	0.005–0.001 μg/L	Highly expensive and energy-demanding	Has a complex preparatory phase	Low	Saadati et al. (2013)
LC–MS	0.01–0.1 μg/L	High cost with minimum energy consumption	Difficult to operate	Moderate	Omotola and Olatunji (2020)
LC–MS^2	0.15–10 ng/L	High initial and operating costs, but lower energy consumption as compared to GC–MS	Very difficult	The highest waste emission	Borrull et al. (2020)

8.3 ADVANCED EC DETECTION TECHNIQUES

Advances in conventional detection techniques for ECs have enabled the detection of contaminants with particle sizes as small as 100 nm. These detection techniques are, however, costly, sophisticated, sensitive, and slow in analysis. The alternatives for these are sensor-based detection techniques.

8.3.1 Sensor-Based Detection

Sensor-based analytical techniques are used to detect MPs, pharmacological substances, and PFAS. These are used because they are cost-effective and rapid detection methods, in contrast to laboratory techniques that are costly, time-consuming, and highly sensitive and selective. Sensors are basically of two types, electrical and non-electrical (Mahlein, 2016).

8.3.1.1 Electrochemical Sensors

Electrochemical sensing has emerged as a promising technique for detecting MPs, nanoparticles, and pharmacological substances owing to their faster detection ability, cost-effectiveness, high sensitivity and selectivity, simple operational procedure, and real-time monitoring perspective, making them suitable for in situ analysis (Mohamed, 2016; Kim et al., 2026; Costa-Rama and Fernández-Abedul, 2021). The detection principle of electrochemical sensors is based on a potentiostat, which measures voltage fluctuations caused by the oxidation or reduction of contaminant particles as they interact with the working electrode, which is maintained at a constant potential and submerged in the sample. When charged particles collide with the electrode, a signal is generated by electron transfer between them (Cheng and Compton, 2014). Electrochemical sensors, widely used for EC detection, are described below.

8.3.1.1.1 Metal Oxide Sensors

Metal oxide sensors are envisaged to show high performance in detecting MPs, specifically among all other ECs, even at very low concentrations and in complex wastewater, owing to their surface modification by weathering (Tang et al., 2020; Turner and Holmes, 2015). Several research publications have demonstrated that different MPs, such as polyvinyl chloride and polyformaldehyde, can interact with heavy metals such as Cu, Ni, Ag, Hg, Cr, Co, Zn, Cd, and Pb. Surface modifications of MPs can lead to a variety of interactions, such as polarization of the surface, which can induce Coulombic interactions (Zhang et al., 2020). It can also lead to bioaccumulation of metal oxides via biofilm formation or precipitation (Rochman et al., 2014; Ashton et al., 2010).

8.3.1.1.2 Field-Effect Transistor-Based Biosensors

The advent of field-effect transistor (FET)-based sensors for detecting ECs in wastewater monitoring is very recent, driven by advances in sensing channels. The FET sensor has three electrodes (gate, drain, and source), a sensing channel, and an insulating component. The gate electrode is insulated from all other counterparts by a dielectric material. At the same time, the other two electrodes are connected to the semiconducting layer. When a contaminant is adsorbed into the gate layer, an electric field is generated, which sets off the sensor. The newer types of FETs currently being studied are graphene-, transition metal dichalcogenide-, and phosphorene-based FETs (Mao et al., 2017). The pollutants that can be detected include pesticides, metals, heavy metals, and MPs (Elli et al., 2022).

8.3.1.2 Eosin Y-Based Fluorescent Sensor

An eosin Y-based fluorescent sensor is used for the detection of polyfluorinated compounds. Fluorescent sensors work on host–guest interactions, as discussed below in the detection of nanoparticle section. In this case, the host exhibits fluorescence, which, in the presence of polyethyleneimine (PEI), is quenched due to electrostatic interactions between Eosin Y and PEI. If polyfluoroalkyl substances (PFAS) are also present in the reaction mixture with PEI, quenching is reduced by electrostatic interactions between PFAS and PEI. This fluorescence recovery is detected, and the system acts as a sensor (Liang et al., 2015).

8.3.2 Nanoparticle Detection

Nanoparticles have been used for ages, but the idea that they could adversely affect the environment is very recent. The detection techniques that can be used are transmission electron microscopy, dynamic light scattering, fluorescent labeling, high-performance liquid chromatography–inductively coupled plasma mass spectrometry (HPLC–ICP-MS) and field flow fractionation–inductively coupled plasma mass spectrometry (FFF–ICP-MS). Among these, only two can be used for our purposes.

8.3.2.1 Detection by Fluorescent Labeling

Fluorescence makes the detection of labeled NPs highly effective and efficient, owing to the techniques' high sensitivity and selectivity. This allows for real-time monitoring and visualization of target particles in biological and environmental systems. Some nanoparticles form unique complexes through physical interactions with small molecules or proteins, exhibiting host–guest interactions. Host–guest reaction is demonstrated by self-assembled supramolecular amphiphilic fluorescent nanoparticles synthesized by Zhou et al. (2022). The host molecule was a terpyridine-modified arene coordinated to Zn^{2+}, and the guest molecule was polyethylene glycol. Muhammed et al. (2019) prepared arene-capped silver nanoclusters that exhibited host–guest interactions with neutral alkylamines and cationic quaternary ammonium compounds, resulting in remarkable luminescence enhancement.

8.3.2.2 Single-Particle Inductively Coupled Plasma Mass Spectroscopy (ICP-MS)

Long et al. (2016) used ICP-MS to determine the presence of gold nanoparticles in natural water. This can be extended to other nanoparticles in wastewater as well. ICP-MS is a highly sensitive technique that can detect nanoparticles and ionized elements in very low concentrations (effectively one nanoparticle) in a sample. When a water sample is analyzed, a constant, low-diffused signal is generally generated due to both dissolved ions and nanoparticles being present. If nanoparticles are present, then pulse signals are generated.

8.3.3 AI-Assisted Detection Techniques

AI-assisted detection techniques enhance the performance of existing technologies and integrate with them the ability to monitor ECs in real time and assess the information to evaluate the risk they pose. The information assessed can be necessary for early warning systems. Such developments can play an important role in maintaining water quality in accordance with water quality regulatory standards and also protect the ecosystem and its health. Yeliz Durgun (2024) introduced a method for assessing water quality by incorporating advanced machine learning algorithms, including random forest, support vector machines, and neural networks, to classify spectral analysis data and assess the effectiveness of UV disinfection in disinfected water samples. This type of approach can be used to analyze real-world wastewater samples and assess the risk they pose. Targeted screening is difficult because of limited resources for reference standards. Sun et al. (2025) introduced AI/ML-powered pseudo-target screening, using tetracycline (TC) as a base model to detect contaminants, as this antibiotic is among the most commonly detected EC in water systems. The analysis data were applied to the XGBoost model, which differentiated between TC and non-TC, with a correct identification rate of 100. Yue et al. (2025) showcased a graph neural network (GNN) and its ability to create embedded molecular structure vectors to analyze 542 data points from a dataset of 28 EC types and predict the oxidation rate by ozonization. Compilation and analysis of data on these diverse structures and properties of these ECs would have been a time-consuming and laborious process if done traditionally. A similar approach may be extended for detection data analysis for various other types of ECs in complex wastewater.

8.4 INSIGHT AND PROSPECTS

The detection of EPs has evolved from basic spectroscopic methods to highly integrated, multi-modal analytical platforms. Current research highlights a shift toward techniques that offer both high sensitivity and real-time monitoring capabilities. AI algorithms can process large amounts of data from sources such as sensors and satellites to identify changes in EC levels, demonstrating effective real-time monitoring.

Machine learning models can predict the fate and behavior of ECs, thereby enabling the development of effective remediation strategies. In the future, we can expect the use of detection techniques with advanced data processing to detect multiple compounds in a sample. Some of them are also in development. With growing awareness, it is being considered for smart labeling of substances, which could be a potential game-changer for visualizing contaminants in biological ecosystems.

REFERENCES

Adib, D., Mafigholami, R., & Tabeshkia, H. (2021). Identification of microplastics in conventional drinking water treatment plants in Tehran, Iran. *Journal of Environmental Health Science and Engineering*, *19*(2), 1817–1826.

Ahmed, W., Angel, N., Edson, J., Bibby, K., Bivins, A., O'brien, J. W., … & Li, J. (2020). First confirmed detection of SARS-CoV-2 in untreated wastewater in Australia: A proof of concept for the wastewater surveillance of COVID-19 in the community. *The Science of the Total Environment*, *728*, 138764. https://doi.org/10.1016/j.scitotenv.2020.138764.

Altman, J., & Bayer, S. A. (1985). Embryonic development of the rat cerebellum. III. Regional differences in the time of origin, migration, and settling of Purkinje cells. *Journal of Comparative Neurology*, *231*(1), 42–65.

Ashton, K., Holmes, L., & Turner, A. (2010). Association of metals with plastic production pellets in the marine environment. *Marine Pollution Bulletin*, *60*(11), 2050–2055.

Bhuiyan, M. H. R., & Ngadi, M. O. (2023). SR-FTIR microspectroscopy: Emerging 2D structural-chemical analytical technique for food quality and safety monitoring. *Food Bioscience*, *56*, 103392.

Biswas, R. (2025). State of the art detection methods of microplastics as marine litter: A mini review. *Discover Oceans*, *2*(1), 24.

Borrull, J., Colom, A., Fabregas, J., Borrull, F., & Pocurull, E. (2020). Liquid chromatography tandem mass spectrometry determination of 34 priority and emerging pollutants in water from the influent and effluent of a drinking water treatment plant. *Journal of Chromatography A*, *1621*, 461090.

Bradley, E. B., & Frenzel, C. A. (1970). On the exploitation of laser Raman spectroscopy for detection and identification of molecular water pollutants. *Water Research*, *4*(1), 125–128.

Chakraborty, I., Banik, S., Biswas, R., Yamamoto, T., Noothalapati, H., & Mazumder, N. (2023). Raman spectroscopy for microplastic detection in water sources: A systematic review. *International Journal of Environmental Science and Technology*, *20*(9), 10435–10448.

Chen, H., Hu, Q., Li, W., Cai, X., Mao, L., & Li, R. (2023). Approaches to nanoparticle labeling: A review of fluorescent, radiological, and metallic techniques. *Environment & Health*, *1*(2), 75–89.

Cheng, W., & Compton, R. G. (2014). Electrochemical detection of nanoparticles by 'nano-impact' methods. *TrAC Trends in Analytical Chemistry*, *58*, 79–89.

Conis, E. (2016). DDT disbelievers: Health and the new ECONOMIC poisons in Georgia after World War II. *Southern Spaces*, *28*. https://southernspaces.org/2016/ddt-disbelievers-health-and-new-economic-poisons-georgia-after-world-war-ii/

Costa-Rama, E., & Fernández-Abedul, M. T. (2021). Based screen-printed electrodes: A new generation of low-cost electroanalytical platforms. *Biosensors*, *11*(2), 51.

Doskocz, N., Affek, K., Matczuk, M., Drozd, M., & Załęska-Radziwiłł, M. (2025). Nanoparticles in wastewater: A comprehensive approach to understanding their ecotoxicity and genotoxicity. *Desalination and Water Treatment*, *321*, 100988.

Dou, N., Zhang, S., & Qu, J. (2019). Simultaneous detection of acetaminophen and 4-aminophenol with an electrochemical sensor based on silver–palladium bimetal nanoparticles and reduced graphene oxide. *RSC Advances*, *9*(54), 31440–31446.

Durgun, Y. (2024). Real-time water quality monitoring using AI-enabled sensors: Detection of contaminants and UV disinfection analysis in smart urban water systems. *Journal of King Saud University-Science*, *36*(9), 103409.

Elli, G., Hamed, S., Petrelli, M., Ibba, P., Ciocca, M., Lugli, P., & Petti, L. (2022). Field-effect transistor-based biosensors for environmental and agricultural monitoring. *Sensors*, *22*(11), 4178.

Fatta, D., Achilleos, A., Nikolaou, A., & Meric, S. (2007). Analytical methods for tracing pharmaceutical residues in water and wastewater. *TrAC Trends in Analytical Chemistry*, *26*(6), 515–533.

Fdez-Sanromán, A., Messaoud, N. B., Pazos, M., Rosales, E., & Queirós, R. B. (2025). Development of eco-friendly and cost-effective electrochemical sensor for the simultaneous detection of 4-aminophenol and paracetamol in water. *Sensing and Bio-Sensing Research*, *48*, 100782.

Gan, F., Wu, K., Ma, F., Wei, C., & Du, C. (2022). In-situ monitoring of nitrate in industrial wastewater using Fourier transform infrared attenuated total reflectance spectroscopy (FTIR-ATR) coupled with chemometrics methods. *Heliyon*, *8*(12), e12423.

Gautam, R. K., Mottaghipisheh, J., Verma, S., Singh, R. P., Muthukumaran, S., Navaratna, D., & Ahrens, L. (2025). PFAS contamination in key Indian states: A critical review of environmental impacts, regulatory challenges and predictive exposure. *Journal of Hazardous Materials Advances*, *18*, 100748.

Gómez-Pertusa, C., Grindlay, G., García, M. J. V., Yáñez, M. A., & Gras, L. (2025). Determination of nanoparticles in wastewaters by means spICP-MS after a separation/preconcentration treatment with nylon filters. *Talanta*, *298*, 128805.

Gros, M., Petrović, M., & Barceló, D. (2006). Multi-residue analytical methods using LC-tandem MS for the determination of pharmaceuticals in environmental and wastewater samples: A review. *Analytical and Bioanalytical Chemistry*, *386*(4), 941–952.

Guo, S., Beleites, C., Neugebauer, U., Abalde-Cela, S., Afseth, N. K., Alsamad, F., … & Bocklitz, T. (2020). Comparability of Raman spectroscopic configurations: A large scale cross-laboratory study. *Analytical Chemistry*, *92*(24), 15745–15756.

Hogenboom, A. C., Niessen, W. M. A., & Brinkman, U. T. (2000). Characterization of photodegradation products of alachlor in water by on-line solid-phase extraction liquid chromatography combined with tandem mass spectrometry and orthogonal-acceleration time-of-flight mass spectrometry. *Rapid Communications in Mass Spectrometry*, *14*(20), 1914–1924.

Hosomi, K., Takahiro, K., Nishiyama, F., & Yokoyama, S. (2019). Plasma-induced recovery of plasmonic sensitivity of aged silver nanoparticles to ethanol vapor and plasma exposure-time dependence. *Thin Solid Films*, *673*, 52–56.

Huang, T., McClelland, A., & Zeng, T. H. (2022, July). Trace PFAS detection in water sources using silver nanoparticles for surface-enhanced Raman spectroscopy (SERS). In *2022 IEEE 22nd International Conference on Nanotechnology (NANO)* (pp. 342–345). IEEE.

Kaur, H., Rana, B., Tomar, D., Kaur, S., & Jena, K. C. (2021). Fundamentals of ATR-FTIR spectroscopy and its role for probing in-situ molecular-level interactions. In *Modern Techniques of Spectroscopy: Basics, Instrumentation, and Applications*, ed. D.K. Singh, M. Pradhan, A. Materny (pp. 3–37). Springer Singapore.

Kim, E. B., Akhtar, M. S., Ameen, S., & Li, H. (2026). Metal Oxide Electrodes: A New Frontier in Electrochemical Microplastic Detection. *Trends in Environmental Analytical Chemistry*, *49*, e00289.

Krishnan, R. Y., Manikandan, S., Subbaiya, R., Karmegam, N., Kim, W., & Govarthanan, M. (2023). Recent approaches and advanced wastewater treatment technologies for mitigating emerging microplastics contamination–A critical review. *Science of the Total Environment*, *858*, 159681.

Kumar, M., Patel, A. K., Shah, A. V., Raval, J., Rajpara, N., Joshi, M., & Joshi, C. G. (2020). First proof of the capability of wastewater surveillance for COVID-19 in India through the detection of genetic material of SARS-CoV-2. *Science of The Total Environment*, *746*, 141326.

Lal, S., Link, S., & Halas, N. J. (2007). Nano-optics from sensing to waveguiding. *Nature Photonics*, *1*(11), 641–648.

Li, X., He, S., Xiao, H., He, T. T., Zhang, J. D., Luo, Z. R., ... & Cao, L. Y. (2022a). Neonicotinoid insecticides promote breast cancer progression via G protein-coupled estrogen receptor: In vivo, in vitro and in silico studies. *Environment International*, *170*, 107568.

Li, M., Hou, Z., Meng, R., Hao, S., & Wang, B. (2022b). Unraveling the potential human health risks from used disposable face mask-derived micro/nanoplastics during the COVID-19 pandemic scenario: A critical review. *Environment International*, *170*, 107644.

Liang, J., Deng, X., & Tan, K. (2015). An eosin Y-based “turn-on” fluorescent sensor for detection of perfluorooctane sulfonate. *Spectrochimica Acta Part A: Molecular and Biomolecular Spectroscopy*, *150*, 772–777.

Liu, Y., Lüttjohann, S., Vianello, A., Lorenz, C., Liu, F., & Vollertsen, J. (2024). Detecting small microplastics down to 1.3 μm using large area ATR-FTIR. *Marine Pollution Bulletin*, *198*, 115795.

Long, C. L., Yang, Z. G., Yang, Y., Li, H. P., & Wang, Q. (2016). Determination of gold nanoparticles in natural water using single particle-ICP-MS. *Journal of Central South University*, *23*(7), 1611–1617.

Luo, Y., Su, W., Rabbi, M. F., Wan, Q., Xu, D., Wang, Z., … & Wu, J. (2024). Quantitative analysis of microplastics in water environments based on Raman spectroscopy and convolutional neural network. *Science of The Total Environment*, *926*, 171925.

Mahlein, A. K. (2016). Plant disease detection by imaging sensors–parallels and specific demands for precision agriculture and plant phenotyping. *Plant Disease*, *100*(2), 241–251.

Mao, S., Chang, J., Pu, H., Lu, G., He, Q., Zhang, H., & Chen, J. (2017). Two-dimensional nanomaterial-based field-effect transistors for chemical and biological sensing. *Chemical Society Reviews*, *46*(22), 6872–6904.

Mayr, A. V., Weinhold, A., Nolzen, A., Keller, A., & Ayasse, M. (2025). The neonicotinoid Acetamiprid alters the chemical profile of the primitive eusocial bee Lasioglossum malachurum. *Ecotoxicology and Environmental Safety*, *306*, 119311.

Medema, G., Heijnen, L., Elsinga, G., Italiaander, R., & Brouwer, A. (2020). Presence of SARS-Coronavirus-2 RNA in sewage and correlation with reported COVID-19 prevalence in the early stage of the epidemic in the Netherlands. *Environmental science & technology letters*, *7*(7), 511–516.

Menger, R. F., Funk, E., Henry, C. S., & Borch, T. (2021). Sensors for detecting per-and polyfluoroalkyl substances (PFAS): A critical review of development challenges, current sensors, and commercialization obstacles. *Chemical Engineering Journal*, *417*, 129133.

Meyers, N., Kopke, K., Buhhalko, N., Mattsson, K., Janssen, C. R., Everaert, G., & De Witte, B. (2024). Value for money: A cost-effectiveness analysis of microplastic analytics in seawater. *Microplastics and Nanoplastics*, *4*(1), 4.

Mikac, L., Rigó, I., Himics, L., Tolić, A., Ivanda, M., & Veres, M. (2023). Surface-enhanced Raman spectroscopy for the detection of microplastics. *Applied Surface Science*, *608*, 155239.

Mohamed, H. M. (2016). Screen-printed disposable electrodes: Pharmaceutical applications and recent developments. *TrAC Trends in Analytical Chemistry*, *82*, 1–11.

Muhammed, M. A. H., Cruz, L. K., Emwas, A. H., El-Zohry, A. M., Moosa, B., Mohammed, O. F., & Khashab, N. M. (2019). Pillar [5] arene-stabilized silver nanoclusters: Extraordinary stability and luminescence enhancement induced by host–guest interactions. *Angewandte Chemie International Edition*, *58*(44), 15665–15670.

Muthulakshmi, L., Mohan, S., & Tatarchuk, T. (2023). Microplastics in water: Types, detection, and removal strategies. *Environmental Science and Pollution Research*, *30*(36), 84933–84948.

Oaks, J. L., Gilbert, M., Virani, M. Z., Watson, R. T., Meteyer, C. U., Rideout, B. A., ... & Ahmed Khan, A. (2004). Diclofenac residues as the cause of vulture population decline in Pakistan. *Nature*, *427*(6975), 630–633.

Obinaju, B. E., & Martin, F. L. (2016). ATR-FTIR spectroscopy reveals polycyclic aromatic hydrocarbon contamination despite relatively pristine site characteristics: Results of a field study in the Niger Delta. *Environment International*, *89*, 93–101.

Omotola, E. O., & Olatunji, O. S. (2020). Quantification of selected pharmaceutical compounds in water using liquid chromatography-electrospray ionisation mass spectrometry (LC-ESI-MS). *Heliyon*, *6*(12), e05787.

Ortúzar, M., Esterhuizen, M., Olicón-Hernández, D. R., González-López, J., & Aranda, E. (2022). Pharmaceutical pollution in aquatic environments: A concise review of environmental impacts and bioremediation systems. *Frontiers in Microbiology*, *13*, 869332.

Osman, A. I., Hosny, M., Eltaweil, A. S., Omar, S., Elgarahy, A. M., Farghali, M., Yap, P. S., Wu, Y. S., Nagandran, S., Batumalaie, K., Gopinath, S. C. B., John, O. D., Sekar, M., Saikia, T., Karunanithi, P., Hatta, M. H. M., & Akinyede, K. A. (2023). Microplastic sources, formation, toxicity and remediation: A review. *Environmental Chemistry Letters*, *21*, 1–41. Advance online publication.

Picó, Y., Blasco, C., & Font, G. (2004). Environmental and food applications of LC–tandem mass spectrometry in pesticide-residue analysis: An overview. *Mass Spectrometry Reviews*, *23*(1), 45–85.

Prakash, V., Pain, D. J., Cunningham, A. A., Donald, P. F., Prakash, N., Verma, A., ... & Rahmani, A. R. (2003). Catastrophic collapse of Indian white-backed Gyps bengalensis and long-billed Gyps indicus vulture populations. *Biological Conservation*, *109*(3), 381–390.

Quintelas, C., Melo, A., Costa, M., Mesquita, D. P., Ferreira, E. C., & Amaral, A. L. (2020). Environmentally-friendly technology for rapid identification and quantification of emerging pollutants from wastewater using infrared spectroscopy. *Environmental Toxicology and Pharmacology*, *80*, 103458.

Ray, S. S., Lee, H. K., Huyen, D. T. T., Chen, S. S., & Kwon, Y. N. (2022). Microplastics waste in environment: A perspective on recycling issues from PPE kits and face masks during the COVID-19 pandemic. *Environmental Technology & Innovation*, *26*, 102290.

Rehman, A. U., Crimi, M., & Andreescu, S. (2023). Current and emerging analytical techniques for the determination of PFAS in environmental samples. *Trends in Environmental Analytical Chemistry*, *37*, e00198.

Rochman, C. M., Hentschel, B. T., & Teh, S. J. (2014). Long-term sorption of metals is similar among plastic types: Implications for plastic debris in aquatic environments. *PloS One*, *9*(1), e85433.

Ruan, X., Xie, L., Liu, J., Ge, Q., Liu, Y., Li, K., ... & Zhang, L. (2024). Rapid detection of nanoplastics down to 20 nm in water by surface-enhanced raman spectroscopy. *Journal of Hazardous Materials*, *462*, 132702.

Saadati, N., Abdullah, M. P., Zakaria, Z., Sany, S. B. T., Rezayi, M., & Hassonizadeh, H. (2013). Limit of detection and limit of quantification development procedures for organochlorine pesticides analysis in water and sediment matrices. *Chemistry Central Journal*, *7*(1), 63.

Sandra, P., Nogueira, J., Sandra, T., & David, F. (2002). Recent Applications in LC-MS: Environmental Analysis. *LC-GC Europe*, *15*(11), 6–11.

Schymanski, D., Goldbeck, C., Humpf, H. U., & Fürst, P. (2018). Analysis of microplastics in water by micro-Raman spectroscopy: Release of plastic particles from different packaging into mineral water. *Water Research*, *129*, 154–162.

Sefiloglu, F. Ö., Stratmann, C. N., Brits, M., van Velzen, M. J., Groenewoud, Q., Vethaak, A. D., ... & Lamoree, M. H. (2024). Comparative microplastic analysis in urban waters using μ-FTIR and Py-GC-MS: A case study in Amsterdam. *Environmental Pollution*, *351*, 124088.

Shi, J., Zhang, R., Pei, Y., Liao, C., & Wu, X. (2020). Exposure to acetamiprid influences the development and survival ability of worker bees (Apis mellifera L.) from larvae to adults. *Environmental Pollution*, *266*, 115345.

Singh, G., Thakur, N., & Kumar, R. (2024). Nanoparticles in drinking water: Assessing health risks and regulatory challenges. *Science of the Total Environment*, *949*, 174940.

Sohrabi, H., Dezhakam, E., Khataee, A., Nozohouri, E., Majidi, M. R., Mohseni, N., ... & Yoon, Y. (2022). Recent trends in layered double hydroxides based electrochemical and optical (bio) sensors for screening of emerging pharmaceutical compounds. *Environmental Research*, *211*, 113068.

Sun, Y., Wu, B., Dong, H., Zhu, J., Ren, N., Ma, J., & You, S. (2025). Machine learning-powered pseudo-target screening of emerging contaminants in water: A case study on tetracyclines. *Water Research*, *274*, 123039.

Swan, G. E., Cuthbert, R., Quevedo, M., Green, R. E., Pain, D. J., Bartels, P., … & Wolter, K. (2006). Toxicity of diclofenac to Gyps vultures. *Biology Letters*, *2*(2), 279–282.

Talukdar, P. J., Bharti, K., Banerjee, S., Basu, S., Das, S. K., Paul, R. R., ... & Lahiri, B. (2024). A comprehensive FTIR micro-spectroscopic analysis and classification of precancerous human oral tissue aided by machine learning. *Sensors & Diagnostics*, *3*(11), 1854–1865.

Tang, S., Lin, L., Wang, X., Feng, A., & Yu, A. (2020). Pb (II) uptake onto nylon microplastics: Interaction mechanism and adsorption performance. *Journal of Hazardous Materials*, *386*, 121960.

Thompson, R. C., Olsen, Y., Mitchell, R. P., Davis, A., Rowland, S. J., John, A. W., … & Russell, A. E. (2004). Lost at sea: Where is all the plastic? *Science*, *304*(5672), 838–838.

Turner, A., & Holmes, L. A. (2015). Adsorption of trace metals by microplastic pellets in fresh water. *Environmental Chemistry*, *12*(5), 600–610.

United Nations Environment Programme. (2019). Global chemicals outlook II - From legacies to innovative solutions: Implementing the 2030 agenda for sustainable development.

Wang, T., Li, S., Mu, R., Lu, Z., Su, J., Chen, J., & Zhan, J. (2024). Size-resolved SERS detection of trace polystyrene nanoplastics via selective electrosorption. *Analytical Chemistry*, *96*(49), 19545–19552.

Wang, Y., Jang, H. J., Topel, M., Dasetty, S., Liu, Y., Ateia, M., … & Chen, J. (2025). Reversible parts-per-trillion-level detection of perfluorooctane sulfonic acid in tap water using field-effect transistor sensors. *Nature Water*, *3*(10), 1187–1197.

Wei, D., Chen, S., & Liu, Q. (2015). Review of fluorescence suppression techniques in Raman spectroscopy. *Applied Spectroscopy Reviews*, *50*(5), 387–406.

Wu, F., Zhang, J., Xiao, A., Gu, X., Lee, W. L., Armas, F., ... & Alm, E. J. (2020). SARS-CoV-2 titers in wastewater are higher than expected from clinically confirmed cases. *Msystems*, *5*(4), 10–1128.

Wurster, C. F. (2015). *DDT wars: Rescuing our national bird, preventing cancer, and creating the environmental defense fund*. Oxford University Press.

Xia, X., Mu, H., Li, Y., Hou, Y., Li, J., Zhao, Z., … & Wei, L. (2024). Which emerging micropollutants deserve more attention in wastewater in the post-COVID-19 pandemic period? Based on distribution, risk, and exposure analysis. *Science of The Total Environment*, *951*, 175511.

Xu, X., Li, H., Hasan, D., Ruoff, R. S., Wang, A. X., & Fan, D. L. (2013). Near-field enhanced plasmonic-magnetic bifunctional nanotubes for single cell bioanalysis. *Advanced Functional Materials*, *23*(35), 4332–4338.

Yang, Y., Chen, B., Hower, J., Schindler, M., Winkler, C., Brandt, J., ... & Hochella, M. F. (2017). Discovery and ramifications of incidental magnéli phase generation and release from industrial coal-burning. *Nature Communications*, *8*(1), 194.

Yue, J., Pang, H., Wei, R., Hu, C., & Qu, J. (2025). Machine learning-assisted molecular structure embedding for accurate prediction of emerging contaminant removal by ozonation oxidation. *Environmental Science & Technology*, *59*(18), 9298–9311.

Zhang, H., Pap, S., Taggart, M. A., Boyd, K. G., James, N. A., & Gibb, S. W. (2020). A review of the potential utilisation of plastic waste as adsorbent for removal of hazardous priority contaminants from aqueous environments. *Environmental Pollution*, *258*, 113698.

Zhang, X., Qiu, C., Ullah, R., Pohl, C., & Liu, Y. (2021). Determination of perfluorinated and polyfluorinated alkyl substances in drinking water samples using automated solid phase extraction and liquid chromatography–tandem mass spectrometry. *Separation Science Plus*, *4*(1), 24–35.

Zhang, Z., Geng, Y., Zhou, W., Shao, X., Lin, H., & Zhou, Y. (2024). Development of a multi-spectroscopy method coupling μ-FTIR and μ-Raman analysis for one-stop detection of microplastics in environmental and biological samples. *Science of the Total Environment*, *917*, 170396.

Zhou, Y., Yang, L., Ma, L., Han, Y., Yan, C. G., & Yao, Y. (2022). Nano-theranostics constructed from terpyridine-modified pillar [5] arene-based supramolecular amphiphile and its application in both cell imaging and cancer therapy. *Molecules*, *27*(19), 6428.

Carbon Dots and Other Nanomaterials for Advanced Water Quality Monitoring

9

Kuppusamy Kanagaraj, Tharmaraj Vairaperumal, and Arumugam Thangamani

9.1 INTRODUCTION

One of the most important factors in the protection of the population's health and environmental sustenance is water quality surveillance. Clean water is a vital component of human life, agriculture, and industry, but safe drinking water is getting more and more endangered by urbanization, industrialization, and pollution of the environment. Quality of water is affected by many factors such as the existence of chemical pollutants (heavy metals and organic substances), physical pollutants (turbidity and suspended solids), and biological pollutants (bacteria and viruses). The quality of the water is not only a question of its detection but also a question of its origin, its flow, and the evaluation of the possible threat to human health and the ecosystem [1]. Conventional water quality testing techniques, albeit efficient in a number of aspects, usually fare poorly in sensitivity, speed, and cost-effectiveness, particularly when real-time or high-volume monitoring is needed.

The inherent weaknesses of traditional water quality surveillance, including chemical analysis in laboratories, manual sampling, and time-intensive biological analysis, have led to the development of new technologies that can provide more precise, sensitive, and effective detection. Recent breakthroughs in nanotechnology, especially the creation of carbon nanotubes (carbon dots (CDs) and other carbon-based nanomaterials, have offered new possibilities to improve the quality of water testing. With their peculiar optical, electrical, and chemical characteristics, these materials have unprecedented opportunities in the detection of pollutants at significantly lower levels than conventional methods can [2]. Nanomaterials in water monitoring have been considered a breakthrough in the science of the environment, as they provide real-time, on-site, and highly sensitive testing that can be conducted at a fraction of the cost of traditional testing.

CDs are one of the most perspective nanomaterials to use in water quality recognition because they are small, highly biocompatible, their fluorescence can be tuned, and they can be easily manufactured.

DOI: 10.1201/9781003741626-9

These are zero-dimensional nanomaterials mainly composed of carbon and, in most cases, of heteroatoms such as nitrogen, sulfur, or phosphorus to make them more optical and chemical [3,4]. The fluorescence properties of CDs can be utilized to identify diverse waterborne pollutants such as toxic heavy metals, organic contaminants, and micro-organisms. CDs are highly sensitive to the fluorescence response of their various analytes, thus making them excellent sensors of water quality that require real-time operation or operation in different environments. Their capacity to adjust their surface chemistry and functionalize them with sites of certain binding molecules also enhances their selectivity toward certain pollutants, which allows targeted detection.

Along with CDs, other nanomaterials, including metal nanoparticles (e.g., gold and silver), quantum dots (QDs), graphene-based materials, and metal oxides, have also proven to have enormous potential in water quality monitoring. The surface plasmon resonance (SPR) effects of gold nanoparticles provide a surface attribute that other materials do not have the ability to exhibit; similarly, titanium dioxide (TiO_2) nanoparticles' photocatalytic effects are surface characteristics not accessible to other materials. These mixtures of nanomaterials can result in very versatile, multifunctional water sensing systems that are able to sense a very wide range of contaminants in water [5]. As an example, QDs, whose optical characteristics are determined by their sizes, have the potential to offer multiplexed sensing of multiple pollutants at once. Graphene and graphene oxide (GO), which possess a large surface area and electrical conductivity, can be incorporated in sensors to sense minute quantities of organic and inorganic pollutants. Sensors that not only detect but also eliminate the harmful pollutants of water can be made using metal oxide nanoparticles, which are reactive and which can be used in solving the problem of water contamination in two ways.

Combining all these different nanomaterials into miniature, easy-to-use sensing systems represents a promising future in water quality monitoring. Through integration of the merits of CDs and other nanomaterials, it can be formulated to come up with a sensor that is not only highly sensitive but also affordable to install, easily deployable, and capable of giving real-time feedback about the water quality [6,7]. The possibility to replace traditional laboratory analysis with real-time on-site water quality testing can potentially transform the current water management strategies, especially in areas that lack access to sophisticated infrastructures or laboratories. Moreover, the possibility to quickly detect particular pollutants in water bodies may help to greatly minimize the risks of waterborne illnesses, industrial contamination, and environmental damage [8]. This chapter will discuss the synthesis, properties, and applications of CDs and other nanomaterials in water quality monitoring, besides noting how they could be used to change the existing practice of water quality testing (Figure 9.1).

9.2 OVERVIEW OF WATER QUALITY MONITORING

The role of water quality monitoring improves the quality of the aquatic environment, the potability of drinking water, and the health of the population (Figure 9.2). Water quality is any physical, chemical, and biological characteristic of water that is subjected to natural processes and human activities. Routine monitoring of water quality is necessary in order to identify contamination, safeguard the environment, and adhere to the regulations. Water quality monitoring allows one to determine the presence of pollutants, assess their effects on human health, and come up with mitigation measures [9,10].

Monitoring of water quality usually implies measuring a great variety of physical, chemical, and biological parameters. The parameters are a measure of water status, possible risks, and water appropriateness to various purposes, such as drinking, farming, leisure, and manufacturing. The conventional means of water quality testing have been an excellent tool over the years, but these methods have major drawbacks concerning time, cost, and sensitivity. Nanotechnology and, in particular, the application of carbon-based and other nanoscale materials offer new answers to these constraints, and these include more rapid, more precise, and cost-effective ways to measure water quality [11,12].

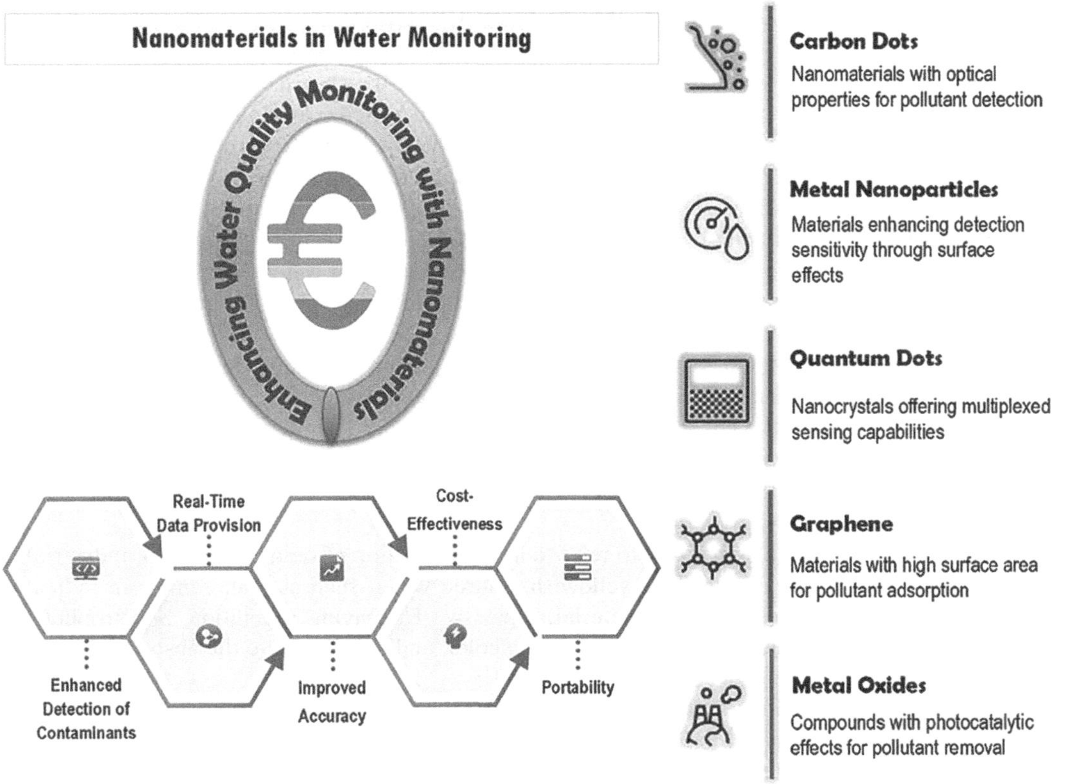

FIGURE 9.1 An overview of the chapter detailing the use of nanomaterials like carbon dots and graphene for enhanced water monitoring.

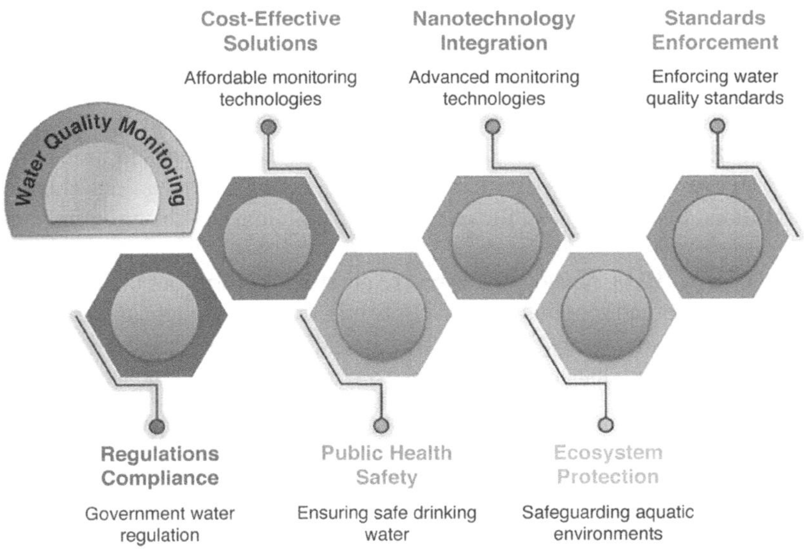

FIGURE 9.2 An overview of water quality monitoring, highlighting cost-effective solutions, regulation compliance, and ecosystem protection strategies.

9.2.1 Physical Parameters in Water Quality

The physical parameters of water quality are the appearances, which affect the look, the feel, and the behavior of the water. These parameters are essential in the study of the overall state of water and most of the time give immediate information on the presence of pollutants. The most important physical parameters that are observed during water quality measurements are the following:

Temperature: Water temperature influences the solubility of gases, the speed of chemical reactions, and the metabolic activity of aquatic organisms in general. Change in temperature can be an indicator of possible contamination, since industrial discharges or runoffs in warm water can result in a change in the local temperature, which affects aquatic life [13]. The temperature is normally quantified through thermometers, thermocouples, or digital probes.

Turbidity: Turbidity is the measure of the turbidity or haziness of water due to the presence of suspended particles, i.e., sediment, algae, or pollutants. When turbidity is high, it usually shows that there is pollution or runoff from the construction sites, farms, or urban locations. The water quality may be determined by turbidity because it decreases light penetration, influences the growth of aquatic plants, and disturbs the natural equilibrium of the ecosystems. The nephelometers, or turbidimeters, are widely used to measure turbidity [14].

Color: Water color can also be used to refer to a contamination of organic material, industrial effluents, or agricultural runoff. Brownish or yellowish-colored water, such as that, can be an indication of humic material or of dissolved organic material, typically of decaying vegetation. Spectrophotometric techniques are the common methods used to measure color, and they measure the absorbance of light on the sample of water [15].

Conductivity: Water conductivity is a measure of its conductivity to electricity, which is determined by the concentration of dissolved ions. Higher conductivity may suggest the dissolved salts that could be chlorides, sulfates, and nitrates that are usually byproducts of either industrial, agricultural, or wastewater discharge. In measuring conductivity, an electrode-based sensor is used to measure the capability of water to carry an electrical charge [16].

These physical parameters give a sense of the water quality, and they may act as the first line of defense in detecting contamination. But at the same time, they lack specifics of the nature of pollutants found in the water, and here chemical and biological parameters are involved.

9.2.2 Chemical Parameters in Water Quality

The chemical parameters are necessary to comprehend the types and concentrations of various types of pollutants in the water sample. These parameters assist in locating the hazardous chemicals, the aptness of water to be consumed by humans, and the effects of the pollutants on the environment [17,18].

- **Dissolved Oxygen (DO)**: Aquatic organisms require DO to carry out their respiration. Reduced DO may result in organic pollution, which utilizes oxygen through microbial decomposition, resulting in hypoxic or anoxic regions. Electrochemical sensors, e.g., the Clark electrode, or optical sensors that measure the change in fluorescence are used to quantify the DO concentration [19].
- **Chemical Oxygen Demand (COD)**: COD is the quantity of oxygen needed to oxidize organic matter chemically in a water sample. High COD values indicate the existence of huge amounts of organic contaminants, which can be brought about by industrial effluents, sewage, or agricultural runoffs. The colorimetric or titrimetric methods are used to measure COD, which gives an indicator of the ability of water to purify itself [20].
- **Total Dissolved Solids (TDS)**: TDS is the total sum of dissolved ions, i.e., salts, minerals, and metals. High TDS can be the result of urban runoff, industrial effluent, or sewage effluent. Conductivity-based sensors are used to measure TDS, and they detect the amount of ionic content present in water [21].

- **Specific Pollutants**: Some of these toxic chemicals and pollutants fall under this category: heavy metals (lead, mercury, cadmium), pesticides, herbicides, and nitrates. Such pollutants may be very harmful to human health and the environment and usually are collected in the food chain. Determining and quantifying these pollutants is usually done using sophisticated analytical methods such as spectrophotometry, atomic absorption spectroscopy (AAS), inductively coupled plasma mass spectrometry (ICP-MS), and high-performance liquid chromatography (HPLC) [22].
- Chemical parameters allow obtaining a detailed image of the chemical composition of the water and possible health hazards. But they need special apparatus and are frequently both time-consuming and expensive to determine. Owing to these reasons, fast and handheld sensors, particularly ones that involve nanomaterials, are becoming even more useful in offering real-time monitoring of chemicals [23–27].

9.2.3 Nanomaterials: A New Paradigm for Water Quality Monitoring

Nanotechnology, especially the application of carbon-based and other nanoscale materials, can help overcome most of the limitations of conventional methods of water quality monitoring. Nanomaterials are designed to have special optical, electrical, and chemical characteristics that allow them to detect a large number of contaminants with high sensitivity and selectivity. Compared to larger materials, nanomaterials have a high surface area, which can be tuned, and their small size makes them ideal in applications of water quality sensing, as they are cheaper, faster, and enable more accurate measurements [28–33].

- **Enhanced Sensitivity**: Nanomaterials are able to pick up pollutants at a much lower concentration than the conventional processes, allowing the detection of trace contaminants [34]. Nanomaterials are most sensitive to changes in their environment due to the increased surface-area-to-volume ratios, which increases their reactivity.
- **Portability and Cost-Effectiveness**: Nanomaterial sensors can be reduced to a smaller size and built into devices that are portable and easy to use by the user in conducting an on-site test of water. These sensors can be relatively inexpensive to produce and run compared with the more traditional methods that require a laboratory and thus can be used in remote locations or at scale [35].
- **Real-Time and Continuous Monitoring**: Nanomaterials facilitate real-time and continuous water quality monitoring, giving instant feedback of the contamination levels. This is especially applicable in the use of monitoring industrial effluents, the detection of waterborne pathogens, or tracing heavy metal pollution [32].

Examples of nanomaterials that have demonstrated potential in water quality monitoring include CDs, QDs, metal nanoparticles, and graphene-based materials, to mention a few. These substances have the potential to provide quick, sensitive, and economical detection of a wide variety of pollutants, which provides a paradigm shift in our method to monitor and control water quality [36].

9.3 NANOTECHNOLOGY IN WATER QUALITY MONITORING

Nanotechnology is the science and engineering of the handling of matter on a nanoscale, or in the range of 1–100 nm (Figure 9.3). This area has transformed the science of materials and has provided new and

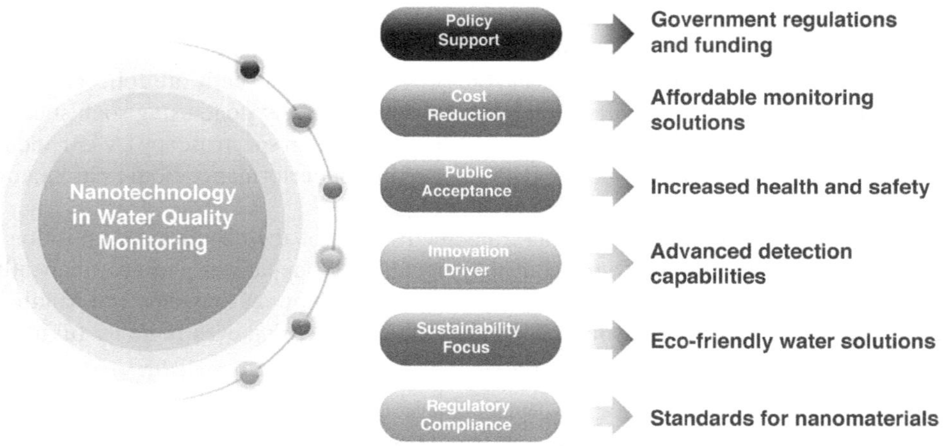

FIGURE 9.3 A conceptual overview showing how government regulations and funding drive innovation and cost reduction in nano-enhanced water monitoring.

improved properties that are not available in bulk materials. Materials with structures or dimensions that are on the order of nanometer size are known as nanomaterials and have several distinct properties: high surface-area-to-volume ratio, high chemical reactivity, and new optical, electrical, and mechanical properties. The mentioned characteristics are what make them especially applicable in a broad range of applications, specifically, environmental monitoring and water quality testing [37–39].

9.3.1 Promising Nanomaterials for Water Quality Monitoring

Among the different kinds of nanomaterial available, a few have shown a substantial promise for water quality monitoring purposes—each with different properties that make them suitable for different kinds of contaminants. These include CDs, metal nanoparticles, QDs, graphene-made materials, and metal oxide nanoparticles in this list. Some of these nanomaterials are explored in more detail below [39–48].

9.3.1.1 Carbon Dots

CDs are zero-dimensional nanomaterials that received much interest due to the outstanding optical characteristics, specifically fluorescence. CDs are tiny (usually between 1 and 10 nm across) and consist of carbon as their primary constituent and can be prepared by either a top-down or bottom-up approach. Their emission of fluorescence can be optimized through the change of their sizes, functionalization of the surface, or the variety of heteroatoms, including nitrogen, sulfur, or phosphorus, which are doped. Such properties render CDs suitable for use in water quality measurements, specifically in the detection of low concentrations of particular pollutants [49].

Carbon dots can be functionalized so as to identify heavy metals (e.g., lead, mercury), organic pollutants (e.g., pesticides, pharmaceuticals), or pathogens (e.g., bacteria, viruses), triggering a change in their fluorescence properties. This transformation is measurable to generate real-time and high-sensitivity contaminant detection. Furthermore, carbon dots are biocompatible, biodegradable, and not toxic, thus vital when using them in monitoring the environment [50].

9.3.1.2 Metal Nanoparticles

Other optical and chemical properties of metal nanoparticles, including gold (Au), silver (Ag), and copper (Cu), have been widely studied in terms of water quality monitoring. These nanoparticles also possess some exceptional properties, including SPR, which enables them to couple with electromagnetic radiation in a manner that cannot be achieved with bulk materials. One example would be the gold nanoparticles, which have been extensively used to carry out colorimetric sensing since they can change color when in the presence of a particular analyte. This renders them the best choice for identifying contaminants on the ground, especially where the monitoring system must be cheap and portable. Silver nanoparticles, however, are antimicrobial in nature and can therefore be utilized in the detection and removal of microbial contaminants in water [51,52]. The primary strength of metal nanoparticles in water quality monitoring is the versatility that they possess. Researchers have been able to customize the nanoparticles to detect a whole array of pollutants by changing the size, shape, and surface chemistry of the structures to detect heavy metals, organic compounds, and biological agents.

9.3.1.3 Quantum Dots

QDs are optical nanocrystals which are semiconductor crystallites that depend on size. Because QDs are so small, quantum confinement effects are seen in the QDs, which result in rare electronic and optical properties, including tunable fluorescence and the possibility to emit light of a certain wavelength depending on their size. This renders QDs suitable for multiplexed detection of various water contaminants since they may be customized to produce various wavelengths when exposed to various analytes [53,54].

9.3.1.4 Graphene-Based Materials

Graphene and its derivatives, including GO and reduced graphene oxide (rGO), have received much attention regarding environmental sensing because of its high electrical conductivity, high surface area, and mechanical stability. Graphene materials are suitable in the detection of heavy metals, organic pollutants, and gases in water. An example of such is GO, which has a high surface area and can be readily functionalized with chemical groups which specifically target pollutants. It has been employed in electrochemical sensors in detecting heavy metals such as lead and mercury as well as organic pollutants such as dyes and pesticides. Graphene is also highly conductive, and therefore, its transfer of electrons may be rapid, increasing the sensitivity of electrochemical sensors [55,56]. The materials based on graphene are also very versatile and can be used as sensing devices in different applications: wearable sensors, field deployable devices, and integrated monitoring devices.

9.3.1.5 Metal Oxide Nanoparticles

Titanium dioxide (TiO_2), zinc oxide (ZnO), and iron oxide (Fe_2O_3) metal oxide nanoparticles have found extensive application in the environment, especially in organic pollutant detection and degradation. These materials are photocatalytic in nature, implying that they are able to decompose pollutants in UV light. Moreover, the metal oxides are very reactive and could be used in monitoring water quality, as they could react with a number of contaminants. Metal oxide nanoparticles are commonly incorporated as part of a composite with other nanomaterials, including carbon-based materials or QDs, to increase the sensitivity and specificity of water quality sensors. As an example, TiO_2 nanoparticles can be functionalized to specifically target heavy metals or organic compounds and offer a potent instrument to monitor trace contaminants in water [57].

Although nanomaterials have a high potential of monitoring water quality, some challenges still exist before such technologies can be adopted very broadly. Such problems like the stability of complex water matrices, the interference of other substances, and the scalability of sensor technologies should be discussed. Also, the environmental effects of nanomaterials should be viewed long-term to make sure that the advantages of adopting this kind of material in water quality surveillance do not surpass the risks involved. Future studies on nanotechnology in water quality monitoring will probably involve enhancing the stability, sensitivity, and selectivity of nanomaterial sensors. Moreover, the combination of such sensors into real-time, portable, and automated monitoring devices will improve their practical usage in developed and developing regions [58].

Nanotechnology has suggestive potential in transforming water quality measurement by offering highly sensitive, cost-effective, and portable devices for measuring various forms of contaminants. Some of the most promising nanomaterials under development in water quality testing include CDs, metal nanoparticles, QDs, and graphene-based materials as well as metal oxide nanoparticles. These materials are all unique in their own way, with high surface area, tunable optical properties, and chemical reactivity, making them the best choice to develop next-generation sensors capable of delivering real-time, on-site water quality monitoring. Nanomaterials will be more and more significant as research develops to ensure the security of our water resources and human health [59].

9.4 CARBON DOTS

9.4.1 What Are Carbon Dots?

CDs are a special kind of fluorescent nanoparticle, which has gained popularity in such areas as bioimaging, biosensing, and environmental sensing (Figure 9.4). These are nanoscale materials that are usually between 1 and 10 nm in size, and as such, their size is within the domain of 0D nanomaterials. The distinguishing feature of the CDs among the other nanoparticles is that they have innate fluorescence and photoluminescence properties, which have made the use of them the most suitable in a broad spectrum of detection applications. Contrary to conventional QDs, which contain heavy metals, which are toxic, CDs are entirely made of carbon atoms, thus non-toxic and biocompatible, which is an important trait of environmental and biological applications [60].

The size, functionalization of the surface, and doping components of the CDs can be manipulated to tune the fluorescence, allowing specific and sensitive detection of the analyte. Heteroatoms, including

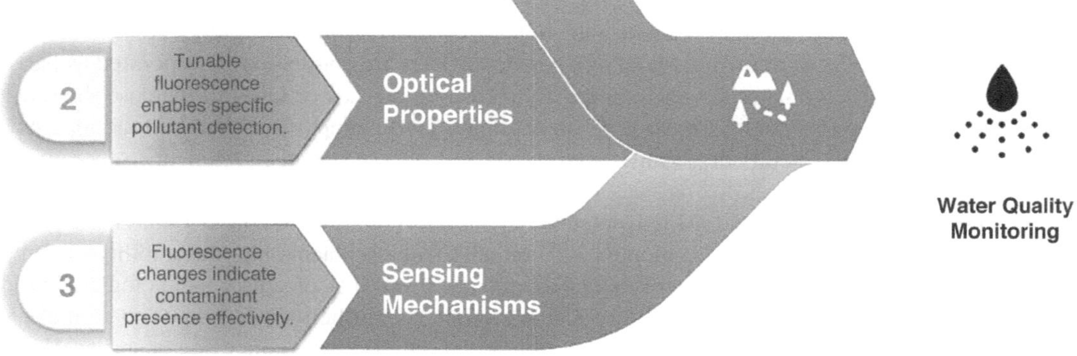

FIGURE 9.4 An overview of carbon dots in sensing, illustrating how synthesis approaches dictate their optical properties and subsequent mechanisms for detecting pollutants in water.

nitrogen, sulfur, or phosphorus, are also commonly doped into CDs to increase their fluorescence intensity or change their emission characteristics. The synthesis procedures used to produce CDs can also be used to produce nanoparticles with specialized functional groups on their surfaces, which can react with contaminants in a specific manner, thereby allowing targeted sensing [61,62].

9.4.2 Synthesis of Carbon Dots

The production of CDs could be divided into two general strategies: top-down and bottom-up strategies. Both methods can be used to produce tunable sizes, surface chemistry, and optical properties of CDs, though each possesses its own advantages and limitations. These approaches are important to understand in order to apply CDs to a specific application, including water quality monitoring, in which the detailed properties of the nanomaterial are important in determining the sensitivity and selectivity in detection [2].

9.4.2.1 Top-Down Methods

The methods used on a top-down basis include the fragmentation of large carbonaceous materials to nanoscale. Such procedures usually start with bulk materials like graphite, carbon nanotubes, GO, or other carbon-based materials. The carbon source is then either chemically oxidized, laser ablated, or electrochemically reduced to nanoscale particles [63].

Chemical Oxidation: In this approach, carbon-rich substances are oxidized with strong acids, and the process results in the creation of smaller structures of carbon, such as CDs. The process enables the manufacture of CDs whose size distribution and surface functional groups can be controlled.

Laser Ablation: In this method, bulk carbon material is reduced to tiny pieces of carbon by high-energy laser pulses. Laser ablation provides a very high level of accuracy with regard to size and surface properties of CDs. Nonetheless, the technique may be costly and may not be easy to apply on a large-scale manufacturing.

Electrochemical Techniques: CDs can also be obtained in the process of electrochemical oxidation of graphite or carbon nanotubes. This process may be greener than chemical oxidation because it does not use any severe chemicals. Adjusting the electrochemical parameters, e.g., voltage and current density, the size and surface properties of the CDs can be tuned.

9.4.2.2 Bottom-Up Methods

Bottom-up techniques, however, consist in the synthesis of CDs by use of small molecule precursors by chemical reactions. Such methods usually incorporate organic substances such as sugars, citric acid, or amino acids that are polymerized or undergo pyrolysis in controlled conditions to produce CDs.

- **Hydrothermal Treatment**: In this procedure organic precursors undergo heating in a closed system (an autoclave is usually lined with Teflon) at high pressure and temperature. The process is what makes the precursors decompose to result in carbon nanoparticles. Hydrothermal synthesis is one of the widely used processes of making CDs because it is a simple and scalable process [64].
- **Pyrolysis**: Pyrolysis is a process whereby the organic matter is heated at a high temperature in the absence of oxygen because it decomposes the material into smaller carbon structures, such as CDs [65].
- **Solvothermal Synthesis**: It resembles hydrothermal synthesis except that organic solvents are used in place of water. Solvothermal may produce CDs with particular surface functionalization, which may also increase their interaction with pollutants and enhance the performance of sensors [66].

Both the top-down and bottom-up techniques present the option of controlling the size, surface chemistry, and optical properties of CDs. The fact that either of these methods can be selected depending on the desired application, e.g., water quality monitoring, is of great benefit when it comes to creating functional and efficient sensors.

9.4.3 Optical Properties of Carbon Dots

The most useful property of CDs is their fluorescence and photoluminescence, especially in the water quality sensor industry. CDs produce fluorescence at certain wavelengths that can be readily measured when excited by the light of a specific wavelength. The size of CDs, their surface functionalization, and the presence of dopants, e.g., nitrogen or sulfur can affect the emission wavelength, intensity, and decay time of CDs [67].

Environmental factors like the addition of metal ions, organic molecules, or even biological contaminants can also affect the photoluminescence intensity of CDs. Such responsiveness to environmental fluctuation makes CDs the best choice for detecting pollutants in water since it is capable of responding to the presence of certain contaminants by altering its fluorescence. Furthermore, the emission of the fluorescence of the CDs can either be quenched or enhanced, based on the kind of analyte in presence; this offers a simple and easily measurable signal to sensors [68].

9.4.4 Mechanism of Sensing with Carbon Dots

The application of CDs in the monitoring of water quality can be carried out in different ways using different mechanisms that capitalize on their fluorescence characteristics (Figure 9.5). The sensing mechanisms enable CDs to identify a wide variety of contaminants, and, as such, they become very useful in real-time monitoring of the environment.

Detection of the contaminants by the CDs is carried out using fluorescence-based processes like quenching, enhancement, and Förster resonance energy transfer (FRET). In fluorescence quenching, pollutants react with surface functional groups of CDs, which lowers fluorescence intensity by absorbing photons or through non-radiative energy transfer. For example, metal ions such as Hg^{2+} or Pb^{2+} bind to the surface and quench the emission, which makes them detectable quantitatively. Conversely, fluorescence enhancement is a phenomenon involving the interaction of some contaminant or biomolecules with CDs and can enhance more efficient radiative recombination, leading to high fluorescence signal generation to detect it with high sensitivity. Another of these mechanisms is FRET, which is the transfer of energy of excited CDs (donor) to acceptor molecules around them, where the presence of contaminants like metal ions or pesticides changes fluorescence reactions based on the distance between the donor and acceptor to selectively sense target analytes [69,70].

9.4.5 Applications of Carbon Dots in Water Quality Monitoring

CDs are highly biocompatible with excellent fluorescence characteristics and are good nanomaterials in water quality monitoring. The tunable optical properties allow them to detect sensitive, different contaminants in drinking water, wastewater, and natural water systems. Toxic heavy metal ions, including Hg^{2+}, Pb^{2+}, As^{3+}, and Cd^{2+}, can be selectively sensed by functionalized CDs via fluorescence on–off responses and can therefore be used to monitor metal contamination by heavy metals in industrial or mining areas. On the same note, CDs that have been surface-modified can detect organic contaminants such as pesticides, herbicides, and industrial chemicals by generating detectable changes in intensity or

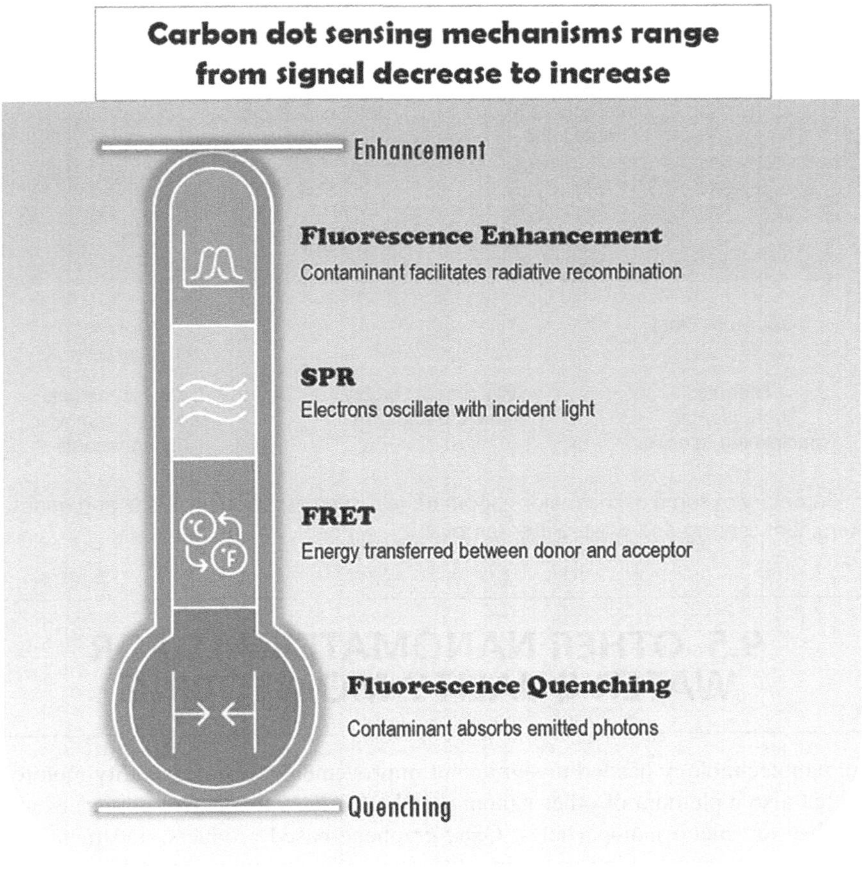

FIGURE 9.5 An overview of carbon dot sensing, showing how mechanisms like FRET, SPR, and fluorescence changes enable sensitive contaminant detection in environmental monitoring.

color of fluorescence. Moreover, CDs may be modified by using particular ligands or antibodies that identify microbial pathogenic agents, including *Escherichia coli* and *Salmonella*, which allows one to detect microbial contamination very sensitively and promote the effective monitoring of the safety of drinking water and waterborne infections [71].

9.4.6 Challenges and Future Directions

The CDs have a high potential in monitoring water quality, but a number of challenges need to be met to be applied in practice. Their stability over complex water matrices with dissolved organic matter, salts, and ions has the potential to influence sensor performance, necessitating better stability in order to be useful in the field. To detect the small concentration of pollutants in large water bodies is also not easy, as these are sensitive in nature, although the use of CDs can be combined with other nanomaterials to detect the pollutants. Also, massive production with high reproducibility and quality is a significant challenge with high optimization of synthesis and manufacturing strategies required. It can be concluded that future studies ought to aim at improving the precision, selectivity, and stability of carbon dot sensors and incorporating them into portable, user-friendly devices, which allow quick, inexpensive, and very sensitive water contaminants monitoring [72].

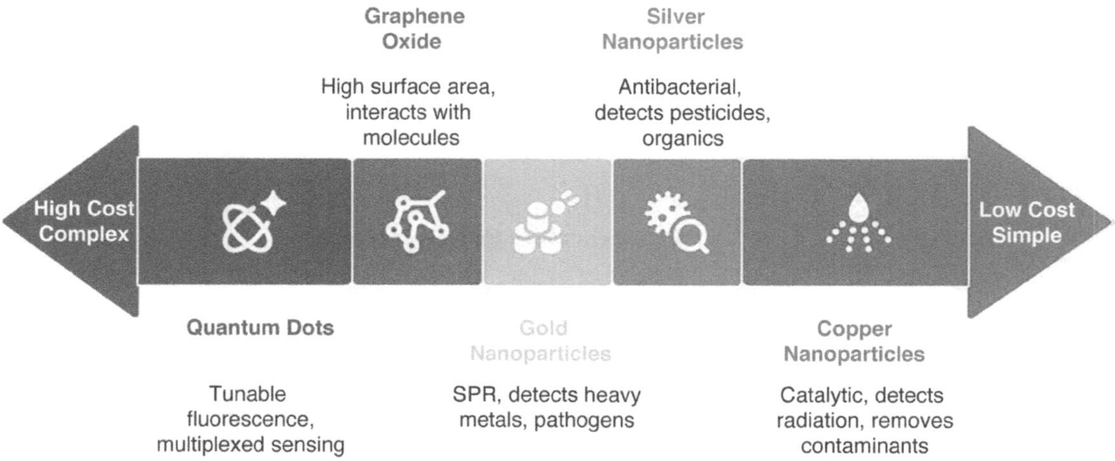

FIGURE 9.6 Summary of some of the major nanomaterials, such as graphene oxide and quantum dots, used in water contaminant sensing and wastewater removal.

9.5 OTHER NANOMATERIALS FOR WATER QUALITY MONITORING

As explained, nanotechnology has led to significant improvements in water quality monitoring by using not only CDs but also a plethora of other nanomaterials that have distinctive properties and capabilities (Figure 9.6). They are metal nanoparticles, QDs, graphene-based products, and metal oxide nanoparticles that are especially practical in identifying different pollutants in water, including heavy metals and organic pollutants and even microbial pathogens. Their optical, electrical, and chemical characteristics are unique, therefore making them sensitive, versatile, and effective in real-time, in situ testing of water quality. This chapter explores the different nanomaterials, which have demonstrated a lot of promise in improving the water quality monitoring and their properties and applications, as well as the issues.

9.5.1 Metal Nanoparticles

Among the most popular nanomaterials investigated to date in terms of environmental monitoring, including gold (Au), silver (Ag), and copper (Cu), are metal nanoparticles that have distinct optical properties and can be readily reacted and functionalized. These nanoparticles have enhanced surface areas, and hence, they are more reactive and responsive to external stimuli. Moreover, their optical characteristics, particularly the capacity to show powerful SPR, make them very beneficial in detecting pollution and contaminants in water.

9.5.1.1 Gold Nanoparticles

In environmental monitoring, gold nanoparticles are one of the most widely used metal nanoparticles. The SPR is the main characteristic of their collective oscillation of the conduction electrons on the surface of gold particles when they are illuminated. It is this effect that enables gold nanoparticles to have strong and tunable absorption and scattering properties. In water quality assay, heavy metals, pathogens, and organic pollutants can be detected using gold nanoparticles. The application of gold nanoparticles in sensor uses is extensive and encompasses colorimetric and spectroscopic detection systems by use of SPR. For instance, when gold nanoparticles are aggregated in the presence of target molecules, e.g., certain metal ions, the color of the solution may change, and this change can be easily observed with the naked eye or spectrophotometrically. This renders gold nanoparticle-based sensors very efficient in real-time detection of water contamination [73].

It is also possible to functionalize gold nanoparticles with different ligands, antibodies, or molecules that identically bind a certain contaminant. This selective binding enhances the selectivity of the sensors, making them very sensitive to detect pollutants at traces. More specifically, gold nanoparticles are also finding application in the detection of heavy metals like mercury (Hg), lead (Pb), and arsenic as a common contaminant of water supplies, particularly in industrial areas [74].

This functionalization enhances the application of these molecules in microbial detection systems for water quality monitoring. Silver nanoparticles also have many mechanical behaviors that are not very common; for example, the closer they get to malachite ovals or alizarin antennae, respectively, silk proteins will undergo molecular conformation changes and appear purple or red. Silver nanoparticles also exhibit strong optical properties, such as SPR, which have been utilized to develop sensors. For example, silver nanoparticles have been used in analysis test items where their performance is dependent on the state they are squeezed into, and release increases an analyte in general red or green caryophyllene. With this way of detection, sensitive and very fast results have gently but unevenly developed [75].

9.5.1.2 Copper Nanoparticles

A less common subject of study is copper nanoparticles, which, owing to their unique combination of reactivity and cost efficiency, could eventually become too important a tool for monitoring water quality on the ground; in hot and dry regions, however, many times there's no power available to treat contaminated pond waters (often associated with agriculture). Copper nanoparticles are especially convenient for water purification and pollution control. They have a strong catalytic effect and are thus most suited to being used in sensors that detect (through reflection or transmission) radiation waves from polluted sources, organic compounds, oils, and locally produced heavy metals [76].

9.5.2 Quantum Dots

Semiconductor nanocrystals, also known as QDs, have size-dependent optical characteristics, which can be exploited to enable multiplexed sensing. These compounds' unique fluorescence features are due to the quantum confinement effects, and consequently these nanomaterials have quite a range of emission spectra. The ability to tune emissions is indeed an important benefit: for any given nanodot, QDs allow light to be emitted at different wavelengths depending on its size, which makes them extremely useful tools in the detection of many pollutants [75].

Fluorescence Properties of Quantum Dots: The bright and stable fluorescence of QDs is essential for the development of sensors that are sensitive and reliable. The size of QDs can be modified to a great extent, which eventually affects their emission spectrum: for instance, smaller QDs emit light closer to shorter wavelengths (blue), while larger ones go out even further into longer wavelengths (red). Depending on the size of QDs, their fluorescence is size-dependent. This makes QDs particularly suitable for applications that require multiplex detection capability in a single experiment [67].

Water Quality In Water Quality Monitoring The QDs have been applied to produce sensors which can identify a wide variety of water pollutants. As an example, by functionalizing QDs with particular antibodies, it is possible to detect pathogens such as Escherichia coli or Salmonella in water sources, which is essential to keep the water source safe in drinking and other recreational activities [77]. Also, QDs have been employed to detect heavy metals, including cadmium (Cd^{2-}), lead (Pb^{2-}), and mercury (Hg^{2-}), based on their fluorescent properties. Also, in QDs, multiplexed sensors have a potential, as different QDs are set to detect various contaminants at the same time. This capability to detect multiple pollutants at once is extremely beneficial in monitoring water quality analysis since a very small amount of samples is needed to provide a comprehensive analysis.

9.5.3 Graphene and Graphene Oxide

Graphene is a single layer of carbon atoms, a hexagonal structure, which has received considerable interest in recent years because of its very high electrical conductivity, high surface area, and mechanical strength. GO, which is a graphene derivative with oxygen functional groups, possesses some distinctive characteristics that make it very useful in the application of water quality monitoring. These characteristics are high surface area, hydrophilicity, and the possibility of interaction with other molecules and ions that make GO the best choice in sensing applications [78].

9.5.4 Metal Oxide Nanoparticles

Nanoparticles of metal oxide, like titanium dioxide (TiO_2), zinc oxide (ZnO), and iron oxide (Fe_2O_3), have been a long-standing subject of study regarding their photocatalytic activity and their capability to degrade organic pollutants. This renders these nanoparticles highly reactive at their surfaces, rendering them useful in detection and elimination of contaminants in water. They are also appropriate to use in sensors that detect different toxic substances in water because of their electrical and optical characteristics. *Photocatalysis*: Photocatalytic metal oxide nanoparticles, specifically TiO_2 and ZnO, have been reported as possessing photocatalytic properties that allow them to degrade organic pollutants when exposed to ultraviolet (UV) light. This photocatalytic action is now very useful in environmental cleanup implementation, where these nanoparticles can be exploited to decompose the hard-to-break organic pollutants, including pesticides and industrial chemicals, to less harmful compounds. Detection and removal of pollutants can be done at the same time by incorporating metal oxide nanoparticles in the water treatment systems [79].

Sensing Applications: On top of photocatalytic properties, metal oxide nanoparticles can also be used in sensors to sense heavy metals, gases, and other harmful substances. As an example, TiO_2 and ZnO nanoparticles are commonly employed in electrochemical sensors to detect heavy metal ions in water. A sensitive and quick technique to monitor water quality can be obtained by alterations in the electrical or optical characteristics of these nanoparticles on exposure to contaminants. Water Quality: Existing uses of metal oxide nanoparticles include the detection and removal of heavy metals (such as lead (Pb), arsenic (As), and cadmium (Cd)) and other common contaminants of industrial wastewater. Also, organic pollutants like petroleum products and dyes, and gases, including ammonia and carbon dioxide, are also detected using metal oxide-based sensors [80].

Nanomaterials, such as metal nanoparticles, QDs, graphene and its oxide, and metal oxide nanoparticles, mean that there are enough choices in terms of water quality testing instruments out there. Every single one of these materials has its own special benefits, for example, the high sensitivity or selectivity they provide and their ability to detect a wide spectrum of pollutants at low concentrations. Their multifunctional properties make them highly suitable for use in portable, in situ, cost-effective sensor platforms that can be applied across a variety of environments. With further advances in research in the future, we'll see these nanomaterials used more and more in applied water quality monitoring systems, leading to a progressive improvement in our capacity for checking public health hazards that threaten environmental integrity.

9.6 CONCLUSION

The development of CDs and other nanoparticles as part of water quality sensors can be called an important breakthrough in environmental science and technology. These materials are sensitive, selective, and versatile; hence, they are able to detect a great variety of contaminants in real time and on site. Their use in safe and clean water will just expand as more studies are refining the properties and uses of these nanomaterials. The future of water quality monitoring will most probably be a mix of various nanomaterials embedded into portable and easy-to-use equipment. As nanotechnology is improving and the requirement of environmental sustainability heightens, CDs and other nanomaterials will transform the concept of water quality monitoring into an accurate, affordable, and real-time approach to water safety and human health.

REFERENCES

[1] D.V. Chapman, T. Sullivan, The role of water quality monitoring in the sustainable use of ambient waters, *One Earth* 5(2) (2022) 132–137.

[2] H.B. Glasgow, J.M. Burkholder, R.E. Reed, A.J. Lewitus, J.E. Kleinman, Real-time remote monitoring of water quality: a review of current applications, and advancements in sensor, telemetry, and computing technologies, *Journal of Experimental Marine Biology and Ecology* 300(1–2) (2004) 409–448.

[3] S. Nangan, K. Kanagaraj, G. Kaarthikeyan, A. Kumar, M. Ubaidullah, B. Pandit, R. Govindasamy, T. Natesan, Sustainable preparation of luminescent carbon dots from syringe waste and hyaluronic acid for cellular imaging and antimicrobial applications, *Environmental Research* 237 (2023) 116990.

[4] Y. Sahu, A. Hashmi, R. Patel, A.K. Singh, M.A.B.H. Susan, S.A. Carabineiro, Potential development of N-doped carbon dots and metal-oxide carbon dot composites for chemical and biosensing, *Nanomaterials* 12(19) (2022) 3434.

[5] A. Thakur, P. Devi, A comprehensive review on water quality monitoring devices: materials advances, current status, and future perspective, *Critical Reviews in Analytical Chemistry* 54(2) (2024) 193–218.

[6] A. Nair, K. Kuppusamy, S. Nangan, T. Natesan, J.T. Haponiuk, S. Thomas, R. Ramasubburayan, L. Gnanasekaran, M. Selvaraj, S. Gopi, Multifunctional natural derived carbon quantum dots from Withania somnifera (L.)–Antiviral activities against SARS-CoV-2 pseudoviron, *Environmental Research* 239 (2023) 117366.

[7] S.A. Jaywant, K.M. Arif, A comprehensive review of microfluidic water quality monitoring sensors, *Sensors* 19(21) (2019) 4781.

[8] K. Kanagaraj, J. Rebek Jr., Y. Yu, Control of reactivity and selectivity in isomerization and rearrangement reactions inside confined spaces, Chemical Communications 59 (2023) 9056–9065.

[9] P. Singh, L. Kumar Singh. Carbon quantum dots as a promising tool for heavy metal sensing and removal in wastewater, *Desalination and Water Treatment* 324 (2025) 101435. https://doi.org/10.1016/j.dwt.2025.101435

[10] P.J. Newman, S.C. Nixon, Y.J. Rees, Surface water quality monitoring, classification, biological assessment and standards, *Water Science and Technology* 30(10) (1994) 1.

[11] G. Gallareta-Olivares et al. Metal-doped carbon dots as robust nanomaterials for the monitoring and degradation of water pollutants, *Chemosphere* 312 (2023) 137190.

[12] S. Gorde, M. Jadhav, Assessment of water quality parameters: a review, *International Journal of Engineering Research and Applications* 3(6) (2013) 2029–2035.

[13] K. Kanagaraj, R. Wang, M.-K. Zhao, P. Ballester, J. Rebek, Jr., Y. Yu, Selective binding and isomerization of oximes in a self-assembled capsule, *Journal of the American Chemical Society* 145(10) (2023) 5816–5823.

[14] T. Matos, M. Martins, R. Henriques, L. Goncalves, A review of methods and instruments to monitor turbidity and suspended sediment concentration, *Journal of Water Process Engineering* 64 (2024) 105624.

[15] G. Chamanee, M. Sewwandi, H. Wijesekara, M. Vithanage, Occurrence and abundance of microplastics and plasticizers in landfill leachate from open dumpsites in Sri Lanka, *Environmental Pollution* 350 (2024) 123944.

[16] J. Ji, X. Wei, W. Wu, C. Fan, D. Zhou, K. Kanagaraj, G. Cheng, K. Luo, X. Meng, C. Yang, The more the slower: self-inhibition in supramolecular chirality induction, memory, erasure, and reversion, *Journal of the American Chemical Society* 144(3) (2022) 1455–1463.

[17] H. Rezaei-Sedehi, O. Ashkani. The future of water purification with carbon and graphene quantum dots: a comprehensive review. *Journal of Water and Environmental Nanotechnology* 10(4) (2025) 392–416.

[18] X. Yang, J. Zhang, K.M. Mostofa, M. Mohinuzzaman, H.H. Teng, N. Senesi, G.S. Senesi, J. Yuan, Y. Liu, S.-L. Li, Solubility characteristics of soil humic substances as a function of pH: mechanisms and biogeochemical perspectives, *Biogeosciences* 22(7) (2025) 1745–1765.

[19] M. Shellaiah, K.W. Sun, N. Thirumalaivasan, M. Bhushan, A. Murugan, Sensing utilities of cesium lead halide perovskites and composites: a comprehensive review, *Sensors* 24(8) (2024) 2504.

[20] C. Kiran, P. Patel, S. Sahoo, A. Rahaman, V. Verma, Photocatalytic remediation of river ganga waters using TiO_2 nanoparticles: linking structural and magnetic properties to aquatic water quality improvement, *Journal of Contaminant Hydrology* 276 (2025) 104723. https://doi.org/10.1016/j.jconhyd.2025.104723

[21] C. Peng, W. Liang, J. Ji, C. Fan, K. Kanagaraj, W. Wu, G. Cheng, D. Su, Z. Zhong, C. Yang, Pyrene-tiaraed pillar[5]arene: strong intramolecular excimer emission applicable for photo-writing, *Chinese Chemical Letters* 32(1) (2021) 345–348.

[22] D. Kurre, S. Samantaray, Sources and causes of emerging contaminants, *Emerging Contaminants in Water*, Springer 2025, pp. 49–88.

[23] K. Kanagaraj, C. Xiao, M. Rao, C. Fan, V. Borovkov, G. Cheng, D. Zhou, Z. Zhong, D. Su, X. Yu, J. Yao, T. Hao, W. Wu, *J.J. Chruma, C. Yang,* A quinoline-appended cyclodextrin derivative as a highly selective receptor and colorimetric probe for nucleotides, *iScience* 23(3) (2020) 100927.

[24] A.C. Estrada et al., Colloidal nanomaterials for water quality improvement and monitoring. *Frontiers in Chemistry* 10 (2022) 1011186.

[25] S. Verma, S. Verma, R. Ramakant, V. Pandey, A. Verma, Comprehensive assessment of physico-chemical and biological parameters in water quality monitoring: a review of contaminants, indicators, and health impacts, *International Journal of Horticulture, Agriculture and Food science* 9(2) (2025) 611019.

[26] S. Mei, K. Wang, H. Schmitt, N. Hofstra, Modelling Escherichia coli concentrations: 45.6%–78.1% of China's rivers show poor microbial water quality☆, *Environmental Pollution* 374(12618) (2025) 4.

[27] N. Thirumalaivasan, S. Nangan, K. Kanagaraj, S. Rajendran, Assessment of sustainability and environmental impacts of renewable energies: focusing on biogas and biohydrogen (Biofuels) production, *Process Safety and Environmental Protection* 189 (2024) 467–485.

[28] P. Venkatesan, R. Govindasamy, G. Periyasami, M. Rahaman, S. Pandiaraj, M. Thiruvengadam, N. Thirumalaivasan, S.-P. Wu, Eco-friendly, bright luminescent carbon dots and their potential applications for detecting hypochlorous acid in water and live cell imaging, *Journal of Materials Research and Technology* 24 (2023) 6522–6532.

[29] S. Das, K.R. Khondakar, H. Mazumdar, A. Kaushik, Y.K. Mishra, AI and IoT: supported sixth generation sensing for water quality assessment to empower sustainable ecosystems, *ACS ES&T Water* 5(2) (2025) 490–510.

[30] M. Farré, R. Brix, D. Barceló, Screening water for pollutants using biological techniques under European Union funding during the last 10 years, *TrAC Trends in Analytical Chemistry* 24(6) (2005) 532–545.

[31] R. Ramasubburayan, K. Kanagaraj, L. Gnanasekaran, N. Thirumalaivasan, N. Senthilkumar, Greener approach supported nitrogen-infused carbon dots for biocompatible cellular markers and fluorescent ink based spray-assisted fingerprint analysis, *Waste and Biomass Valorization* 15(11) (2024) 6431–6440.

[32] I. Yaroshenko, D. Kirsanov, M. Marjanovic, P.A. Lieberzeit, O. Korostynska, A. Mason, I. Frau, A. Legin, Real-time water quality monitoring with chemical sensors, *Sensors* 20(12) (2020) 3432.

[33] K. Kanagaraj, K. Bavanidevi, T.J. Chow, K. Pitchumani, Selective "turn-off" fluorescent sensing of mercury ion using aminocyclodextrin: 3-hydroxy-N-phenyl-2-naphthamide complex in aqueous solution, *RSC Advances* 4 (2014) 11714–11722..

[34] T.A. Saleh, Trends in the sample preparation and analysis of nanomaterials as environmental contaminants, *Trends in Environmental Analytical Chemistry* 28 (2020) e00101.

[35] K. Kanagaraj, K. Pitchumani, Highly selective "turn-on" fluorescent and colorimetric sensing of fluoride ion using 2-(2-hydroxyphenyl)-2,3-dihydroquinolin-4(1H)-one based on excited-state proton transfer, *Chemistry – An Asian Journal* 9 (2014) 146–152.

[36] X.-Y. Xue et al., Nanomaterials for water pollution monitoring and remediation. *Environmental Chemistry Letters* 15(1) (2017) 23–27.

[37] K. Kanagaraj, A. Affrose, S. Sivakolunthu, K. Pitchumani, Highly selective fluorescent sensing of fenitrothion using per-6-amino-β-cyclodextrin:Eu(III) complex, *Biosensors and Bioelectronics* 35 (2012) 452–455.

[38] S. Malik, K. Muhammad, Y. Waheed, Nanotechnology: a revolution in modern industry, *Molecules* 28(2) (2023) 661.

[39] G. Jiang, Y. Li, J. Liu, L. Liu, F. Pi, Progress on aptamer-based SERS sensors for food safety and quality assessment: methodology, current applications and future trends, *Critical Reviews in Food Science and Nutrition* 64(3) (2024) 783–800.

[40] N. Thirumalaivasan, S. Gopi, K. Karthik, S. Nangan, K. Kanagaraj, S. Rajendran, Nano-PCM materials: bridging the gap in energy storage under fluctuating environmental conditions, *Process Safety and Environmental Protection* 189 (2024) 1003–1021.

[41] P.J. Vikesland, Nanosensors for water quality monitoring, *Nature Nanotechnology* 13(8) (2018) 651–660.

[42] C. Xiao, W. Wu, W. Liang, D. Zhou, K. Kanagaraj, G. Cheng, D. Su, Z. Zhong, J.J. Chruma, C. Yang, Redox-triggered chirality switching and guest-capture/release with a pillar[6]arene-based molecular universal joint, *Angewandte Chemie International Edition* 59(21) (2020) 8094–8098.
[43] B.A. Samy, K. Raman, S. Velayutham, N. Senthilkumar, N. Thirumalaivasan, K. Kanagaraj, R. Pothu, R. Boddula, A.B. Radwan, N. Al-Qahtani, Natural product extract fractions as potential arthritis treatments: a detailed analysis using in-silico, in-vivo, and in-vitro methods, *International Immunopharmacology* 144 (2025) 113595.
[44] L.N. Thwala, S.C. Ndlovu, K.T. Mpofu, M.Y. Lugongolo, P. Mthunzi-Kufa, Nanotechnology-based diagnostics for diseases prevalent in developing countries: current advances in point-of-care tests, *Nanomaterials* 13(7) (2023) 1247.
[45] K. Akbar, E. Moretti, A. Vomiero, Carbon dots for photocatalytic degradation of aqueous pollutants: recent advancements. *Advanced Optical Materials* 9(17) (2021) 2100532.
[46] A. Roy, A. Sharma, S. Yadav, L.T. Jule, R. Krishnaraj, Nanomaterials for remediation of environmental pollutants, *Bioinorganic Chemistry and Applications* 2021(1) (2021) 1764647.
[47] M.A. Darwish, W. Abd-Elaziem, A. Elsheikh, A.A. Zayed, Advancements in nanomaterials for nanosensors: a comprehensive review, *Nanoscale Advances* 6(16) (2024) 4015–4046.
[48] H.-B. Zhang, K. Kanagaraj, J. Rebek, Jr., Y. Yu, Hydrophobic and metal-coordinated confinement effects trigger recognition and selectivity, *The Journal of Organic Chemistry* 86(13) (2021) 8873–8881.
[49] Z. Yang, T. Xu, H. Li, M. She, J. Chen, Z. Wang, S. Zhang, J. Li, Zero-dimensional carbon nanomaterials for fluorescent sensing and imaging, *Chemical Reviews* 123(18) (2023) 11047–11136.
[50] Z.A. Qureshi et al., Carbon dots as versatile nanomaterials in sensing and imaging: efficiency and beyond. *Heliyon* 10(11) (2024) e31634. https://doi.org/10.1016/j.heliyon.2024.e31634
[51] N. Thirumalaivasan, S. Nangan, D. Verma, M. Shellaiah, S. Ali, S. Rajendran, K. Kanagaraj, R. Pothu, R. Boddula, A.B. Radwan, Exploring the diverse nanomaterials employed in dental prosthesis and implant techniques: an overview, *Nanotechnology Reviews* 14(1) (2025) 20250140.
[52] Y.-S. Lin, Y.-F. Lin, A. Nain, Y.-F. Huang, H.-T. Chang, A critical review of copper nanoclusters for monitoring of water quality, *Sensors and Actuators Reports* 3 (2021) 100026.
[53] H.-Y. Lee, N. Thirumalaivasan, S.-P. Wu, H_2O_2-responsive boronic ester-modified mesoporous silica nanocarrier for TfR mediated tumor-specific drug delivery applications, *ACS Applied Bio Materials* 8(7) (2025) 6079–6087. https://doi.org/10.1021/acsabm.5c00645.
[54] A.D. Yoffe, Semiconductor quantum dots and related systems: electronic, optical, luminescence and related properties of low dimensional systems, *Advances in Physics* 50(1) (2001) 1–208.
[55] R. Tian, Z. Wang, P. Zhao, Smart carbon nanomaterial-based electrochemical sensor network with edge computing for real-time water quality monitoring. *Microchemical Journal* 221 (2026) 116795. https://doi.org/10.1016/j.microc.2025.116795
[56] A. Razaq, F. Bibi, X. Zheng, R. Papadakis, S.H.M. Jafri, H. Li, Review on graphene-, graphene oxide-, reduced graphene oxide-based flexible composites: from fabrication to applications, *Materials* 15(3) (2022) 1012.
[57] B. Senthil Rathi, L.S. Ewe, S. Sanjay, S. Sujatha, W.K. Yew, R. Baskaran, S.K. Tiong, Recent trends and advancement in metal oxide nanoparticles for the degradation of dyes: synthesis, mechanism, types and its application, *Nanotoxicology* 18(3) (2024) 272–298.
[58] M. Ateia, H. Wei, S. Andreescu, Sensors for emerging water contaminants: overcoming roadblocks to innovation, *Environmental Science & Technology* 58(6) (2024) 2636–2651.
[59] N. Thirumalaivasan, Collagen-composite scaffolds for alveolar bone and dental tissue regeneration: advances in material development and clinical applications—a narrative review, *Dentistry Journal* 13(9) (2025) 396.
[60] F.U. Nisa, M. Tahir, S. Khalid, N. Amin, H. Yin, Y. Long, H. Tang, K. Iiaz, A.U. Khan, M. Naseem, Revolutionizing micro-scale energy storage by 0D carbon nanostructures: synthesis, integration, performance optimization mechanisms and sustainable applications, *Advanced Functional Materials* 35(13) (2025) 2418053.
[61] M. Pradeep, A. Cecil, N. Suresh, R. Govindasamy, N. Thirumalaivasan, Exploring plant-derived carbon dots: dual-functional nanomaterials for endodontic infections and enhanced bio-imaging, *Journal of Oral Biology and Craniofacial Research* 15(6) (2025) 1408–1414.
[62] M. Yang, Y. Shi, Q. Song, Z. Wei, X. Dun, Z. Wang, Z. Wang, C.-W. Qiu, H. Zhang, X. Cheng, Optical sorting: past, present and future, *Light: Science & Applications* 14(1) (2025) 103.
[63] A. Anwar, T.-P. Chang, C.-T. Chen, Graphene oxide synthesis using a top–down approach and discrete characterization techniques: a holistic review, *Carbon Letters* 32(1) (2022) 1–38.
[64] L. Ndlwana, N. Raleie, K.M. Dimpe, H.F. Ogutu, E.O. Oseghe, M.M. Motsa, T.A. Msagati, B.B. Mamba, Sustainable hydrothermal and solvothermal synthesis of advanced carbon materials in multidimensional applications: a review, *Materials* 14(17) (2021) 5094.

[65] H. Luo, L. Lari, H. Kim, S. Hérou, L.C. Tanase, V.K. Lazarov, M.-M. Titirici, Structural evolution of carbon dots during low temperature pyrolysis, *Nanoscale* 14(3) (2022) 910–918.

[66] P. Praserthdam, M. Inoue, O. Mekasuvandumrong, W. Thanakulrangsan, S. Phatanasri, Effect of organic solvents on the thermal stability of porous silica-modified alumina powders prepared via one pot solvothermal synthesis, *Inorganic Chemistry Communications* 3(11) (2000) 671–676.

[67] A.S. Sharma, S. Ali, D. Sabarinathan, M. Murugavelu, H. Li, Q. Chen, Recent progress on graphene quantum dots-based fluorescence sensors for food safety and quality assessment applications, *Comprehensive Reviews in Food Science and Food Safety* 20(6) (2021) 5765–5801.

[68] A. Priyadharshini, K.S. Velu, A.A. Napoleon, P. Roy, N. Ahmad, S. Palanisamy, S. You, S. Mohandoss, S.-C. Kim, N, P, S-doped carbon dots as photoluminescent probe for highly sensitive detection of Cu2+ and Fe3+ ions: applications in cancer cell imaging and real samples, *Journal of Photochemistry and Photobiology A: Chemistry* 462 (2025) 116267.

[69] A. MP, S. Pardhiya, P. Rajamani, Carbon dots: an excellent fluorescent probe for contaminant sensing and remediation, *Small* 18(15) (2022) 2105579.

[70] S. Miao, K. Liang, B. Kong, Förster resonance energy transfer (FRET) paired carbon dot-based complex nanoprobes: versatile platforms for sensing and imaging applications, *Materials Chemistry Frontiers* 4(1) (2020) 128–139.

[71] K. Tandey, K. Shrivas, Green carbon dots-based dluorescence spectrometry for netal ion sensing, *Green Carbon Dots: Sustainable Analytical Approaches*, ACS Publications 2025, pp. 161–189.

[72] H.F. Etefa, A.A. Tessema, F.B. Dejene, Carbon dots for future prospects: synthesis, characterizations and recent applications: a review (2019–2023), *C* 10(3) (2024) 60.

[73] K. Khurana, N. Jaggi, Localized surface plasmonic properties of Au and Ag nanoparticles for sensors: a review, *Plasmonics* 16(4) (2021) 981–999.

[74] V. Rama, K. Kanagaraj, T. Subramanian, P. Suresh, K. Pitchumani, Pyridinium ylide-assisted KY zeolite catalyzed tandem synthesis of polysubstituted cyclopropanes, *Catalysis Communications* 26 (2012) 39–43.

[75] P. Prosposito, L. Burratti, I. Venditti, Silver nanoparticles as colorimetric sensors for water pollutants, *Chemosensors* 8(2) (2020) 26.

[76] K. Kanagaraj, K. Lin, W. Wu, G. Gao, Z. Zhong, D. Su, C. Yang, Chiral buckybowl molecules, *Symmetry* 9(9) (2017) 174 (1–25).

[77] G. Zheng et al., Water pollution control and treatment based on quantum dot chemical and biological high-sensitivity sensing, *Journal of Sensors* 2021(1) (2021) 8704363.

[78] P. Brisebois, M. Siaj, Harvesting graphene oxide–years 1859 to 2019: a review of its structure, synthesis, properties and exfoliation, *Journal of Materials Chemistry C* 8(5) (2020) 1517–1547.

[79] Y.-Q. Chen, H.-W. Guan, K. Kanagaraj, J. Rebek, Jr., Y. Yu, Metal coordination to a deep cavitand promotes binding selectivities in water, *Chinese Chemical Letters*, 2022, 33(11), 4908–4911.

[80] A. Singh, R. Chauhan, V.D. Rajput, T. Minkina, R. Prasad, A. Goel, Exploring the insights of bioslurry—nanoparticle amalgam for soil amelioration, *Environmental Science and Pollution Research* 31(48) (2024) 58310–58323.

10 Biodegradation of Emerging Contaminants in Urban and Rural Water Treatment Systems

Nimish Khandelwal, Mohd. Zafar, Biswanath Mahanty, and Shishir Kumar Behera

10.1 INTRODUCTION

Emerging contaminants (ECs) are a diverse group of natural or synthetic chemicals that have either been recently discovered or identified as potentially toxic to ecosystems and public health (Chaber-Jarlachowicz, Gworek, and Kalinowski 2025). The spread of ECs in aquatic environments poses a global challenge, necessitating the development of effective detection and elimination strategies (Stefanakis and Becker 2020). Pharmaceuticals and personal care products (PPCPs), microplastics (MPs), and industrial organics are arguably the most prominent classes of ECs (Sultan, Anik, and Rahman 2024; Li et al. 2024; R. Kumar et al. 2022). PPCPs comprise a wide range of pharmaceuticals (e.g., antibiotics and anti-inflammatory agents) and personal care products (e.g., cosmetics and disinfectants). These may enter wastewater streams through discharges from hospitals, industries, and households, as well as contributions from agricultural runoff and landfill leachates (Hena, Gutierrez, and Croué 2021; Anand et al. 2022). MPs, i.e., particles <5 mm, generated from the breakdown of larger plastics, synthetic fibers, and microbeads, are another important group of ECs, increasingly detected in both treated and untreated wastewater. Remaining largely undetected, these MPs pose a greater threat to ecosystems (Morin-Crini et al. 2022; Talukdar et al. 2024). Industrial organics, or persistent organic pollutants (POPs), include chemicals such as polychlorinated biphenyls, polybrominated

DOI: 10.1201/9781003741626-10

TABLE 10.1 Major emerging contaminants in urban and rural wastewater

SOURCE TYPE	*MAJOR CLASSES OF ECs*	*EXAMPLES*
Urban WWTPs	Pharmaceuticals	Carbamazepine, Diclofenac, Ibuprofen
	Antibiotics	Ciprofloxacin, Sulfamethoxazole
	Hormones	Estradiol, Estrone
	Personal care products	Triclosan, Parabens
	Industrial chemicals	Bisphenol A, Phthalates
Rural/Decentralized	Veterinary drugs	Tetracycline
	Pesticides	Atrazine, Imidacloprid
	Agricultural runoff contaminants	Nutrient-bound micropollutants

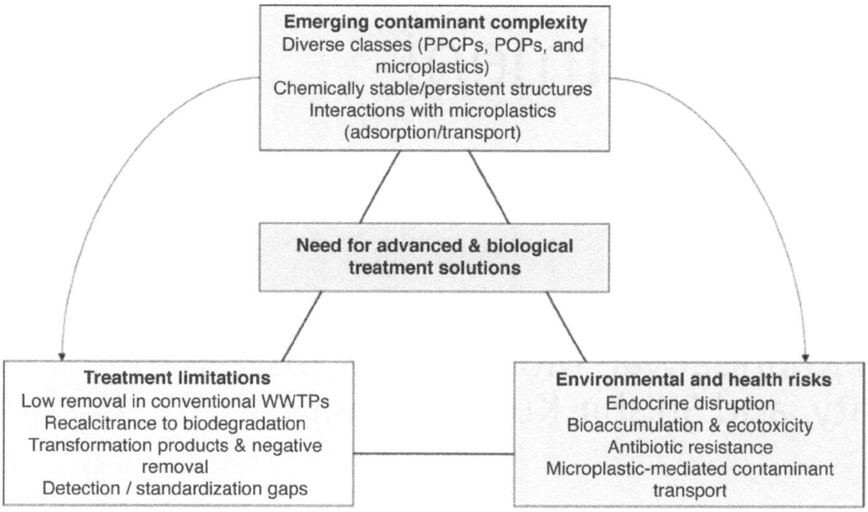

FIGURE 10.1 Drivers of the shift toward advanced and biological treatment solutions for emerging contaminants (PPCPs: pharmaceuticals and personal care products; POPs: persistent organic pollutants; WWTPs: wastewater treatment plants).

diphenyl ethers, and other industrial chemicals (Wilkinson et al. 2017). A comprehensive list of ECs is provided in Table 10.1 (Nishmitha et al. 2025; Boahen, Owusu, and Adjei-Anim 2025).

A significant proportion of ECs are chemically stable, recalcitrant, and highly resistant to degradation using conventional wastewater treatment plants (WWTPs) (Loganathan et al. 2023). As standard physicochemical processes used in WWTPs are not designed for PPCPs, MPs, and industrial organics, these ECs get released into aquatic environments (Hena, Gutierrez, and Croué 2021; McDougall et al. 2022). Centralized collection systems in urban WWTPs act as hotspots for ECs, whereas agricultural runoff and less advanced wastewater treatment infrastructure in rural areas make it diffuse pollution (Morin-Crini et al. 2022). PPCPs and industrial organics have also shown a tendency to bioaccumulate, disrupt endocrine function, and induce antibiotic resistance (Figure 10.1) (Anand et al. 2022; McDougall et al. 2022).

MPs with increasingly large hydrophobic surface areas serve as carriers for ECs, particularly PPCPs. The extent of adsorption, however, may be influenced by polymer type, environmental conditions (e.g., pH, salinity) (Ho et al. 2026; Atugoda et al. 2021), and alterations in surface properties during photodegradation or aging (Y. Xu et al. 2024). These interactions in environmental matrices can increase the spread of antibiotic resistance genes (ARGs), disrupting the nutrient cycles and microbial dynamics in the aquatic ecosystems (Bodzek, Pohl, and Rosik-Dulewska 2024; Kelly et al. 2021) (Figure 10.1). This, in turn, creates a need for leveraging microbial communities that are capable of biodegrading and/or bio-transforming PPCPs and mitigating the environmental risks associated with MPs (Demaria et al. 2025).

Treatment technologies, such as membrane bioreactors (MBRs) and advanced oxidation processes (AOPs), are certainly promising, but scalability and financial and operational challenges do exist (Loganathan et al. 2023; Boleydei and Vaneeckhaute 2024). Lack of standardized detection and analytical methods for PPCPs and MPs further complicates the risk assessment and regulatory interventions (Morin-Crini et al. 2022; Talukdar et al. 2024). Methods focusing on biodegradation of ECs utilizing bacteria, fungi, and microalgae are sustainable, cost-effective, and promising (Sesay et al. 2025). The efficiency of these methods is highly dependent on the physicochemical properties of the contaminant, microbial composition, and operational parameters. Recent technological developments in microalgae–bacteria consortia, constructed wetlands (CWs), and biological–physical–chemical hybrid systems have been effective in treating a broader spectrum of contaminants (Khalidi-Idrissi et al. 2023).

Despite the potential of biological treatment, lack of standardized protocol, limited insights on environmental interactions, and evidence on economic and operational scalability are some of the serious limitations. This chapter bridges these fundamental gaps by providing an overview on parameters governing the fate of ECs, the mechanism of transformation, and the challenges associated with transformation products (TPs). Subsequently, the enzyme-mediated microbial degradation pathways and aerobic and anaerobic degradation mechanisms have been illustrated. Finally, the comparative analysis has attempted to focus beyond localized treatment successes and toward a comprehensive, sustainable framework for global water detoxification.

10.2 SOURCES AND FATE OF EMERGING POLLUTANTS IN WWTPS

10.2.1 The Sources

Domestic, hospital, and industrial wastewater are the main sources for ECs arriving at WWTPs. PPCPs (such as analgesics, antibiotics, beta-blockers, fragrances, and detergents), and endocrine-disrupting compounds (EDCs) are primarily detected in domestic wastewater (Tran, Reinhard, and Gin 2018; Jiang et al. 2024; Sun et al. 2019).

PPCPs are typically released through human excretion and bathing or washing activities (Parashar and Hait 2023) as major point sources (Stefanakis and Becker 2020). A significant portion of prescribed medications, i.e., between 25% and 30% of pharmaceuticals sold in parts of Europe, are discarded or improperly disposed in sewage. Hospital effluents contribute higher concentrations of toxic substances, including antibiotics, cytotoxic drugs, contrast agents, specialized medical substances, and disinfectants (Tran, Reinhard, and Gin 2018). Hospital sources constitute a highly concentrated point source of ECs (Stefanakis and Becker 2020). While the volumetric flow from hospitals is far smaller than that of municipal systems, the concentrations of prescription medicines, disinfectants, and solvents are significantly higher in the former (Stefanakis and Becker 2020). Discharge of X-ray contrast media (e.g., iopromide, iopamidol, and diatrizoate) is the main source of organic iodine in aquatic systems. However, many of these hospital-derived compounds are highly persistent and increasingly being detected in both groundwater and artificial recharge wells (Labad et al. 2023; Hrkal and Pastuszek 2023), indicating the ineffectiveness of conventional WWTPs.

In contrast, a diverse array of complex organic pollutants, including surfactants, additives, and plasticizers, are discharged through industrial wastewater and stormwater runoff in manufacturing sites (Stefanakis and Becker 2020). Industrial effluents containing dyes, solvents, per- and polyfluoroalkyl substances (PFAS), flame retardants, surfactants, and plasticizers further complicate this mixture (Lenka, Kah, and Padhye 2021; Y. Zhou et al. 2019; Tawfik et al. 2022). EDCs, such as bisphenol A and phthalates, originate from plastic products and packaging (Deblonde, Cossu-Leguille, and Hartemann 2011). These industrial pathways are particularly notable for releasing highly persistent and mobile substances, such as PFAS (Jia et al. 2025).

10.2.2 Parameters Governing Fate of Emerging Contaminants

The environmental fate of the ECs in WWTPs is driven by three primary physicochemical parameters, (i) the octanol–water partition coefficient (log K_{ow}), (ii) the acid dissociation constant (pK_a), and (iii) the diffusion coefficient (D). The K_{ow} value determines partitioning of ECs between the aqueous and solid phases. Compounds with log $K_{ow} > 4$ indicate strongly hydrophobic compounds, likely to be adsorbed by sludge and sediment (Wiest et al. 2021; Rout et al. 2021). For example, moderately hydrophilic carbamazepine (log $K_{ow} \approx 2.45$) remains in the aqueous phase and is unlikely to be removed in sludge sorption (Deblonde, Cossu-Leguille, and Hartemann 2011; Wiest et al. 2021).

The pK_a value determines the contaminant's degree of ionization at natural pH levels. This, in turn, helps understand the effects on its solubility, mobility, and biodegradability (Tran, Reinhard, and Gin 2018; Rout et al. 2021). Ionized species tend to be more water-soluble with limited sorption into sludge. Pharmaceutical compounds, such as diclofenac ($pK_a \approx 4.2$), are ionized at neutral pH, resulting in moderate sorption accompanied by varying rates of biodegradability (Deblonde, Cossu-Leguille, and Hartemann 2011). The speciation of EDCs and pharmaceuticals can, therefore, dictate their removal efficiency and environmental persistence.

Diffusion coefficient (D) of contaminants reflects their mobility in water and its accessibility to microbial degradation and/or physical removal processes (Rout et al. 2021). High D values facilitate rapid transport by diffusion through various treatment barriers and enhance their susceptibility to microbial biodegradation and favorable parameters to transport. For example, PFAS, with high mobility and low sorption, are poorly removed by conventional processes and persist in effluents (Lenka, Kah, and Padhye 2021).

10.2.3 Fate of Emerging Contaminants in Wastewater Treatment Plants

The traversal of ECs in primary, secondary, and tertiary stages of a WWTP involves complex chemical and biological transformations (Figure 10.2). The fate of these ECs is determined by their physicochemical properties, treatment technologies, and the potential for transformation into byproducts (Q. Zhou et al. 2024; Mishra et al. 2023). The main techniques involved in primary treatment include sedimentation and skimming for the removal of suspended solids, i.e., physical removal. This is largely ineffective for most ECs, achieving less than 50% removal. Hydrophobic ECs, such as triclosan, tend to adsorb onto these solids, allowing partial removal along with primary sludge. Most hydrophobic and low-sorbing ECs, however, remain in the aqueous phase (Parashar and Hait 2023; Rout et al. 2021).

Biological treatments, such as activated sludge and MBRs, serve as the main stage for sorption and removal of ECs. The removal rates of some biodegradable PPCPs vary between 70% and 90%. However, some recalcitrant ECs, such as carbamazepine, antibiotics, and PFAS, persist through this stage (Rout et al. 2021). This stage successfully reduces MPs further, especially with advanced biological processes. Nevertheless, some of them still escape into the effluent (Parashar and Hait 2023). Desorption, deconjugation of metabolites, and abiotic artifacts may reflect a negative removal rate – i.e., increase in concentration after treatment (Baalbaki et al. 2016; Astuti et al. 2022).

Tertiary treatment methods, including AOPs (such as ozonation and UV), activated carbon adsorption, and membrane filtration, target the residual ECs in wastewater. These advanced technologies can

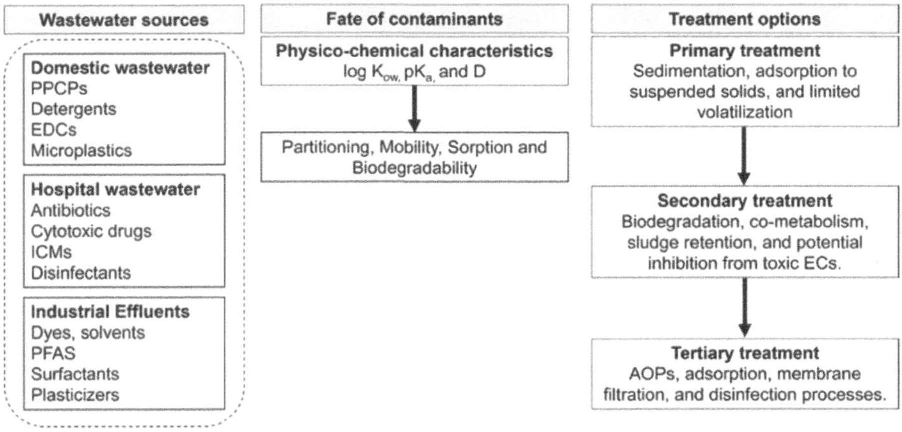

FIGURE 10.2 Schematic of sources, fate of emerging contaminants, and treatment options in wastewater treatment plants (PPCPs: pharmaceuticals and personal care products; EDCs: endocrine-disrupting chemicals; ICMs: iodinated contrast media; PFAS: per- and polyfluoroalkyl substances; ECs: emerging contaminants; AOPs: advanced oxidation processes; UF: ultrafiltration; RO: reverse osmosis; log K_{ow}: octanol–water partition coefficient; pK_a: acid dissociation constant; D: diffusion coefficient).

achieve removal efficiencies exceeding 90% (Morin-Crini et al. 2022; Rout et al. 2021). Complete removal of ECs, especially the polar and persistent compounds, is infrequent, and they remain detectable in the final effluents (Rout et al. 2021; Meng et al. 2021). For MPs in particular, advanced filtration and coagulation–ozonation chains can achieve removal efficiencies up to 99% (Parashar and Hait 2023).

10.2.4 Transformation Products as a Key Challenge

Generation of TPs is a significant challenge in wastewater treatment. TPs generated in biodegradation, oxidation, and deconjugation of metabolites (Wilkinson et al. 2017; Hashem et al. 2025) may be more toxic or persistent than their parent compounds. Biological treatment often leads to partial breakdown of ECs through hydroxylation, hydrolysis, methylation, and formylation (Qian et al. 2024, 2025; R. Kumar et al. 2022). Tertiary treatment often leads to fragmentation and/or oxidation of ECs, creating newer byproducts. For example, oxidation of pharmaceuticals generates hydroxylated TPs, while chlorination may produce halogenated byproducts (Morin-Crini et al. 2022; Qian et al. 2024; R. Kumar et al. 2022).

It is important to note that the distribution and fate of ECs in WWTPs are highly compound-specific and dependent on the design and operational parameters. The physicochemical parameters (log K_{ow}, pK_a, and D) govern the partitioning and biodegradation behavior throughout subsequent treatment stages. While primary and secondary treatments remove a significant fraction of hydrophobic and biodegradable ECs, many hydrophilic, persistent compounds persist into the effluent. Tertiary treatments enhance removal but can also generate new TPs, possibly with additional risks. Comprehensive monitoring, including both parent ECs and their TPs, is essential for accurate risk assessment and effective management (Astuti et al. 2022).

10.3 MECHANISMS OF BIODEGRADATION IN SECONDARY TREATMENT SYSTEMS

Technological innovations in gene manipulation, design of enzymatic cocktails, novel bioreactor systems, and advanced monitoring and control of microbial degradation of ECs are of particular importance

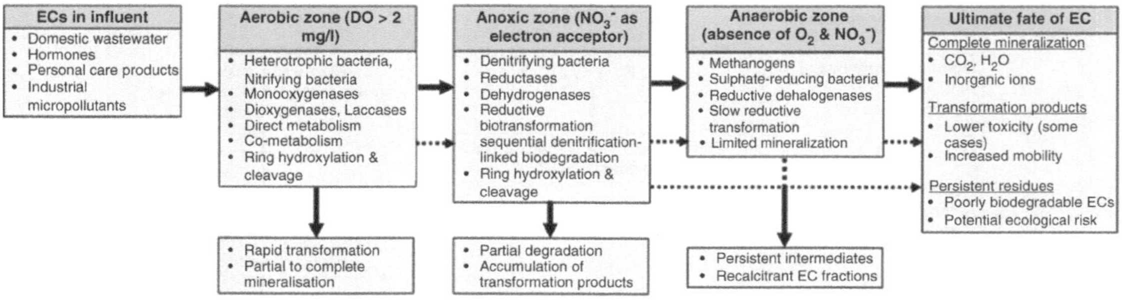

FIGURE 10.3 Integrated biodegradation pathways of emerging contaminants in secondary wastewater treatment (EC: emerging contaminant).

(Figure 10.3). The presence of ARGs in WWTPs has also drawn considerable attention due to their adverse impact on the efficacy of antibiotic therapy for pathogenic bacterial infections (Qin et al. 2020).

PPCPs, antibiotics, and hormones undergo distinct transformation and degradation processes across aerobic, anoxic, and anaerobic zones in wastewater treatment systems. In the aerobic zone (DO > 2 mg/L), the heterotrophs and nitrifiers use O_2-dependent monooxygenases and dioxygenases to hydroxylate and co-metabolize the ECs, leading to rapid degradation and mineralization (Akay et al. 2024; Ruas et al. 2022; Sánchez, Ruiz, and Soto 2022). In anoxic zones, the denitrifiers perform reductive biotransformation via reductases for sequential denitrification-linked degradation of ECs, while nitrates serve as the electron acceptor. However, incomplete mineralization results in accumulation of TPs for some ECs (Dubey, Vellanki, and Kazmi 2023). Under anaerobic environments, methanogens and sulfate-reducing bacteria thrive that use reductive dehalogenases and dehydrogenase systems for contaminant removal, however, with persistent intermediates in sludge or digestate (H. Liu et al. 2025; Barka et al. 2025).

While contaminants are rapidly degraded under aerobic conditions, formation of stable TPs under anoxic/anaerobic conditions limits complete mineralization (Akay et al. 2024). The residues in anoxic/anaerobic treatments may require further processing or pose ecological risks (Khalidi-Idrissi et al. 2023; Akay et al. 2024; Dubey, Vellanki, and Kazmi 2023). Integrated biological treatment systems combining aerobic, anoxic, and anaerobic zones may leverage EC removal by complementing microbial processes and enzymatic pathways across redox gradients (Khalidi-Idrissi et al. 2023; Ruas et al. 2022; Sánchez, Ruiz, and Soto 2022).

10.3.1 Enzyme-Mediated Microbial Pathways

ECs can be degraded by microorganisms, either as primary substrates or via co-metabolism processes, in which the presence of an additional carbon source can enhance degradation rates. Figure 10.4 illustrates the degradation of various ECs to their end products by a suitable microbial enzyme system. For common ECs of pharmaceutical origin, such as ibuprofen, diclofenac, and sulfamethoxazole, removal efficiencies between 70% and 99% are very common (T. Liu et al. 2023; Suleiman et al. 2023). Bioaugmentation with an engineered strain of *Pseudomonas stutzeri* has been reported to achieve 96% removal of alprazolam in about 20 days (Shahriarinour et al. 2021). The effectiveness of biodegradation depends on environmental factors such as temperature, pH, and substrate concentration. EC degradation via various metabolic pathways and/or co-metabolism is affected by both chemical and biological factors, including nutrient and O_2 availability, heavy metal concentrations, and the types of microorganisms. Biodegradation of nonsteroidal anti-inflammatory drugs is favored in mild acidic conditions. However, the presence of heavy metals inhibits the biodegradation of ibuprofen by a *Bacillus thuringiensis* B1 strain (Marchlewicz et al. 2017).

Indigenous microorganisms, including bacteria and fungi, engage in various metabolic reactions to facilitate the degradation of ECs *via* different microbial enzymes that can transform these ECs into simpler molecules, i.e., CO_2 and H_2O (Tong et al. 2022). The bacterial heterotrophic and autotrophic

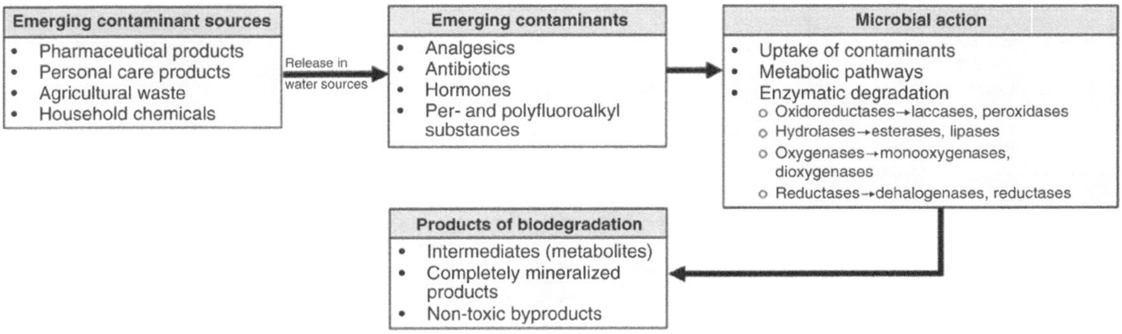

FIGURE 10.4 Enzyme-mediated microbial degradation of emerging contaminants.

metabolisms utilize different strategies to metabolize and co-metabolize the ECs. The autotrophic bacteria degrade ECs using a nonspecific monooxygenase enzyme system. Different ECs (such as diclofenac, naproxen, clofibric acid, and bezafibrate) can be degraded by the ammonium monooxygenase enzymatic system under both aerobic and anaerobic conditions. However, autotrophic sulfur-reducing bacteria couple sulfate reduction to ECs metabolism anaerobically using nitrate and sulfate as alternative electron acceptors. These autotrophic bacteria can co-metabolize organic compounds while reducing nitrate and sulfate and can simultaneously remediate the contaminants and cycle the essential nutrients (Wang et al. 2023). Several benzene compounds, such as benzyl alcohol, benzaldehyde, and 1,2,4-trihydroxybenzene, have been reported to be oxidized by *Desulfovibrio* using sulfate as an electron acceptor (Keen and Patrick 2013). Under aerobic conditions, oxidoreductase (e.g., oxidases, peroxidases, oxygenases, and dehydrogenases) and hydrolase (e.g., amidases and esterases) enzymes oxidize ECs using simple carbon substrates. Anew, a methanotrophic bacterial consortium consisting of *pmo*A genes encoding methane monooxygenase has been observed to facilitate the biodegradation of trichloroethylene with methane (Ahmed et al., M. N. Iqbal, and Dhama 2017). Under anaerobic conditions, reduction of ECs is initiated by the reductive enzyme system, including acetate kinase, reductive dehalogenase, and fumarate reductase.

Heterotrophic bacterial strains, such as *Sphingomonas* Ibu-2 and *Rhizorhabdus wittichii* MPO18, degrade pharmaceutical compounds (as the primary carbon source) under aerobic conditions (Mqambalala et al. 2025). In contrast, heterotrophic *Pseudomonas sp.* degrades phenylacetic acid (ibuprofen) under anaerobic conditions. However, co-metabolic degradation by heterotrophic microorganisms has a threshold for efficient degradation without toxic effects on microorganism growth. A persistent insecticide, imidacloprid, is efficiently co-metabolized by the heterotroph *Stenotrophomonas maltophilia* using various carbon substrates (Zhonghua Liu et al. 2013). Facultative microorganisms, such as *Pseudomonas* sp., *Bacillus* spp., and *Azoarcus evansii,* can effectively degrade ECs under both aerobic and anaerobic conditions by shifting their respiratory mechanisms.

Fungi are well known for aerobic degradation of ECs and their genomic capabilities for the efficient degradation of very complex pollutants such as hydrocarbons, polycyclic aromatic hydrocarbons (PAHs), and pesticides (Vaksmaa et al. 2023). Degradation of ECs by fungi is facilitated by specific enzymes, such as laccase, dioxygenase, and peroxidase, which act aerobically to degrade pollutants. Oxidation of PAHs was reported using lignin peroxidase and manganese peroxidase under aerobic conditions by *Aspergillus sydowii* (Ganesh Kumar et al. 2021).

10.3.2 Molecular Mechanisms in Biodegradation of ECs

Microbial and enzyme-based approaches for degrading ECs require a thorough understanding of the associated microbial enzymatic systems. Advanced molecular approaches and design of novel enzyme cocktails can improve bioremediation technologies and mitigate the impacts of diverse ECs. Molecular approaches, such as genomics, proteomics, transcriptomics, and metabolomics, can reveal microbial

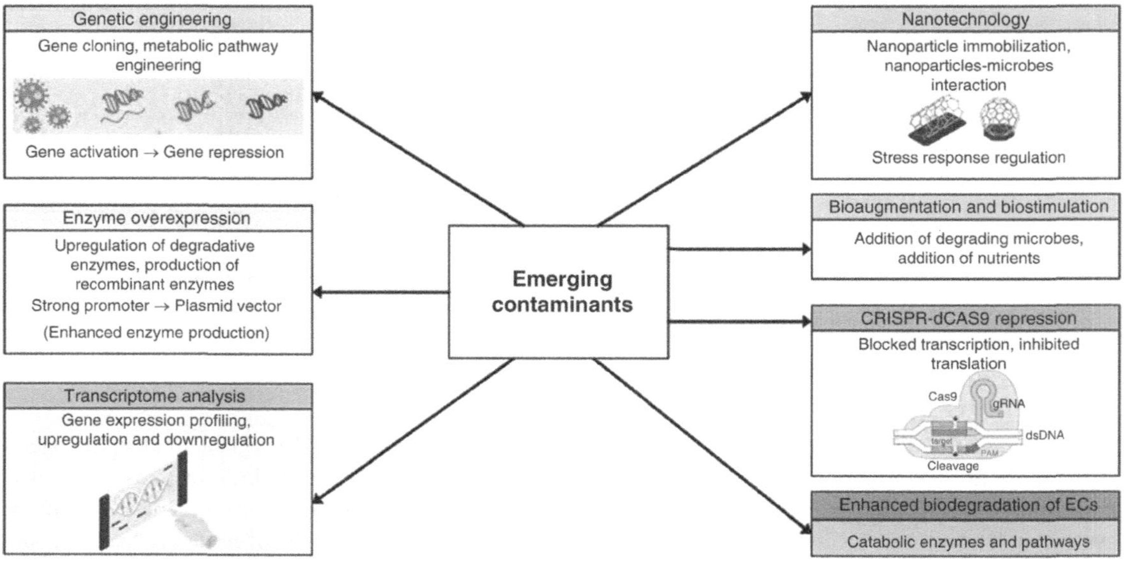

FIGURE 10.5 Molecular mechanisms related to biodegradation of emerging contaminants.

metabolism and provide in-depth information on soil–microbial interactions and functions, key metabolic genes, and the mechanisms of toxicity of ECs (Alidoosti et al. 2024). Figure 10.5 illustrates the various molecular mechanisms to enhance the biodegradation of ECs.

Using these omics techniques, i.e., changes in microbial gene expression profiles (proteomics), identification and quantification of metabolites (metabolomics) associated with EC degradation can be explored. Genetic manipulation is a critical mechanism for enhancing the biodegradation of recalcitrant contaminants. A CRISPR-Cas9-engineered *Pseudomonas putida* strain is reported to degrade 90% of tetracycline in wastewater through regulation of efflux pump expression (Mamatha Bhanu, Kataki, and Chatterjee 2024). This approach enables manipulation of microbial metabolic pathways. Metagenomics revealed the hidden potential of microbes toward bioremediation through genome-based identification of new promoters, genes, and degradative pathways of ECs. The whole-genome sequences of well-known bacterial genera such as *Pseudomonas, Shewanella, Deinococcus*, and *Dehalococcoides* reveal novel genes that could provide better insights into their biodegradation ability and substrate selectivity (Bhardwaj et al. 2020). The whole genome of *Streptomyces nigra* LM01 revealed its potential to degrade and dechlorinate atrazine to hydroxyatrazine or convert it to cyanuric acid (Zhao and Wang 2024). Transcriptomics captures the gene expression profile of a single organism at a particular growth stage; metatranscriptomics can examine gene expression across the microbial community by extracting RNA from an environmental sample (Bokade et al. 2023). The effect of bioaugmentation and biostimulation on the transcriptomic profiles of some bacterial strains of the genera *Rhodococcus, Mycolicibacterium, Dietzia, Pseudomonas, Arthrobacter*, and *Gordonia* was investigated (Brzeszcz et al. 2020). Metaproteomics techniques, e.g., mass spectrometry (MS) and two-dimensional electrophoresis (2-DE), can reveal key catabolic enzymes involved in microbial degradation of ECs. To understand the process of phenanthrene degradation, a proteomic investigation of *S. paucimobilis* 20006FA was performed using 2-DE, and the resulting peptides were analyzed by MALDI-TOF/TOF (Macchi et al. 2018). These proteomic investigations revealed an enzymatic system that transforms phenanthrene into various degradative products. The presence of dioxygenase enzyme-coding genes suggested the potential for aromatic biodegradation *via* the salicylate and protocatechuate pathways for phenanthrene degradation. Metabolomics is an approach to identify metabolites involved in the degradation of ECs. Several studies have examined metabolites produced under natural biodegradation of ECs. A comparative intracellular metabolome (e.g., fatty acids, polyhydroxyalkanoates, and polar metabolites) of *Sinorhizobium* sp. was analyzed during phenanthrene biodegradation (Keum et al. 2006). Fluxomics is a method of measuring metabolic reaction rates and involves studying the complete set of metabolic fluxes within a microbial cell (Alidoosti et al. 2024). The full collection of metabolic fluxes (fluxomics) is a unique characteristic of the microbial cell. Fluxomics offers several benefits over proteomics and genomic profiling because it depends on the expression profile (metabolite data).

10.3.3 Influence of Operational Parameters in Aerobic, Anoxic, and Anaerobic Degradation

ECs can be effectively removed from wastewater treatment processes through sorption combined with enzyme-mediated biotransformation. The bioremediation of ECs of pharmaceutical origin (e.g., ibuprofen, diclofenac), personal care products, and hormones, including EDCs, is influenced by different redox conditions, including aerobic (O_2 as electron acceptor and dissolved oxygen (DO) >2 mg/L), anoxic (nitrate/denitrification, DO ~ 0 mg/L), and anaerobic (methanogenesis/sulfate reduction, DO = 0 mg/L). Other operational parameters, such as solid retention time (SRT), hydraulic retention time (HRT), pH, and temperature, affect the bioavailability of compounds impacting the removal efficiency (Ruas et al. 2022). Aerobic processes yield a high biodegradation rate by diverse autotrophic and heterotrophic microbial consortia, followed by anoxic processes that include denitrifying bacteria. Anaerobic processes are often limited to sorption and result in a slow biodegradation of ECs.

SRT is an important parameter to control biomass age and microbial diversity. A longer SRT enriches slow-growing nitrifiers (i.e., suitable for co-metabolism) and enhances biotransformation through increasing sorption sites. An optimal SRT of 10–20 days favored high removal (>95%) of EDCs (e.g., ibuprofen and estrogen) (Clara et al. 2005). In contrast, a low SRT of one to five days limits the biodegradation efficiencies of these compounds below 60%. A longer HRT in aerobic bioremediation usually provides sufficient time for enzymatic biodegradation and equilibrium sorption. A laboratory study to assess the effect of HRT on biodegradation of PPCPs (Fernandez-Fontaina et al. 2012) suggested limited biodegradation at shorter HRT and/or higher micropollutant loading rates. In order to cope with these situations, an increased loading rate of these micropollutants was required. Similarly, pH affects ionization and enhances pK_a-dependent sorption and bioavailability of ECs and microbial enzyme activities during bioremediation processes. The pH between 6.5 and 8 has been shown to increase the sorption under aerobic conditions, and more than 90% removal of ibuprofen was observed at pH 7. Similarly, temperature is one of the influential parameters controlling the biodegradation rate of ECs. An increase in temperature between 20°C and 30°C accelerates the degradation of estrogens by >90%. The anoxic process showed stability in the temperature range of 15–25°C. However, mesophilic (30–37°C) and thermophilic (55°C) processes have shown robustness for biodegradation of some antibiotics in the waste stream (Carballa et al. 2004; Luo et al. 2014).

Secondary treatment is a core component of WWTPs, employing processes such as activated sludge, MBRs, trickling filters, and CWs to degrade organic compounds through microbial metabolism under aerobic, anoxic, or anaerobic conditions. The biodegradation mechanism of ECs is challenging and varies, with 25–99% efficiency depending on the compounds and system (Sesay et al. 2025). The primary treatment relies mainly on physical settling, while the secondary treatment is facilitated by sorption of compounds onto sludge and biotransformation *via* enzymatic and microbial processes. Microbial consortia leverage synergistic mechanisms such as cross-feeding and co-metabolism to enhance the degradation of complex pharmaceuticals, including ibuprofen, diclofenac, and sulfamethoxazole, achieving 70–99% removal efficiencies (Sesay et al. 2025).

MBRs have been successfully employed to treat ECs by combining biological treatment with membrane filtration. An immobilized MBR system, including *Pseudomonas aeruginosa* encapsulated in alginate/gellan gum microbeads, has been shown to protect against the toxic effects of ECs during continuous biodegradation of these compounds (Suleiman et al. 2023). The predominant mechanisms of EC removal in the MBR system included size exclusion, electrostatic sorption onto the membrane surface and biofilm, and subsequent biodegradation *via* hydrophobic interactions with the membrane (Rout et al. 2021). A vertical electroactive biofilter is known for its efficient removal of pharmaceutical compounds (80%) and herbicides (50–70%) through the incorporation of electric fields into a conventional biofilter system. These bioelectrochemical reactors were able to efficiently enhance the wastewater detoxification by altering the chirality of the contaminants (Sesay et al. 2025). Similarly, anaerobic MBRs (AnMBRs) are well known for recovering sustainable energy (methane) while significantly reducing COD (up to 85%) from mixed pharmaceutical effluents (Khalidi-Idrissi et al. 2023).

Integrated microalgae-based WWTPs comprising an anoxic reactor, an aerobic photobioreactor, and an anaerobic digester are considered an advanced, sustainable technology for removing various ECs (Ruas et al. 2022). An integrated anoxic–aerobic bioreactor system has been proven to be an efficient technology for removing at least 48% of all tested ECs, including enrofloxacin, ciprofloxacin, progesterone, and propylparaben. Similarly, an anaerobic/anoxic/aerobic system was used to effectively remove organophosphates, and different removal rates were reported under anaerobic, anoxic, and aerobic conditions (L. Zhou et al. 2024).

10.4 CASE STUDIES AND PROCESS COMPARISONS

Conventional activated sludge systems act as the baseline for wastewater treatment, as they are designed primarily to remove organic matter and nutrients through microbial activity in aerated tanks. However, these systems are often inadequate when it comes to the removal of ECs due to their complex chemical structures and low concentrations (Rout et al. 2021; Ulucan-Altuntas et al. 2023). Several advanced treatment processes have demonstrated enhanced removal efficiency by optimizing the microbial environment and integrating chemical and physical separation methods, such as redox-optimized systems, biofilm reactors, and hybrid physical–biological–membrane treatment systems (Figure 10.6).

One of the remarkable examples of this ingenuity can be found in the redox-optimized systems utilized in East Asian contexts. Rather than treating wastewater as a uniform stream, processes such as the Daewoo nutrient removal (DNR) process create a sophisticated metropolis for microbes. By dividing reactors into distinct anaerobic, anoxic, and aerobic zones, engineers can cultivate specialized microbial communities within a single system. This multi-zone configuration achieves high nitrogen and phosphorus removal while fostering specific microbial activity to degrade the pharmaceutical compounds. Similarly, in the Symbio process, engineers create microbial flocs with an aerobic outer region and an anoxic inner core utilizing low aeration. This allows simultaneous nitrification and denitrification within the same tank (Behera et al. 2011).

The use of inert carriers to increase attached biomass concentration in biofilm reactors, e.g., moving bed biofilm reactors (MBBRs) and integrated fixed-film activated sludge (IFAS) systems (Tang 2018), has been revolutionary. At longer SRT, these systems give microbes the necessary time to break down

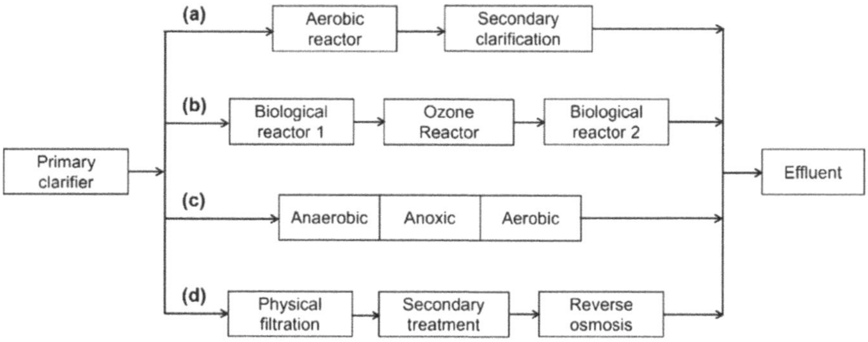

FIGURE 10.6 Schematic representation of discussed treatment schemes: (a) conventional activated sludge process; (b) redox-optimized system (BO_3B process), (c) modified biological systems, and (d) hybrid physical–biological–membrane treatment system.

organics that would otherwise evade conventional treatment (Oulton, Kohn, and Cwiertny 2010; Faria et al. 2020). For the most recalcitrant pollutants, where biodegradation by microbes is not adequate, the biological–ozone–biological (BO_3B) hybrid system has emerged as a new normal (de Wilt et al. 2018):

i. Preliminary biological step: Clears the bulk organic load to unmask the more difficult contaminants.
ii. Ozonation: Uses targeted oxidation to shatter the hard-to-degrade chemical bonds of POPs.
iii. Final biological step: Meticulously mineralizes any potentially toxic byproducts generated during the ozone stage, ensuring the final effluent is safe for the environment.

The efficacy of these innovative designs can be observed across case studies from around the world, which show how operational parameters and complementary treatment technologies influence the removal of ECs (Table 10.2). A comprehensive study of municipal WWTPs in Ulsan, Korea, showcased the critical

TABLE 10.2 Key plant/process features summary

REGION	*PROCESS TYPE*	*CONFIGURATION/ KEY FEATURES*	*TARGETED POLLUTANTS*	*KEY OPERATIONAL INSIGHT*	*REFERENCE*
Australia/ USA	AWRDP (RO, NF, Ozonation)	Multi-barrier physical separation; RO membranes; Ozonation dose 17.5 mg/L	Hormones (Estradiol), analgesics (CBZ, Ketoprofen)	RO is the most effective barrier, achieving high removal regardless of biodegradability	Khan et al. (2004)
Netherlands	BO_3B Pilot	Sequential system: BR1→ O_3R→ BR2; sand filter media; HRT 1.46 h	Caffeine, Ibuprofen, CBZ, Diclofenac, Sulfamethoxazole	BR1 achieves TOC reduction (38%) to spare ozone; O_3R targets CBZ (>99% at high dose)	de Wilt et al. (2018)
Korea	DNR, Symbio	Multi-zone redox control (anoxic–oxic), varying SRT (16.8–43 d)	Antibiotics, hormones, analgesics (CBZ, Metoprolol)	An optimized redox structure is critical for removing recalcitrant compounds (e.g., CBZ)	Behera et al. (2011)
Colombia/ Brazil	CWSHF	Horizontal subsurface flow; 72 h HRT; planted polyculture (Colombia)	Carbamazepine, Sildenafil, Methylparaben	High sensitivity to biological activity/ climate; efficiency depends on compound structure (MP removed, CBZ recalcitrant)	Delgado et al. (2020)
UK/ Denmark	MBBR (intermittent feeding)	Polishing MBBR; intermittent feeding with raw wastewater	Diclofenac, Metoprolol, Atenolol	Strategic feeding promotes specialized degraders, drastically cutting Diclofenac half-life (2.1 h)	Tang (2018)

Acronyms: CBZ = carbamazepine; BO_3B = bio-ozone-bio; CWSHF = subsurface horizontal flow constructed wetlands; AWRDP = advanced water recycling demonstration plant; NF = nanofiltration; RO = reverse osmosis; TOC = total organic carbon; O_3R = ozone reactor; MP = methylparaben; MBBR = moving bed biofilm reactor.

role of engineered biological conditions. Across the systems studied, HRTs ranged from 7.9 to 15 h, while SRTs ranged from 16.8 to 43 days. While more than 99% removal was observed for easily biodegradable analgesics (e.g., acetaminophen) and hormones (e.g., estradiol and estriol) across all treatment types, the removal of pharmaceuticals was highly dependent on the treatment processes. For example, carbamazepine exhibited an average removal of 23.1%; however, the modified DNR process achieved the highest removal of 51%, 55%, and 90%, respectively, for carbamazepine, metoprolol, and triclosan. Coexistence of anoxic and oxic zones was observed to facilitate the biodegradation of highly refractory compounds (Behera et al. 2011).

The pilot-scale BO_3B process from Netherlands had been efficient in treating high-organic-carbon effluents prior to targeted chemical oxidation (de Wilt et al. 2018). The initial biological reactor (BR1) removed 38% of total organic carbon (TOC), achieving >95% removal of readily biodegradable compounds such as caffeine and ibuprofen. The subsequent ozonation step (O_3R) effectively targeted the recalcitrant compounds, achieving complete removal of carbamazepine at an optimal dose of 0.4 g O_3/g TOC. This was significantly higher than the removal (<35%) from conventional biological methods. The final biological step (BR2) proved crucial, displaying a 17% TOC reduction, demonstrating successful biodegradation of potentially toxic ozonation byproducts.

CWs represent passive, low-cost options studied in tropical and subtropical regions. However, their performance relies heavily on substrate interaction and seasonal climate. In a Colombian horizontal subsurface flow CW, >97% of methylparaben was removed primarily through microbial degradation, while carbamazepine showed <10% removal. Conversely, sildenafil achieved up to 97% removal in planted systems, suggesting that specific plant/microbial metabolism can overcome structural resistance for certain compounds (Delgado et al. 2020). In Brazil, a vertical flow CW demonstrated acute sensitivity to seasonal climate. In winter, removal efficiencies were positive, with up to 99% removal for the antibiotic erythromycin and 80% for the analgesic diclofenac. However, in warmer summer conditions, several compounds exhibited negative removal efficiencies. This phenomenon is attributed to biological deconjugation processes, where excreted metabolites revert back to the parent compound, increasing the measured concentration in the effluent (M. Kumar et al., 2022; De Vargas et al. 2021).

The advanced water recycling demonstration plant in Australia, with a focus on high-level water reuse, tested various physical and chemical polishing methods (Khan et al. 2004). The reverse osmosis (RO) process was demonstrated to be the most consistent for the removal of estradiol, estrone, ethinylestradiol, and carbamazepine. RO acts as a near-universal physical barrier, regardless of the biodegradability or chemical class of compound. Ozonation was observed to degrade the salicylic acid partially. However, for high-level water polishing, RO consistently outperformed the other methods.

The removal performance of ECs varies widely across systems and generally correlates with the compound's susceptibility to biological activity and ability to maximize microbial diversity or employ targeted oxidation (Table 10.3, Figure 10.7). In optimized biological environments, such as MBBR, DNR, and Symbio, an increased SRT value facilitates the microorganisms to adapt and engage in co-metabolism. In this manner, the transformation of the PPCPs is supported by the metabolism of readily available primary substrates. For example, in Denmark, a radical improvement in diclofenac removal was observed in intermittently fed MBBRs, showing how the promotion of specialized microbial function increases the biodegradation rates (Tang 2018).

While not the ultimate elimination mechanism, sorption onto biomass or media, such as gravel in CWs, is essential for retaining hydrophobic compounds and increasing their contact time with microbial communities for subsequent biodegradation (Delgado et al. 2020). For highly recalcitrant compounds, such as carbamazepine and certain contrast media, which are known to resist biological attack because of their stable aromatic structures, ozonation has been demonstrated to be inevitable. This non-selective chemical process targets double bonds and aromatic groups, effectively breaking down persistent structures, which is not possible by biodegradation (Oulton, Kohn, and Cwiertny 2010; de Wilt et al. 2018). RO fundamentally bypasses this stable structure and microbial activity by relying on physical size exclusion.

TABLE 10.3 Removal of pollutants, including antibiotics, hormones, and analgesics

COMPOUND (CLASS)	*TREATMENT TYPE*	*REGION*	*INFLUENT CONCENTRATION*	*TREATMENT VOLUME/ CAPACITY*	*RE*	*DOMINANT MECHANISM*	*REFERENCE*
Estradiol (hormone)	Modified systems	Korea	4 ng/L	19.5–247.3 × 10^3 m^3/d (plant flow)	~100%	Biodegradation	Behera et al. (2011)
Ibuprofen (analgesic)	BO_3B (BR1)	Europe	200 μg/L (spiked)	1.4 L (effective reactor volume)	>95%	Biodegradation of easily degradable organic matter	de Wilt et al. (2018)
Acetaminophen (analgesic)	Conventional/ Modified	Korea	3.54–10.23 μg/L	19.5–247.3 × 10^3 m^3/d (plant flow)	>99%	Biodegradation/ Biological	Behera et al. (2011)
Methylparaben (analgesic/PCP)	CWHSF	Colombia	200 μg/L	15 mL/min (operating flow)	>97%	Bacterial degradation of simple molecular structure	Delgado et al. (2020)
Diclofenac (analgesic)	MBBR (Intermittent feed)	UK and Denmark	3–20 μg/L (spiked)	3 L (lab)/500–900 L (pilot reactors)	>80% total removal	Enhanced biodegradation/ co-metabolism	Tang (2018)
Carbamazepine (analgesic/ recalcitrant)	BO_3B (O_3R)	Europe	200 μg/L (spiked)	1.6 L (liquid volume of O3R)	>99% (at 0.4 g O_3/g TOC)	Chemical oxidation (ozonation)	de Wilt et al. (2018)
Carbamazepine (analgesic/ recalcitrant)	RO (AWRDP)	Australia and USA	Not specified (spiked)	Pilot demonstration plant scale	Significant removal (below detection limits)	Physical rejection	Khan et al. (2004)
Erythromycin (antibiotic)	VF-CW (Summer)	Brazil	0.1359 μg/g	1100 L/d (average input flow)	Negative RE	Deconjugation of metabolites	De Vargas et al. (2021)
Sulfamethoxazole (antibiotic)	BO_3B (BR1)	Europe	200 μg/L (spiked)	1.4 L (effective reactor volume)	>99%	Biodegradation	de Wilt et al. (2018)

Acronyms: AWRDP = advanced water recycling demonstration plant; BO_3B = bio-ozone-bio; BR1 = initial biological reactor; CWSHF = subsurface horizontal flow constructed wetlands; RO = reverse osmosis; O_3R = ozone reactor; MP = methylparaben; MBBR = moving bed biofilm reactor; VF-CW = vertical flow constructed wetland; PCP = personal care product; RE = removal efficiency.

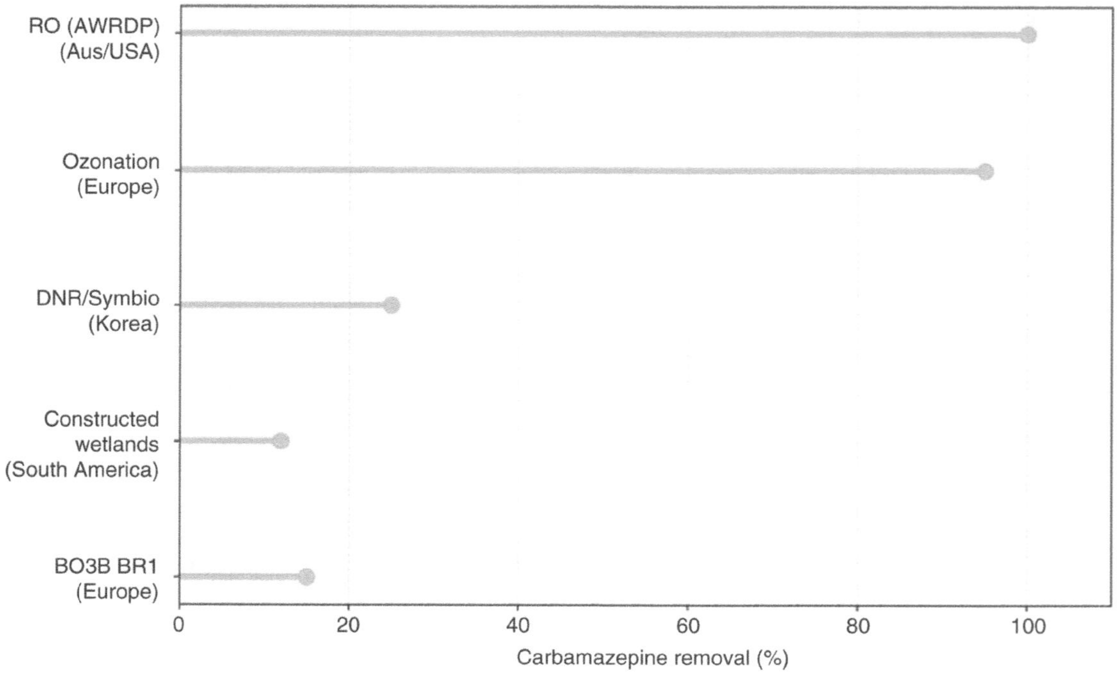

FIGURE 10.7 Comparative removal performance of carbamazepine among treatment types (AWRDP = advanced water recycling demonstration plant; BO_3B = bio-ozone-bio; BR1 = initial biological reactor; DNR: Daewoo nutrient removal; RO: reverse osmosis).

This has been noted as the most robust mechanism for consistent, near-total removal of all ECs, including biologically inert compounds such as carbamazepine and iodinated contrast media.

Among these systems, structural complexity of ECs (e.g., aromaticity and halogenation) plays a more decisive role in removal efficiency. The biological systems consistently underperform for carbamazepine, indicating intrinsic biodegradation resistance rather than operational limitation.

10.5 BIODEGRADATION MODELING APPROACHES

In order to correctly understand, predict, and optimize the complex behavior of ECs in urban and rural wastewater systems, which cannot be captured by experiments alone, modeling is a necessary approach (Sulaiman et al. 2024). Models such as the activated sludge model (ASM) and the plant-wide model are helpful in capturing complex mechanisms, such as sorption, biodegradation, volatilization, and photolysis, while studying the higher-order interactions between recalcitrant ECs (Dubey, Vellanki, and Kazmi 2023; Gabrielli et al. 2023; Bencsik et al. 2024). This is done by coupling hydrodynamics, gas transfer, biokinetics, and sludge line processes, providing an overall mass balance for ECs, which is impossible to measure through experimental measurement approaches alone (Bencsik et al. 2024; Moya-Llamas, Pacazocchi, and Trapote 2023).

ASMs are a set of models categorized as ASM1 (i.e., included carbon oxidation, nitrification, and denitrification), ASM2 (i.e., included additional biological/chemical phosphorus removal), and ASM3, which incorporated improved carbon storage and endogenous respiration (Sin and Al 2021) and are extensively applicable to WWTPs for the removal of macro-pollutants. ASMs are generally focused on COD, N, and P and omit trace levels of PPCPs and EDCs. However, to model these ECs, ASMs can be extended to incorporate additional state variables and processes specific to these compounds. The integrated processes may include sorption and biodegradation/biotransformation as dominant removal mechanisms in the aerobic activated sludge process. The most widely used and established extended ASM framework for xenobiotics (Plósz, Langford, and Thomas 2012) integrates these trace elements into ASM structures to assess biodegradation and sorption. This extended ASM structure (mainly integrated in ASM1 or ASM2) for xenobiotics has been successfully validated for pharmaceutical compounds. Besides, other extensions (e.g., pseudo-first-order or co-metabolic kinetics) use similar approaches and are applicable for the indirect decomposition of these compounds, depending on SRT, temperature, and biomass concentration. These extended ASMs can be successfully implemented in suitable software packages such as GPS-X, BioWin, and MATLAB®. In addition to ASM, anaerobic digestion models (ADMs) can also be used in an extended form for the EC removal in AD processes. The ADM extension for ECs is compound-specific and less standardized than for ASM.

In these models, sorption and biodegradation are added as new state variables for each EC (e.g., dissolved S_{EC} and sorbed X_{EC}). It is assumed that sorption is at instantaneous equilibrium, faster than biodegradation and HRT, and is represented by a linear partition coefficient (K_d) (L/kg SS).

$$X_{EC} = K_d \times X_{SS} \times S_{EC} \tag{10.1}$$

where X_{EC} is sorbed EC concentration (µg/kg TSS or similar trace units), S_{EC} is dissolved EC concentration (µg/L), X_{SS} is total suspended solids concentrations (kg SS/m^3), and K_d is solid–liquid distribution coefficient (L/kg TSS), compound-specific (e.g., high for triclosan and low for ibuprofen) (Fernandez-Fontaina et al. 2014).

Similarly, the total EC balance in a mixed flow reactor includes terms for influent, effluent, and waste sludge. The dynamic equation for dissolved EC can be represented as follows:

$$\frac{dS_{EC}}{dt} = \frac{Q_{in}}{V}\left(S_{EC,in} - S_{EC}\right) - \rho_{bio} + \frac{Q_{waste}}{V}\left(f_{sorbed} \times total_{EC}\right) \tag{10.2}$$

where S_{EC} is dissolved EC concentration (µg/L), $S_{EC,in}$ is influent EC concentration (µg/L), Q_{in} is influent flow rate (m^3/d), Q_{waste} is waste sludge flow rate (m^3/d), V is reactor volume (m^3), and ρ_{bio} is biodegradation rate constant (L/(kg TSS·d).

The biodegradation model is also widely used for the removal of many ECs considering two approaches. The first one assumes pseudo-first-order kinetics under low EC concentration represented as follows:

$$\rho_{bio} = k_{biol} \cdot S_{EC} \cdot X_{hetero} \tag{10.3}$$

or normalized to solids as follows:

$$\rho_{bio} = k'_{biol} \cdot X_{SS} \cdot S_{EC} \tag{10.4}$$

where k_{biol}(L/g biomass·d) or k'_{biol}(L/kg SS·d) is compound-specific, calibrated from batch experiments. This is simple and widely used for persistent or moderately degradable ECs.

The other approach, i.e., co-metabolic biodegradation, assumes indirect degradation of ECs via enzymes meant for primary substrate. In this process, biodegradation of ECs is linked to growth on primary substrates (e.g., acetate, propionate, H_2). At trace concentrations, biodegradation simplifies to pseudo-first-order kinetics:

$$\rho_{bio} = k_{biol} \cdot S_{EC} \cdot \frac{X_{het} + X_{aut}}{X_{TSS}} \cdot X_{TSS} \tag{10.5}$$

or normalized:

$$\rho_{bio} = k'_{biol} \cdot X_{TSS} \cdot S_{EC} \tag{10.6}$$

where k'_{biol} is the biomass-specific biodegradation rate constant (L/g TSS·d), and X_{het} and X_{aut} are heterotrophic and autotrophic biomass (from base ASM), respectively.

Similarly, for co-metabolic degradation, which is common for ECs (e.g., via ammonia monooxygenase in nitrifiers), the process is linked to primary substrate uptake, i.e., ammonia oxidation rate ρ_{NH4} in the ASM nitrification process (Fernandez-Fontaina et al. 2014), represented in Equation 10.7:

$$\rho_{bio,EC} = k_{co} \cdot \rho_{NH4} \cdot S_{EC} \tag{10.7}$$

In this process, temperature correction via Arrhenius law is given below:

$$k_{biol}(T) = k_{biol}(20) \cdot \theta^{(T|20)} \tag{10.8}$$

where $\theta \approx 1.05$–1.12

A longer SRT (i.e., >10–15 days) increases microbial diversity, as indicated by higher k_{biol} for ECs. These processes can be integrated into the ADM1 framework by including new state variables, such as (dissolved), (sorbed), and new biomass (X_{EC-deg}), if direct metabolism occurred.

Traditional mathematical modeling approaches for the fate and transport of ECs are mechanistic and include advection–dispersion–reaction equations and hydrodynamic models such as SWAT and Delft3D (B. Xu et al. 2024). These models simulate processes such as sorption, biodegradation, photolysis, and volatilization to identify ECs, including both soluble and particulate materials (Noutsopoulos, Charalambous, and Koumaki 2020). However, these mechanistic approaches struggle with data scarcity, nonlinear interactions, and uncertainty for predicting the fate and transport of novel ECs.

MBR is also a popular WWT technology that combines activated sludge treatment with microfiltration and ultrafiltration membranes for efficient removal of micropollutants (Besha et al. 2017). Several studies have reported that MBRs can effectively remove biodegradable micropollutants of pharmaceutical origin, such as ibuprofen and 17β-estradiol. The removal mechanisms of these micropollutants from wastewater in MBR systems include biodegradation, air stripping, sorption, and photo-transformation (Sipma et al. 2010).

Sorption in MBR is predicted by the solid water distribution coefficient (K_d), i.e., defined as a ratio of ECs in the solid and in the aqueous phase at equilibrium (K_d), and is represented as follows:

$$K_d = \frac{C_{sorbed}}{X_{SS} S_{soluble}} \tag{10.9}$$

where C_{sorbed} is sorbed ECs in (µg/L), X_{SS} is the concentration of the suspended solids in wastewater (kg/L), and $S_{soluble}$ is the concentration of the soluble part of ECs (µg/L).

Biodegradation is a redox-dependent removal mechanism of micropollutants and is represented by pseudo-first-order degradation kinetics as follows:

$$\frac{dC}{dt} = K_{biol} \times X_{ss} \times S_{soluble} \tag{10.10}$$

where C is the concentration of total micropollutants (µg/L), $S_{soluble}$ is the concentration of the soluble part of micropollutants (µg/L), K_{biol} is the pseudo-first-order reaction rate constant (L/g_{ss}/day), X_{ss} is the concentration of the suspended solids (g_{ss}/L), and t is the time in days.

Advanced hybrid modeling integrates the physical models with artificial intelligence (AI) and machine learning (ML)-based data-driven approaches to leverage the benefits of both physical laws (i.e., interpretability and generalizability) and data-driven approaches for better handling of complex data, prediction, and capturing unknown patterns (Zhuang Liu et al. 2025). The hybrid modeling framework

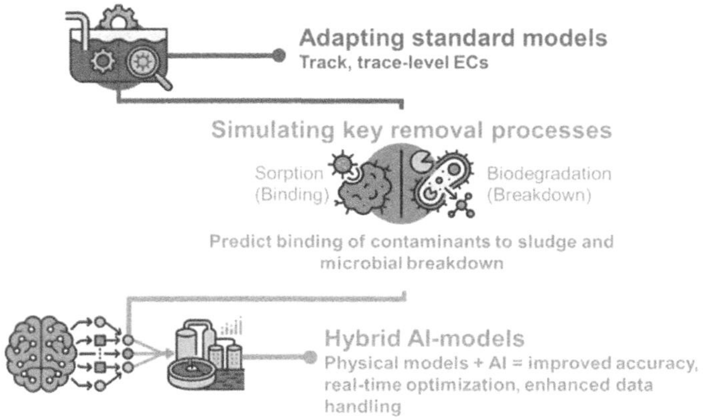

FIGURE 10.8 Modeling emerging contaminant removal using machine learning-based simulations.

enables predictive modeling of fate, transport, toxicity, and removal of ECs, supporting sustainable environmental management. The main benefits of the hybrid modeling approach included improved uncertainty prediction, real-time optimization, and scalability. For example, in the model including diffusion and biodegradation, i.e., the advection–dispersion–reactor (ADR) model, AI can predict the constant (k) using an advanced optimization algorithm (e.g., genetic algorithm or neural network).

$$\frac{\partial C}{\partial t} = D\frac{\partial^2 C}{\partial x^2} - v\frac{\partial C}{\partial x} - kC \tag{10.11}$$

where C is the concentration of ECs, D is the dispersion coefficient, v is velocity, and k is the reaction rate constant.

The ML algorithms such as artificial neural networks, support vector machine, random forest, and XGBoost can handle nonlinear prediction for fate and transport of ECs. The hybrid genetic algorithm–artificial neural network approach can efficiently optimize the removal of ECs in the WWTPs (Zaresefat and Derakhshani 2023). Besides, deep learning approaches, including the Elman Neural Network, can approximate the solutions to the diffusion equations pertaining to ECs using the Levenberg–Marquardt method. Explainable AI can be integrated with these hybrid models to interpret black-box models (Figure 10.8) and improve the PFAS adsorption process (Sulaiman et al. 2024).

10.6 CONCLUSION

Standard physical–chemical processes in WWTPs are not designed to remove ECs such as pharmaceuticals and industrial organics. This limitation leads to the continuous release of recalcitrant pollutants into aquatic environments. These compounds pose significant risks to human health and contribute to the rise of antibiotic resistance. Biodegradation serves as the primary method for EC removal during secondary treatment, the efficiency of which is strongly influenced by physicochemical properties of the compounds and the structure of the microbial community. Additionally, operational parameters such as SRT, HRT, and temperature significantly influence the performance of WWTPs.

To improve removal efficiency, advanced biological configurations utilize redox-optimized systems. These systems use multi-zone environments to encourage specialized microbial activity, which is essential for degrading complex chemical structures. However, it is important to note that for highly recalcitrant compounds such as carbamazepine, biological treatment alone is insufficient. Such compounds require hybrid systems, such as the BO_3B process for enhanced removal. For further improvement, RO systems serve as efficient physical barriers, which are capable of near-total removal. Anew, ozonation is essential for degrading stable aromatic structures that biomass alone cannot mineralize.

The removal of a parent compound does not always result in detoxification. The treatment processes can generate TPs that are, at times, more mobile, persistent, or toxic than the original ECs. Therefore, risk assessments must include comprehensive monitoring of both parent ECs and their metabolites. Hybrid treatment systems must be incorporated to achieve real sustainable water management. Tools such as omics technologies, which include genomics and proteomics, are vital for studying hidden microbial degradation pathways. Furthermore, hybrid modeling, combining traditional physical laws with AI and ML, provides a superior framework for real-time optimization of WWTP performance and prediction of the fate of ECs in complex environments.

REFERENCES

Ahmed, Ishtiaq, Hafiz M. N. Iqbal, and Kuldeep Dhama. 2017. "Enzyme-Based Biodegradation of Hazardous Pollutants – An Overview." *Journal of Experimental Biology and Agricultural Sciences* 5 (4): 402–411. https://doi.org/10.18006/2017.5(4).402.411.

Akay, Caglar, Nadin Ulrich, Ulisses Rocha, Chang Ding, and Lorenz Adrian. 2024. "Sequential Anaerobic–Aerobic Treatment Enhances Sulfamethoxazole Removal: From Batch Cultures to Observations in a Large-Scale Wastewater Treatment Plant." *Environmental Science & Technology* 58 (28): 12609–12620. https://doi.org/10.1021/acs.est.4c00368.

Alidoosti, Fatemeh, Minoo Giyahchi, Shabnam Moien, and Hamid Moghimi. 2024. "Unlocking the Potential of Soil Microbial Communities for Bioremediation of Emerging Organic Contaminants: Omics-Based Approaches." *Microbial Cell Factories* 23 (1): 210. https://doi.org/10.1186/s12934-024-02485-z.

Anand, Uttpal, Bashir Adelodun, Carlo Cabreros, Pankaj Kumar, S. Suresh, Abhijit Dey, Florencio Ballesteros, and Elza Bontempi. 2022. "Occurrence, Transformation, Bioaccumulation, Risk and Analysis of Pharmaceutical and Personal Care Products from Wastewater: A Review." *Environmental Chemistry Letters* 20 (6): 3883–3904. https://doi.org/10.1007/s10311-022-01498-7.

Astuti, Maryani Paramita, Suprihanto Notodarmojo, Cindy Rianti Priadi, and Lokesh P. Padhye. 2022. "Contaminants of Emerging Concerns (CECs) in a Municipal Wastewater Treatment Plant in Indonesia." *Environmental Science and Pollution Research* 30 (8): 21512–21532. https://doi.org/10.1007/s11356-022-23567-8.

Atugoda, Thilakshani, Meththika Vithanage, Hasintha Wijesekara, Nanthi Bolan, Ajit K. Sarmah, Michael S. Bank, Siming You, and Yong Sik Ok. 2021. "Interactions between Microplastics, Pharmaceuticals and Personal Care Products: Implications for Vector Transport." *Environment International* 149 (April): 106367. https://doi.org/10.1016/j.envint.2020.106367.

Baalbaki, Zeina, Tamanna Sultana, Thomas Maere, Peter A. Vanrolleghem, Chris D. Metcalfe, and Viviane Yargeau. 2016. "Fate and Mass Balance of Contaminants of Emerging Concern during Wastewater Treatment Determined Using the Fractionated Approach." *Science of The Total Environment* 573 (December): 1147–1158. https://doi.org/10.1016/j.scitotenv.2016.08.073.

Barka, Evridiki, Asimina Koukoura, Evangelos Statiris, Taxiarchis Seintos, Athanasios S. Stasinakis, Daniel Mamais, Simos Malamis, and Constantinos Noutsopoulos. 2025. "The Fate of Contaminants of Emerging Concern in an Upflow Anaerobic Sludge Blanket Reactor Coupled with Constructed Wetlands for Decentralized Domestic Wastewater Treatment." *Molecules* 30 (13): 2671. https://doi.org/10.3390/molecules30132671.

Behera, Shishir Kumar, Hyeong Woo Kim, Jeong-Eun Oh, and Hung-Suck Park. 2011. "Occurrence and Removal of Antibiotics, Hormones and Several Other Pharmaceuticals in Wastewater Treatment Plants of the Largest Industrial City of Korea." *Science of The Total Environment* 409 (20): 4351–4360. https://doi.org/10.1016/j.scitotenv.2011.07.015.

Bencsik, Dániel, Tanush Wadhawan, Ferenc Házi, and Tamás Karches. 2024. "Plant-Wide Models for Optimizing the Operation and Maintenance of BTEX-Contaminated Wastewater Treatment and Reuse." *Environments* 11 (5): 88. https://doi.org/10.3390/environments11050088.

Besha, Abreham Tesfaye, Abaynesh Yihdego Gebreyohannes, Ramato Ashu Tufa, Dawit Nega Bekele, Efrem Curcio, and Lidietta Giorno. 2017. "Removal of Emerging Micropollutants by Activated Sludge Process and Membrane Bioreactors and the Effects of Micropollutants on Membrane Fouling: A Review." *Journal of Environmental Chemical Engineering* 5 (3): 2395–2414. https://doi.org/10.1016/j.jece.2017.04.027.

Bhardwaj, Pooja, Kunvar Ravendra Singh, Niti B. Jadeja, Prashant S. Phale, and Atya Kapley. 2020. "Atrazine Bioremediation and Its Influence on Soil Microbial Diversity by Metagenomics Analysis." *Indian Journal of Microbiology* 60 (3): 388–391. https://doi.org/10.1007/s12088-020-00877-4.

Boahen, Ebenezer, Lawrencia Owusu, and Stephanie Owusuaa Adjei-Anim. 2025. "A Comprehensive Review of Emerging Environmental Contaminants of Global Concern." *Discover Environment* 3 (1): 144. https://doi.org/10.1007/s44274-025-00259-x.

Bodzek, Michał, Alina Pohl, and Czesława Rosik-Dulewska. 2024. "Microplastics in Wastewater Treatment Plants: Characteristics, Occurrence and Removal Technologies." *Water* 16 (24): 3574. https://doi.org/10.3390/w16243574.

Bokade, Priyanka, Vivek Kumar Gaur, Varsha Tripathi, Shishir Bobate, Natesan Manickam, and Abhay Bajaj. 2023. "Bacterial Remediation of Pesticide Polluted Soils: Exploring the Feasibility of Site Restoration." *Journal of Hazardous Materials* 441 (January): 129906. https://doi.org/10.1016/j.jhazmat.2022.129906.

Boleydei, Hamid, and Céline Vaneeckhaute. 2024. "Advancements and Challenges in Membrane Bioreactor Performance for Decentralized and Low-Temperature Applications: A Comprehensive Review." *Journal of Water Process Engineering* 63 (June): 105523. https://doi.org/10.1016/j.jwpe.2024.105523.

Brzeszcz, Joanna, Piotr Kapusta, Teresa Steliga, and Anna Turkiewicz. 2020. "Hydrocarbon Removal by Two Differently Developed Microbial Inoculants and Comparing Their Actions with Biostimulation Treatment." *Molecules* 25 (3): 661. https://doi.org/10.3390/molecules25030661.

Carballa, Marta, Francisco Omil, Juan M Lema, María Llompart, Carmen García-Jares, Isaac Rodríguez, Mariano Gómez, and Thomas Ternes. 2004. "Behavior of Pharmaceuticals, Cosmetics and Hormones in a Sewage Treatment Plant." *Water Research* 38 (12): 2918–2926. https://doi.org/10.1016/j.watres.2004.03.029.

Chaber-Jarlachowicz, Paulina, Barbara Gworek, and Radosław Kalinowski. 2025. "Removal Efficiency of Pharmaceuticals during the Wastewater Treatment Process: Emission and Environmental Risk Assessment." edited by Muammar Qadafi. *PLOS One* 20 (9): e0331211. https://doi.org/10.1371/journal.pone.0331211.

Clara, M., N. Kreuzinger, B. Strenn, O. Gans, and H. Kroiss. 2005. "The Solids Retention Time—a Suitable Design Parameter to Evaluate the Capacity of Wastewater Treatment Plants to Remove Micropollutants." *Water Research* 39 (1): 97–106. https://doi.org/10.1016/j.watres.2004.08.036.

Deblonde, Tiphanie, Carole Cossu-Leguille, and Philippe Hartemann. 2011. "Emerging Pollutants in Wastewater: A Review of the Literature." *International Journal of Hygiene and Environmental Health* 214 (6): 442–448. https://doi.org/10.1016/j.ijheh.2011.08.002.

Delgado, Nasly, Lizeth Bermeo, Daniel A. Hoyos, Gustavo A. Peñuela, Alberto Capparelli, Damián Marino, Agustín Navarro, and Juan C. Casas-Zapata. 2020. "Occurrence and Removal of Pharmaceutical and Personal Care Products Using Subsurface Horizontal Flow Constructed Wetlands." *Water Research* 187 (December): 116448. https://doi.org/10.1016/j.watres.2020.116448.

Demaria, Francesca, Marcel Suleiman, Philippe Corvini, and Pilar Junier. 2025. "Microbes as Resources to Remove PPCPs and Improve Water Quality." *Microbial Biotechnology* 18 (1): e70084. https://doi.org/10.1111/1751-7915.70084.

Dubey, Monika, Bhanu Prakash Vellanki, and Absar Ahmad Kazmi. 2023. "Emerging Contaminants in Conventional and Advanced Biological Nutrient Removal Based Wastewater Treatment Plants in India- Insights into the Removal Processes." *Science of The Total Environment* 894 (October): 165094. https://doi.org/10.1016/j.scitotenv.2023.165094.

Faria, Clara V., Bárbara C. Ricci, Ana F.R. Silva, Miriam C.S. Amaral, and Fabiana V. Fonseca. 2020. "Removal of Micropollutants in Domestic Wastewater by Expanded Granular Sludge Bed Membrane Bioreactor." *Process Safety and Environmental Protection* 136 (April): 223–233. https://doi.org/10.1016/j.psep.2020.01.033.

Fernandez-Fontaina, E., M. Carballa, F. Omil, and J.M. Lema. 2014. "Modelling Cometabolic Biotransformation of Organic Micropollutants in Nitrifying Reactors." *Water Research* 65 (November): 371–383. https://doi.org/10.1016/j.watres.2014.07.048.

Fernandez-Fontaina, E., F. Omil, J.M. Lema, and M. Carballa. 2012. "Influence of Nitrifying Conditions on the Biodegradation and Sorption of Emerging Micropollutants." *Water Research* 46 (16): 5434–5444. https://doi.org/10.1016/j.watres.2012.07.037.

Gabrielli, Marco, Riccardo Delli Compagni, Lucia Gusmaroli, Francesca Malpei, Fabio Polesel, Gianluigi Buttiglieri, Manuela Antonelli, and Andrea Turolla. 2023. "Modelling and Prediction of the Effect of Operational Parameters on the Fate of Contaminants of Emerging Concern in WWTPs." *Science of The Total Environment* 856 (January): 159200. https://doi.org/10.1016/j.scitotenv.2022.159200.

Ganesh Kumar, A., D. Manisha, K. Sujitha, D. Magesh Peter, R. Kirubagaran, and G. Dharani. 2021. "Genome Sequence Analysis of Deep Sea Aspergillus Sydowii BOBA1 and Effect of High Pressure on Biodegradation of Spent Engine Oil." *Scientific Reports* 11 (1): 9347. https://doi.org/10.1038/s41598-021-88525-9.

Hashem, Jaleela S., Wael Ismail, Yin-Ru Chiang, and Adnan A. Bekhit. 2025. "Diversity of Estrogen Biodegradation Pathways and Application in Environmental Bioremediation." *Frontiers in Microbiology* 16 (September). https://doi.org/10.3389/fmicb.2025.1630636.

Hena, Sufia, Leonardo Gutierrez, and Jean-Philippe Croué. 2021. "Removal of Pharmaceutical and Personal Care Products (PPCPs) from Wastewater Using Microalgae: A Review." *Journal of Hazardous Materials* 403 (February): 124041. https://doi.org/10.1016/j.jhazmat.2020.124041.

Ho, Chia Min, Weiying Feng, Xiaofeng Li, Sundaravelpandian Kalaipandian, Su Kong Ngien, and Xuezheng Yu. 2026. "Exploring the Complex Interactions between Microplastics and Marine Contaminants." *Marine Pollution Bulletin* 222 (January): 118697. https://doi.org/10.1016/j.marpolbul.2025.118697.

Hrkal, Zbynek, and Frantisek Pastuszek. 2023. "Behaviour of PPCP Substances in a Fluvial Aquifer after Infiltration of Treated Wastewater." *Applied Sciences* 13 (16): 9348. https://doi.org/10.3390/app13169348.

Jia, Yonghui, Yuanbo Li, Shuping Kang, Chuanxia Li, Gabriel Munoz, Shilai Hao, Feng Xiao, Yue Zhi, Caihong Liu, and Chuhui Zhang. 2025. "Sources and Occurrence of Per- and Polyfluoroalkyl Substances in Industrial Wastewater and Assessment of Current Treatment Approaches: A Review." *Journal of Hazardous Materials* 499 (November): 140195. https://doi.org/10.1016/j.jhazmat.2025.140195.

Jiang, Tingting, Wenyong Wu, Meng Ma, Yaqi Hu, and Ruoxi Li. 2024. "Occurrence and Distribution of Emerging Contaminants in Wastewater Treatment Plants: A Globally Review over the Past Two Decades." *Science of The Total Environment* 951 (November): 175664. https://doi.org/10.1016/j.scitotenv.2024.175664.

Keen, Patricia, and David Patrick. 2013. "Tracking Change: A Look at the Ecological Footprint of Antibiotics and Antimicrobial Resistance." *Antibiotics* 2 (2): 191–205. https://doi.org/10.3390/antibiotics2020191.

Kelly, John J., Maxwell G. London, Amanda R. McCormick, Miguel Rojas, John W. Scott, and Timothy J. Hoellein. 2021. "Wastewater Treatment Alters Microbial Colonization of Microplastics." edited by Robert Nerenberg. *PLOS ONE* 16 (1): e0244443. https://doi.org/10.1371/journal.pone.0244443.

Keum, Young-Soo, Jong-Su Seo, Yuting Hu, and Qing X. Li. 2006. "Degradation Pathways of Phenanthrene by Sinorhizobium Sp. C4." *Applied Microbiology and Biotechnology* 71 (6): 935–941. https://doi.org/10.1007/s00253-005-0219-z.

Khalidi-Idrissi, A., A. Madinzi, A. Anouzla, A. Pala, L. Mouhir, Y. Kadmi, and S. Souabi. 2023. "Recent Advances in the Biological Treatment of Wastewater Rich in Emerging Pollutants Produced by Pharmaceutical Industrial Discharges." *International Journal of Environmental Science and Technology* 20 (10): 11719–11740. https://doi.org/10.1007/s13762-023-04867-z.

Khan, S.J., T. Wintgens, P. Sherman, J. Zaricky, and A.I. Schäfer. 2004. "Removal of Hormones and Pharmaceuticals in the Advanced Water Recycling Demonstration Plant in Queensland, Australia." *Water Science and Technology* 50 (5): 15–22. https://doi.org/10.2166/wst.2004.0303.

Kumar, Manish, Jayalaxmi Ngasepam, Kiran Dhangar, Jurgen Mahlknecht, and Suvendu Manna. 2022. "Critical Review on Negative Emerging Contaminant Removal Efficiency of Wastewater Treatment Systems: Concept, Consistency and Consequences." *Bioresource Technology* 352 (May): 127054. https://doi.org/10.1016/j.biortech.2022.127054.

Kumar, Rohitashw, Mahrukh Qureshi, Dinesh Kumar Vishwakarma, Nadhir Al-Ansari, Alban Kuriqi, Ahmed Elbeltagi, and Anuj Saraswat. 2022. "A Review on Emerging Water Contaminants and the Application of Sustainable Removal Technologies." *Case Studies in Chemical and Environmental Engineering* 6 (December): 100219. https://doi.org/10.1016/j.cscee.2022.100219.

Labad, Francesc, Antoni Ginebreda, Rotman Criollo, Enric Vázquez-Suñé, Sandra Pérez, and Anna Jurado. 2023. "Occurrence, Data-Based Modelling, and Risk Assessment of Emerging Contaminants in an Alluvial Aquifer Polluted by River Recharge." *Environmental Pollution* 316 (January): 120504. https://doi.org/10.1016/j.envpol.2022.120504.

Lenka, Swadhina Priyadarshini, Melanie Kah, and Lokesh P. Padhye. 2021. "A Review of the Occurrence, Transformation, and Removal of Poly- and Perfluoroalkyl Substances (PFAS) in Wastewater Treatment Plants." *Water Research* 199 (July): 117187. https://doi.org/10.1016/j.watres.2021.117187.

Li, Xingyu, Xiaojing Shen, Weiwei Jiang, Yongkai Xi, and Song Li. 2024. "Comprehensive Review of Emerging Contaminants: Detection Technologies, Environmental Impact, and Management Strategies." *Ecotoxicology and Environmental Safety* 278 (June): 116420. https://doi.org/10.1016/j.ecoenv.2024.116420.

Liu, Hongbo, Xiang Yuan, Yuxuan Yao, Lijin Yao, Junbo Zhang, and Claudia Maurer. 2025. "Microplastics, Antibiotics, and Heavy Metals in Anaerobic Digestion Systems: A Critical Review of Sources, Impacts, and Mitigation Strategies." *Recycling* 10 (3): 116. https://doi.org/10.3390/recycling10030116.

Liu, Tengfei, Lixiao Ning, Chengyu Mei, Shuang Li, Lianbao Zheng, Pei Qiao, Haixia Wang, Tong Hu, and Weihong Zhong. 2023. "Synthetic Bacterial Consortia Enhanced the Degradation of Mixed Priority Phthalate Ester Pollutants." *Environmental Research* 235 (October): 116666. https://doi.org/10.1016/j.envres.2023.116666.

Liu, Zhonghua, Yijun Dai, Yu Huan, Zhenxing Liu, Lei Sun, Qianwen Zhou, Wenjian Zhang, Qi Sang, Hua Wei, and Sheng Yuan. 2013. "Different Utilizable Substrates Have Different Effects on Cometabolic Fate of Imidacloprid in Stenotrophomonas Maltophilia." *Applied Microbiology and Biotechnology* 97 (14): 6537–6547. https://doi.org/10.1007/s00253-012-4444-y.

Liu, Zhuang, Yonghai Gan, Jun Luo, Xiao Luo, Chengcheng Ding, and Yibin Cui. 2025. "Current Status of Emerging Contaminant Models and Their Applications Concerning the Aquatic Environment: A Review." *Water* 17 (1): 85. https://doi.org/10.3390/w17010085.

Loganathan, Paripurnanda, Saravanamuthu Vigneswaran, Jaya Kandasamy, Agnieszka Katarzyna Cuprys, Zakhar Maletskyi, and Harsha Ratnaweera. 2023. "Treatment Trends and Combined Methods in Removing Pharmaceuticals and Personal Care Products from Wastewater—A Review." *Membranes* 13 (2): 158. https://doi.org/10.3390/membranes13020158.

Luo, Jingyang, Leiyu Feng, Wei Zhang, Xiang Li, Hong Chen, Dongbo Wang, and Yinguang Chen. 2014. "Improved Production of Short-Chain Fatty Acids from Waste Activated Sludge Driven by Carbohydrate Addition in Continuous-Flow Reactors: Influence of SRT and Temperature." *Applied Energy* 113 (January): 51–58. https://doi.org/10.1016/j.apenergy.2013.07.006.

Macchi, M., M. Martinez, R. M. Neme Tauil, M. P. Valacco, I. S. Morelli, and B. M. Coppotelli. 2018. "Insights into the Genome and Proteome of Sphingomonas Paucimobilis Strain 20006FA Involved in the Regulation of Polycyclic Aromatic Hydrocarbon Degradation." *World Journal of Microbiology and Biotechnology* 34 (1): 7. https://doi.org/10.1007/s11274-017-2391-6.

Mamatha Bhanu, L.S., Sampriti Kataki, and Soumya Chatterjee. 2024. "CRISPR: New Promising Biotechnological Tool in Wastewater Treatment." *Journal of Microbiological Methods* 227 (December): 107066. https://doi.org/10.1016/j.mimet.2024.107066.

Marchlewicz, Ariel, Urszula Guzik, Katarzyna Hupert-Kocurek, Agnieszka Nowak, Sylwia Wilczyńska, and Danuta Wojcieszyńska. 2017. "Toxicity and Biodegradation of Ibuprofen by Bacillus Thuringiensis B1(2015b)." *Environmental Science and Pollution Research* 24 (8): 7572–7584. https://doi.org/10.1007/s11356-017-8372-3.

McDougall, Louise, Lauren Thomson, Sarah Brand, Antony Wagstaff, Linda A Lawton, and Bruce Petrie. 2022. "Adsorption of a Diverse Range of Pharmaceuticals to Polyethylene Microplastics in Wastewater and Their Desorption in Environmental Matrices." *Science of The Total Environment* 808 (February): 152071. https://doi.org/10.1016/j.scitotenv.2021.152071.

Meng, Yuan, Weiyi Liu, Heidelore Fiedler, Jinlan Zhang, Xinrui Wei, Xiaohui Liu, Meng Peng, and Tingting Zhang. 2021. "Fate and Risk Assessment of Emerging Contaminants in Reclaimed Water Production Processes." *Frontiers of Environmental Science & Engineering* 15 (5): 104. https://doi.org/10.1007/s11783-021-1392-8.

Mishra, Ranjeet Kumar, Spandana Samyukthalakshmi Mentha, Yash Misra, and Naveen Dwivedi. 2023. "Emerging Pollutants of Severe Environmental Concern in Water and Wastewater: A Comprehensive Review on Current Developments and Future Research." *Water-Energy Nexus* 6 (December): 74–95. https://doi.org/10.1016/j.wen.2023.08.002.

Morin-Crini, Nadia, Eric Lichtfouse, Guorui Liu, Vysetti Balaram, Ana Rita Lado Ribeiro, Zhijiang Lu, Friederike Stock, et al. 2022. "Worldwide Cases of Water Pollution by Emerging Contaminants: A Review." *Environmental Chemistry Letters* 20 (4): 2311–2338. https://doi.org/10.1007/s10311-022-01447-4.

Moya-Llamas, M.J., M.G. Pacazocchi, and A Trapote. 2023. "Respirometric Tests in a Combined UASB-MBR System Treating Wastewater Containing Emerging Contaminants at Different OLRs and Temperatures: Biokinetic Analysis." *Journal of Environmental Management* 345 (November): 118643. https://doi.org/10.1016/j.jenvman.2023.118643.

Mqambalala, Avela, Maleke Maleke, Jorge R Osman, and Julio Castillo Hernandez. 2025. "Biodegradation of Emerging Contaminants Controlled by Biological and Chemical Factors." *Microorganisms* 13 (10): 2354. https://doi.org/10.3390/microorganisms13102354.

Nishmitha, P.S., Kochuparambil Ajayaghosh Akhilghosh, Vijayalekshmi Padmachandran Aiswriya, Athira Ramesh, Muthukumar Muthuchamy, and Anbazhagi Muthukumar. 2025. "Understanding Emerging Contaminants in Water and Wastewater: A Comprehensive Review on Detection, Impacts, and Solutions." *Journal of Hazardous Materials Advances* 18 (May): 100755. https://doi.org/10.1016/j.hazadv.2025.100755.

Noutsopoulos, Constantinos, Vera Charalambous, and Elena Koumaki. 2020. "Evaluating the Fate of Emerging Contaminants in Wastewater Treatment Plants through Plant-Wide Mathematical Modelling." *Environmental Processes* 7 (4): 1065–1094. https://doi.org/10.1007/s40710-020-00459-y.

Oulton, Rebekah L, Tamar Kohn, and David M Cwiertny. 2010. "Pharmaceuticals and Personal Care Products in Effluent Matrices: A Survey of Transformation and Removal during Wastewater Treatment and Implications for Wastewater Management." *Journal of Environmental Monitoring* 12 (11): 1956. https://doi.org/10.1039/c0em00068j.

Parashar, Neha, and Subrata Hait. 2023. "Recent Advances on Microplastics Pollution and Removal from Wastewater Systems: A Critical Review." *Journal of Environmental Management* 340 (August): 118014. https://doi.org/10.1016/j.jenvman.2023.118014.

Plósz, Benedek Gy, Katherine H Langford, and Kevin V Thomas. 2012. "An Activated Sludge Modeling Framework for Xenobiotic Trace Chemicals (ASM-X): Assessment of Diclofenac and Carbamazepine." *Biotechnology and Bioengineering* 109 (11): 2757–2769. https://doi.org/10.1002/bit.24553.

Qian, Yuli, Linchang Guan, Yunhao Ke, Liye Wang, Xuebing Wang, Nanyang Yu, Qingmiao Yu, Si Wei, and Jinju Geng. 2024. "Unveiling Intricate Transformation Pathways of Emerging Contaminants during Wastewater Treatment Processes through Simplified Network Analysis." *Water Research* 253 (April): 121299. https://doi.org/10.1016/j.watres.2024.121299.

Qian, Yuli, Yunhao Ke, Liye Wang, Nanyang Yu, Yujie He, Qingmiao Yu, Si Wei, Hongqiang Ren, and Jinju Geng. 2025. "Entropy Similarity-Driven Transformation Reaction Molecular Networking Reveals Transformation Pathways and Potential Risks of Emerging Contaminants in Wastewater: The Example of Sartans." *Environmental Science & Technology* 59 (8): 4153–4164. https://doi.org/10.1021/acs.est.4c13144.

Qin, Kena, Liangliang Wei, Jianju Li, Bo Lai, Fengyi Zhu, Hang Yu, Qingliang Zhao, and Kun Wang. 2020. "A Review of ARGs in WWTPs: Sources, Stressors and Elimination." *Chinese Chemical Letters* 31 (10): 2603–2613. https://doi.org/10.1016/j.cclet.2020.04.057.

Rout, Prangya R, Tian C Zhang, Puspendu Bhunia, and Rao Y Surampalli. 2021. "Treatment Technologies for Emerging Contaminants in Wastewater Treatment Plants: A Review." *Science of The Total Environment* 753 (January): 141990. https://doi.org/10.1016/j.scitotenv.2020.141990.

Ruas, Graziele, Rebeca López-Serna, Priscila Guenka Scarcelli, Mayara Leite Serejo, Marc Árpàd Boncz, and Raúl Muñoz. 2022. "Influence of the Hydraulic Retention Time on the Removal of Emerging Contaminants in an Anoxic-Aerobic Algal-Bacterial Photobioreactor Coupled with Anaerobic Digestion." *Science of The Total Environment* 827 (June): 154262. https://doi.org/10.1016/j.scitotenv.2022.154262.

Sánchez, Marta, Isabel Ruiz, and Manuel Soto. 2022. "The Potential of Constructed Wetland Systems and Photodegradation Processes for the Removal of Emerging Contaminants—A Review." *Environments* 9 (9): 116. https://doi.org/10.3390/environments9090116.

Sesay, Fatmata, Richard Edmond Victor Sesay, Musa Kamara, Xuesong Li, and Chengxin Niu. 2025. "Biodegradation of Pharmaceutical Contaminants in Wastewater Using Microbial Consortia: Mechanisms, Applications, and Challenges." *Journal of Environmental Management* 384 (June): 125564. https://doi.org/10.1016/j.jenvman.2025.125564.

Shahriarinour, Mahdi, Faten Divsar, Fereshteh Kamalpour Dahka, Sharareh Nezamivand Chegini, Mohamad Mahani, Arash Moeini, and Pierfrancesco Cerruti. 2021. "Biodegradation of Alprazolam in Pharmaceutical Wastewater Using Mesoporous Nanoparticles-Adhered Pseudomonas Stutzeri." *Molecules* 27 (1): 237. https://doi.org/10.3390/molecules27010237.

Sin, Gürkan, and Resul Al. 2021. "Activated Sludge Models at the Crossroad of Artificial Intelligence—A Perspective on Advancing Process Modeling." *Npj Clean Water* 4 (1): 16. https://doi.org/10.1038/s41545-021-00106-5.

Sipma, Jan, Begoña Osuna, Neus Collado, Hector Monclús, Giuliana Ferrero, Joaquim Comas, and Ignasi Rodriguez-Roda. 2010. "Comparison of Removal of Pharmaceuticals in MBR and Activated Sludge Systems." *Desalination* 250 (2): 653–659. https://doi.org/10.1016/j.desal.2009.06.073.

Stefanakis, Alexandros I, and Julie A Becker. 2020. "A Review of Emerging Contaminants in Water." In *Waste Management*, 177–202. IGI Global. https://doi.org/10.4018/978-1-7998-1210-4.ch008.

Sulaiman, Muhammad, Osamah Ibrahim Khalaf, Naveed Ahmad Khan, Fahad Sameer Alshammari, and Habib Hamam. 2024. "Mathematical Modeling and Machine Learning-Based Optimization for Enhancing Biofiltration Efficiency of Volatile Organic Compounds." *Scientific Reports* 14 (1): 16908. https://doi.org/10.1038/s41598-024-65153-7.

Suleiman, Marcel, Francesca Demaria, Cristina Zimmardi, Boris Alexander Kolvenbach, and Philippe François-Xavier Corvini. 2023. "Analyzing Microbial Communities and Their Biodegradation of Multiple Pharmaceuticals in Membrane Bioreactors." *Applied Microbiology and Biotechnology* 107 (17): 5545–5554. https://doi.org/10.1007/s00253-023-12677-z.

Sultan, Maisha Binte, Amit Hasan Anik, and Md. Mostafizur Rahman. 2024. "Emerging Contaminants and Their Potential Impacts on Estuarine Ecosystems: Are We Aware of It?" *Marine Pollution Bulletin* 199 (February): 115982. https://doi.org/10.1016/j.marpolbul.2023.115982.

Sun, Jing, Xiaohu Dai, Qilin Wang, Mark C.M. van Loosdrecht, and Bing-Jie Ni. 2019. "Microplastics in Wastewater Treatment Plants: Detection, Occurrence and Removal." *Water Research* 152 (April): 21–37. https://doi.org/10.1016/j.watres.2018.12.050.

Talukdar, Avishek, Pritha Kundu, Sayan Bhattacharya, and Nalok Dutta. 2024. "Microplastic Contamination in Wastewater: Sources, Distribution, Detection and Remediation through Physical and Chemical-Biological Methods." *Science of The Total Environment* 916 (March): 170254. https://doi.org/10.1016/j.scitotenv.2024.170254.

Tang, K. 2018. "Removal of Pharmaceuticals from Hospital Wastewater by Staged Biofilm and Ozone Polishing." Technical University of Denmark. https://backend.orbit.dtu.dk/ws/portalfiles/portal/149731276/Thesis_online_version_Kai_Tang.pdf.

Tawfik, Ahmed, Mohamed Mohsen, Sherif Ismail, Nawaf S Alhajeri, Ahmed I Osman, and David W Rooney. 2022. "Methods to Alleviate the Inhibition of Sludge Anaerobic Digestion by Emerging Contaminants: A Review." *Environmental Chemistry Letters* 20 (6): 3811–3836. https://doi.org/10.1007/s10311-022-01465-2.

Tong, Xuneng, Sanjeeb Mohapatra, Jingjie Zhang, Ngoc Han Tran, Luhua You, Yiliang He, and Karina Yew-Hoong Gin. 2022. "Source, Fate, Transport and Modelling of Selected Emerging Contaminants in the Aquatic Environment: Current Status and Future Perspectives." *Water Research* 217 (June): 118418. https://doi.org/10.1016/j.watres.2022.118418.

Tran, Ngoc Han, Martin Reinhard, and Karina Yew-Hoong Gin. 2018. "Occurrence and Fate of Emerging Contaminants in Municipal Wastewater Treatment Plants from Different Geographical Regions-a Review." *Water Research* 133 (April): 182–207. https://doi.org/10.1016/j.watres.2017.12.029.

Ulucan-Altuntas, Kubra, Neslihan Manav-Demir, Fatih Ilhan, Huseyin Baran Gelgor, Katherine Huddersman, Abhishek Tiwary, and Eyup Debik. 2023. "Emerging Pollutants Removal in Full-Scale Biological Treatment Plants: A Case Study." *Journal of Water Process Engineering* 51 (February): 103336. https://doi.org/10.1016/j.jwpe.2022.103336.

Vaksmaa, Annika, Simon Guerrero-Cruz, Pooja Ghosh, Emna Zeghal, Victor Hernando-Morales, and Helge Niemann. 2023. "Role of Fungi in Bioremediation of Emerging Pollutants." *Frontiers in Marine Science* 10 (March): 1–21. https://doi.org/10.3389/fmars.2023.1070905.

Vargas, Jocelina Paranhos Rosa De, Marília Camotti Bastos, Maha Al Badany, Rolando Gonzalez, Delmira Wolff, Danilo Rheinheimer Dos Santos, and Jérôme Labanowski. 2021. "Pharmaceutical Compound Removal Efficiency by a Small Constructed Wetland Located in South Brazil." *Environmental Science and Pollution Research* 28 (24): 30955–30974. https://doi.org/10.1007/s11356-021-12845-6.

Wang, Jing, Xiaohong Li, Fang Guan, Zhibo Yang, Xiaofan Zhai, Yimeng Zhang, Xuexi Tang, Jizhou Duan, and Hui Xiao. 2023. "The Isolation of Anaerobic and Facultative Anaerobic Sulfate-Reducing Bacteria (SRB) and a Comparison of Related Enzymes in Their Sulfate Reduction Pathways." *Microorganisms* 11 (8): 2019. https://doi.org/10.3390/microorganisms11082019.

Wiest, Laure, Antoine Gosset, Aurélie Fildier, Christine Libert, Matthieu Hervé, Elisabeth Sibeud, Barbara Giroud, et al. 2021. "Occurrence and Removal of Emerging Pollutants in Urban Sewage Treatment Plants Using LC-QToF-MS Suspect Screening and Quantification." *Science of The Total Environment* 774 (June): 145779. https://doi.org/10.1016/j.scitotenv.2021.145779.

Wilkinson, John, Peter S Hooda, James Barker, Stephen Barton, and Julian Swinden. 2017. "Occurrence, Fate and Transformation of Emerging Contaminants in Water: An Overarching Review of the Field." *Environmental Pollution* 231 (December): 954–970. https://doi.org/10.1016/j.envpol.2017.08.032.

Wilt, Arnoud de, Koen van Gijn, Tom Verhoek, Amber Vergnes, Mirit Hoek, Huub Rijnaarts, and Alette Langenhoff. 2018. "Enhanced Pharmaceutical Removal from Water in a Three Step Bio-Ozone-Bio Process." *Water Research* 138 (July): 97–105. https://doi.org/10.1016/j.watres.2018.03.028.

Xu, Boyan, Ching Kwek Pooi, Tsuey Shan Yeap, Kwok Yii Leong, Xi Yee Soh, Shujuan Huang, Xueqing Shi, Giorgio Mannina, and How Yong Ng. 2024. "Hybrid Model Composed of Machine Learning and ASM3 Predicts Performance of Industrial Wastewater Treatment." *Journal of Water Process Engineering* 65 (August): 105888. https://doi.org/10.1016/j.jwpe.2024.105888.

Xu, Yanghui, Qin Ou, Jan Peter van der Hoek, Gang Liu, and Kim Maren Lompe. 2024. "Photo-Oxidation of Micro- and Nanoplastics: Physical, Chemical, and Biological Effects in Environments." *Environmental Science & Technology* 58 (2): 991–1009. https://doi.org/10.1021/acs.est.3c07035.

Zaresefat, Mojtaba, and Reza Derakhshani. 2023. "Revolutionizing Groundwater Management with Hybrid AI Models: A Practical Review." *Water* 15 (9): 1750. https://doi.org/10.3390/w15091750.

Zhao, Shengchen, and Jihong Wang. 2024. "Biodegradation of Atrazine and Nicosulfuron by Streptomyces Nigra LM01: Performance, Degradative Pathway, and Possible Genes Involved." *Journal of Hazardous Materials* 471 (June): 134336. https://doi.org/10.1016/j.jhazmat.2024.134336.

Zhou, Linjun, Xinfeng Yang, Wen Gu, Jie Gu, Weilong Xing, Lei Wang, Lili Shi, Jing Liang, and Tianyun Wang. 2024. "Removal and Prediction of Chemicals in Anaerobic–Anoxic–Aerobic Wastewater Treatment System: Case Study on Organophosphates." *Emerging Contaminants* 10 (4): 100344. https://doi.org/10.1016/j.emcon.2024.100344.

Zhou, Qiqi, Hongfeng Chen, Guijian Liu, and Xiaohui Wang. 2024. "Occurrence, Sustainable Treatment Technologies, Potential Sources, and Future Prospects of Emerging Pollutants in Aquatic Environments: A Review." *Frontiers in Environmental Science* 12 (September): 1455377. https://doi.org/10.3389/fenvs.2024.1455377.

Zhou, Yunqiao, Jing Meng, Meng Zhang, Shuqin Chen, Bo He, Hui Zhao, Qifeng Li, Sheng Zhang, and Tieyu Wang. 2019. "Which Type of Pollutants Need to Be Controlled with Priority in Wastewater Treatment Plants: Traditional or Emerging Pollutants?" *Environment International* 131 (October): 104982. https://doi.org/10.1016/j.envint.2019.104982.

11 Recent Advances in Photocatalytic Degradation of Phenol with Special References to Z-Scheme and S-Scheme Heterojunction

Satyanjib Sahoo, Shaswata Biswal,
Gayatree Sanjibita Sahoo, Biswanath Mahanty,
Tapas Ranjan Sahoo, and Naresh Kumar Sahoo

11.1 INTRODUCTION

The global production of phenol was about 11 million metric tons in 2024, with an estimated demand of 14.58 million tons in 2030, registering a compound annual growth rate (CAGR) of 3.2–3.45% (Ladeia Ramos et al., 2024). Phenol is a major component across various industrial sectors, including the production of phenolic resins used to manufacture adhesives, laminates, and plywood. Further, the market value of phenol is anticipated to increase from USD 36.81 billion in 2025 to USD 50.84 billion in 2033 (Ladeia Ramos et al., 2024; Sahoo et al., 2021; Xu et al., 2019). It is also essential for bisphenol A synthesis, which forms the basis of polycarbonate plastics and epoxy resins, and is widely used in pharmaceutical manufacturing, including the production of aspirin and antiseptics (Grace Pavithra et al., 2023). These

DOI: 10.1201/9781003741626-11

phenolic recalcitrant compounds are released at varying concentrations from sectors such as coke oven plants (28–1,200 mg/L), petrochemical plants (2.8–1,220 mg/L), coal mining (9–6,800 mg/L), and oil refineries (6–500 mg/L). The phenolic effluents also discharge from the pharmaceutical, textile, paint, wood processing, coal gasification, leather, resin synthesis, and pulp and paper industries at concentrations ranging from 0.1 to 1,600 mg/L (Panigrahy et al., 2022). The high solubility of phenol in water (around 83 g/L at 25°C), together with its low pK_a value (10.0), guarantees that it will be found in all pH conditions, making it difficult to use any physical removal process against it and increasing its propensity for bioaccumulation in sediments and biota (Alshabib & Onaizi, 2019). Additionally, among other contaminants, phenol exhibits a high level of bioconcentration in aquatic life due to its log Kow of 1.46, resulting in biomagnification via the food chain or web (Alshabib & Onaizi, 2019). Acute cutaneous exposure leads to corrosive burns owing to protein precipitation and lipid solubilization, while systemic toxicity is caused by uncoupling mitochondrial oxidative phosphorylation and ATP synthesis inhibition, leading to CNS depression, including tremors, coma, and cardiac arrest (Downs & Wills, 2025; Menzikov & Morozov, 2019). Repeated exposure causes various hematology disorders, including anemia, leukocytosis, lymphocytosis, and thrombocytopenia. It can cause reproductive toxicity, fetotoxicity, teratogenicity, and carcinogenicity (Group 2B IARC) (ATSDR, 2008). Phenol has been defined as a priority pollutant by the U.S. EPA, and the level of phenol in drinking water has been regulated to be below 0.001–0.05 mg/L for various forms of phenol. In the European Union Water Framework Directive (2000/60/EC), the phenolic compounds are considered "hazardous substances" (Mhlongo et al., 2023) with environmental quality criteria of 2.8 μg/L for priority pollutants, including pentachlorophenol. The above characteristics emphasize the importance of developing effective techniques to ensure total phenolic mineralization, especially considering the stringent criteria for effluent discharge at sub-mg/L concentration limits.

Conventional methods for phenol removal consist of adsorption, biological treatment, chemical oxidation, and advanced oxidation processes (AOPs). These techniques are ineffective and impractical for large-scale applications. Activated carbon adsorption is a highly efficient technique for phenol removal, achieving up to 90–95% efficiency within 100–500 mg/L of concentration. Still, it results in the disposal of toxic sludge and frequent sorbent regeneration, making it economically unfeasible for high-volume wastewater (Hameed & Rahman, 2008; Yu et al., 2021). Biological processes using bacteria/fungi have 70–90% efficiency but are ineffective in the presence of a multi-pollutant system because of microbial toxicity, require high hydraulic retention times (days to weeks), and are sensitive to temperature and pH variations, which are common in industrial wastewaters (Mumtaz et al., 2024). Chemical precipitation and solvent extraction have 60–80% efficiency but generate secondary chemical waste and are economically not feasible for low concentrations (<100 mg/L). At the same time, traditional AOPs such as Fenton and H_2O_2 reactions require very high reagent dosages (H_2O_2:Fe ratios of 5–10:1) and generate iron sludge. The never-ending shortcomings of conventional treatment processes, namely secondary pollution and operational rigidity, have forced the scientific community to seek innovative, cutting-edge technologies that achieve complete mineralization. Further, all these technologies have been found to exhibit incomplete mineralization (often resulting in recalcitrant byproducts), high operational costs ($1–5/kg of phenol removed), and sensitivity to wastewater matrices, thereby emphasizing the need for a sustainable photocatalysis system (Grace Pavithra et al., 2023). The photocatalytic degradation process has been recognized as a revolutionary technology, distinct from conventional processes, because it can achieve complete mineralization of phenol to CO_2 and H_2O (Hu et al., 2023). Unlike Fenton reactions, which require pH 2.8–3.5, or activated carbon, which demands frequent regeneration at 400–600°C, photocatalysis can operate effectively over a wide pH range (4–9) and harness free solar energy as the primary driving force (Al Miad et al., 2024; Hu et al., 2023). A detailed comparison of phenolic removal using different methods is shown in Table 11.1.

The problem with single-component photocatalysts is that their thermodynamic properties are fundamentally contradictory. It is necessary to have a material with a small bandgap to absorb visible light. At the same time, it is necessary to have a redox potential strong enough to drive difficult reactions (Che et al., 2018; Yang et al., 2024). Yet, this trade-off has been successfully overcome through rational

TABLE 11.1 Comparison of photocatalytic degradation vs. conventional pollutant treatment methods

ASPECT	*MINERALIZATION*	*PH RANGE*	*ENERGY SOURCE*	*SECONDARY POLLUTION*	*REUSABILITY*	*REFERENCES*
Photocatalysis	Complete to CO_2/H_2O	3–9	Solar/UV, low cost	None	High (cycles >5)	(Ahuja et al., 2023)
Fenton	Partial, sludge byproduct	2.8–3.5	Chemicals/heat	Iron sludge	Low (sludge)	(Xu et al., 2020)
Activated Carbon	None (adsorption only)	Neutral	Regeneration at 400–600°C	Spent carbon disposal	Frequent regeneration	(Pet et al., 2024)
Biological	Partial, slow for toxins	6–8	Aeration, long retention	Excess biomass	Microbial acclimation needed	(Narayanan & Narayan, 2019)

heterojunction design, where type II heterojunctions enhance charge separation at the expense of reduced redox ability. Z-scheme heterojunctions preserve the highest redox potential of each semiconductor component while achieving outstanding charge separation via ingenious Z-shaped electron-transfer routes (Zhou et al., 2024). In recent years, the S-scheme heterostructure has been identified as a mechanically robust, structurally simple alternative. The causes of charge-carrier mobility in an S-scheme heterostructure are the internal electrical field, Coulombic interactions between oppositely charged particles, and band-edge bending. Photoinduced electrons and holes with lower potential are removed, boosting the higher-potential charge carrier and thereby enhancing photocatalytic pollutant degradation activity and stability. The operational cost of photocatalysis ($0.5–2/kg phenol removed) is dramatically lower than traditional processes ($1–5/kg), making it the economically sustainable paradigm shift for large-scale remediation (Kumar et al., 2023).

Thus, this chapter summarizes significant progress in Z-scheme and S-scheme heterojunctions, with emphasis on interfacial charge-transfer behavior, band-structure engineering, reactive oxygen species (ROS)-driven transformation pathways, and a thorough, mechanism-based assessment of the photocatalytic degradation of phenol.

11.2 FUNDAMENTAL MECHANISM OF PHOTOCATALYST

Photocatalysis is one of the most promising sustainable technologies for addressing modern energy and environmental problems by harnessing the virtually unlimited power of solar energy. The photocatalytic reaction requires semiconductor materials (SM) that absorb photons and convert solar energy into chemical energy via a series of reactions catalyzed by the designed photocatalytic nanomaterials (NMs). A thorough understanding of the fundamental mechanisms of photocatalysis is essential for designing efficient photocatalytic systems that can facilitate important reactions for the decomposition of environmental pollutants, such as phenol. The overall photocatalytic reaction proceeds through a systematic three-step process that determines its efficiency (Liu et al., 2021). Figure 11.1 illustrates the details of the three-step photocatalytic reaction shown below.

Step 1: *Photon absorption and electron and hole pair formation*: When a photocatalyst is illuminated with photons having energy equal to or greater than the SM's bandgap energy (E_g), the photons are absorbed by the photocatalyst NMs, generating electron–hole pairs. The photoexcited electrons are excited from the valence band (VB) to the conduction band (CB), leaving behind positively charged holes in the VB. This light-absorption process is the basis of photocatalysis, and the

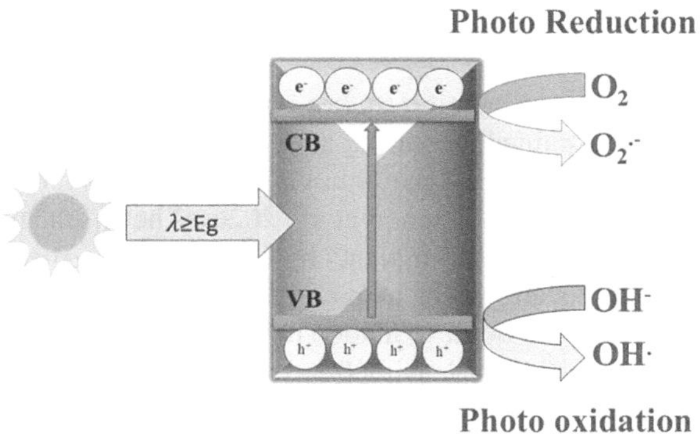

FIGURE 11.1 Fundamental photocatalytic mechanism for pollutant degradation.

bandgap energy directly determines the wavelength of light the material can absorb (Chowdhury et al., 2025; Liu et al., 2021).

Step 2: *Migration and Separation of Charge Carriers:* After photoexcitation, the photoinduced electrons and holes must transfer to the photocatalyst surface to avoid rejoining. In general, the photogenerated electron in the VB of the photocatalyst jumps to its CB, leaving hole pairs in the VB. This is the most difficult step in photocatalysis because of the strong Coulombic attraction between electrons and holes, which leads to rapid rejoining in both the bulk and at the surface, resulting in energy loss as heat. The success of this step depends on the properties of the SM, its crystal structure, defect levels, and most importantly, the heterojunction systems developed to create internal electric fields that facilitate the spatial separation of charge carriers (Chowdhury et al., 2025; Liu et al., 2021).

Step 3: *Surface Redox Reactions:* Finally, the separated electrons and holes participate in oxidation–reduction reactions at the photocatalyst's surface. The photogenerated electrons in the CB reduce the adsorbed oxygen species, forming superoxide radicals, while the photoinduced holes oxidize the adsorbed water, forming hydroxyl radicals (–OH·°). The success of these surface reactions depends on the redox potentials of the CB and VB edges, the concentration of reactive sites, and the affinity of the reactants for the photocatalyst surface (Chowdhury et al., 2025; Liu et al., 2021).

Figure 11.1 shows the band structure of a typical semiconductor photocatalyst, where the CB is at higher energy levels and the VB is at lower energy levels with a bandgap (E_g) in between. When light of sufficient energy hits the material, electrons move from VB to CB, leaving behind holes. The internal electric field at the heterojunction interface pushes electrons to one material and holes to another, separating them in space. This avoids the favorable recombination process and allows the separated carriers to reach the surface, where they can participate in useful chemical reactions. For photocatalytic processes to occur, a greater positive VB potential and a more negative CB potential are required (Wang et al., 2022). This results in an increase in the band gap (E_g), leading to a loss in the capacity to effectively harvest light. It is difficult for a single-component photocatalyst to produce both $\cdot O_2$- and ·OH radicals (Bao et al., 2021). And secondly, due to electrostatic attraction, photogenerated electrons and holes are trapped within single phases and undergo intense recombination (Wang et al., 2024). Figure 11.1 demonstrates the inherent limitation of conventional photocatalysts, where rapid rejoining of photoexcited electron–hole pairs significantly reduces the availability of reactive species required for efficient phenol degradation. Therefore, it is a contradiction to achieve both vast light absorption and strong redox capacity in a semiconductor that is made of a single component at the same time (Wang et al., 2023a). For that reason, extensive research has been conducted on the process of creating heterojunction composite materials. Among the different heterojunctions, the Z-scheme and S-scheme heterostructures have been developed to address the trade-off by conserving strong reducing electrons and highly oxidizing holes, garnering notable attention.

11.2.1 Principle and Design of Z-Scheme Heterostructure

The Z-scheme principle evolved from scientific studies on natural photosynthesis. In the past few years, studies on Z-scheme systems have gained increasing attention due to tremendous progress in their preparation and improvements in charge-carrier separation via direct or indirect charge-transfer mechanisms (Malefane et al., 2023; Mishra et al., 2024; Sahoo et al., 2024). The Z-scheme heterostructure precisely replicates the electron-transport mechanism in plants by combining two distinct photocatalysts (usually an oxidation and a reduction photocatalyst) to increase the efficiency of solar energy conversion (Bian et al., 2021). The electrons from the CB of the oxidation photocatalyst (OP) recombine with the holes from the VB of the reduction photocatalyst (RP), leaving the electrons in the CB of the RP and holes in the VB of the OP. This enables efficient charge separation (Mafa et al., 2021). This photocatalyst combination is called the Z-scheme, and it is attributed to heterostructure carriers moving as they form the letter 'Z' (Liu et al., 2021). Z-scheme photocatalysts are classified into traditional Z-scheme, direct Z-scheme, and all-solid-state (ASS) Z-scheme photocatalysts, based on the different transport media (Ng et al., 2020).

11.2.2 Traditional (Liquid-Phase or Ionic) Z-Scheme

The traditional (liquid-phase or ionic) scheme comprises two semiconductors (SCs), i.e., an oxidation SC (SC-1) and a reduction SC (SC–2). These two SCs are linked via a liquid redox mediator, such as Fe^{3+}/Fe^{2+} or IO_3^-/I^-, via an interfacial charge-transfer route (Low et al., 2020). Their reactions depend on charge transfer via redox-active media. The mechanism of the traditional Z-scheme involves the production of electrons and holes when both the SCs are exposed to solar radiation. Then, the redox liquid electron mediator facilitates the transfer of electrons between SC 1 and SC 2, facilitating charge separation. The mechanism is represented in Figure 11.2(a). From this illustration, it is clear that the Z-scheme system plays a vital role in enhancing charge-carrier separation while keeping high redox potentials that are essential for phenol degradation. There are, however, some limitations associated with the traditional Z-scheme heterojunction system (Molaei, 2024). The conventional Z-scheme photocatalyst has certain drawbacks, such as unfavorable redox shuttle byproducts, competition between the semiconductor and the redox agent for light, and less suitability in liquids (Low et al., 2017). Hence, its application in the solid and gaseous states is highly constrained. To address these problems, the idea of a second-generation Z-scheme heterojunction, which is the ASS heterojunction, was introduced.

11.2.3 All Solid-State Z-Scheme

The all-solid-state Z-scheme aims to eliminate the use of redox pairs and replace them with nanoconductors, expanding their applications. The construction of ASS Z-scheme heterostructures is similar to traditional Z-scheme heterojunctions except that it uses a solid electron mediator like noble metals, graphene, or carbon nanotubes (CNTs) instead of a liquid redox mediator (Chauhan et al., 2024). Consequently, the reverse reaction of the first-generation Z-scheme heterostructure can be suppressed (Ma et al., 2016). The construction and mechanism of the ASS Z-scheme are shown in Figure 11.2(b). Two semiconductors, SC 1 and SC 2, are linked via a conductive material. Electrons from SC 1 recombine with holes from SC 2 via the solid mediator. In general, the heterojunction has advantages over traditional heterojunctions, as it provides greater stability, eliminates backward reactions, and can operate in both solid and gas phases (Xu & Yu, 2022). However, this solid-state Z-scheme heterojunction also has several drawbacks. The application of noble metals is very costly, thereby restricting their use in the ASS Z-scheme heterostructure. Further noble metal NPs are generally strong light harvesters (Tu et al., 2015). Thus, the photocatalyst's capacity to harvest visible light is significantly reduced by the ASS Z-scheme heterostructure design. The preparation of the desired sandwiched structure of the three components is very challenging. To overcome these

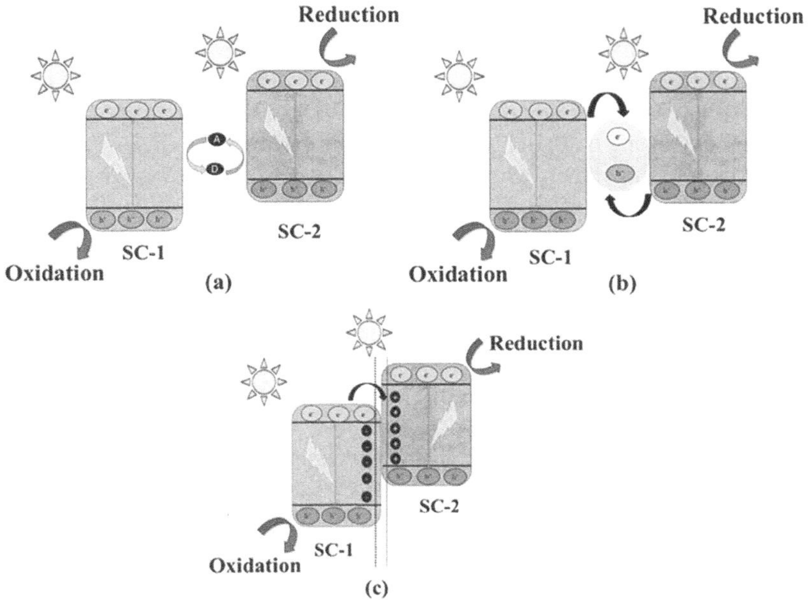

FIGURE 11.2 Schematic representation of all three types of Z-scheme heterojunctions to overcome the photoinduced charge pair recombination (Mohammed et al., 2025).

limitations, Yu et al. (2013) proposed the concept of a third-generation Z-scheme photocatalyst, namely the direct Z-scheme heterostructure.

Direct Z-scheme: In detail, a direct Z-scheme heterostructure comprises only two SMs with direct interfacial contact (Rong et al., 2019). This heterojunction consists of two coupled semiconductors (one with a higher CB) without redox electron mediators, and photo carriers are directly transferred across the interface (Zhang et al., 2021). This reduces the transport distance. The mechanism of the heterojunction is shown in Figure 11.2(c): an electron from the CB of SC 1 recombines with a hole from the VB of SC 2 at the interface. This leaves the remaining electrons and holes with excellent redox ability. Compared to the earlier two generations of Z-scheme photocatalysts, charge-carrier mediators are not needed in the direct Z-scheme heterostructure. The lack of electronic media significantly lowers expenses and simplifies the preparation procedure. Further, the light-shielding effect of the metal-based mediator can be mitigated by constructing a direct Z-scheme photocatalyst; consequently, the direct Z-scheme heterojunction has been extensively investigated for various photocatalytic applications (Mishra et al., 2024; Sahoo et al., 2024; Zhou et al., 2014).

In general, interfacial defect states, built-in electric field (BIEF), and selective facet design control the development of the direct Z-scheme photocatalyst and the mechanism of photoexcited charge transfer. Nonetheless, some issues remain to be addressed, such as selecting SCs with appropriate band gaps and maintaining SC long-term stability under different conditions (Liao et al., 2022). Because of their high redox potential and efficient charge separation, both Z- and S-scheme heterojunctions are advantageous in a variety of energy and environmental applications. However, both have certain limitations that can be addressed by implementing methods such as doping and surface engineering. Over the last several years, Z-scheme heterojunction photocatalysts have advanced dramatically for phenol and nitrophenol degradation, progressing from ternary composites with conductive bridges to advanced quaternary architectures with plasmonic enhancements, and finally to high-surface-area MOF-perovskite hybrids that prioritize stability and scalability. Pioneering this trajectory, Samsudin et al. (2019) constructed a novel Z-scheme g-C_3N_4/CNT/$BiVO_4$ heterostructure photocatalyst for phenol removal under simulated solar light irradiation. The addition of CNT to the heterostructure photocatalyst greatly increases the overall surface area and prevents electron–hole pair rejoining. The Z-scheme g-C_3N_4/CNT/$BiVO_4$ nanocomposite exhibited

80.6% phenol degradation within 120 min, compared to the individual NMs. The photoinduced electrons in the CB of $BiVO_4$ migrate to the VB of g-C_3N_4 via CNT, resulting in the build-up of photoinduced electrons in the CB of g-C_3N_4, which consequently interact with dissolved oxygen to produce superoxide radicals and the photoexcited holes in the VB of $BiVO_4$. The accumulated photoinduced holes in the VB of $BiVO_4$ possess strong oxidation capacity, forming hydroxyl radicals that directly degrade the phenol. Active-species-trapping experiments confirmed the construction of the Z-scheme. Building on this mediated design for visible-light prowess, Jin et al. (2019) developed a Z-scheme heterostructure by using $LaCoO_3$ and g-C_3N_4 via a facile impregnation method. The $LaCoO_3$/g-C_3N_4-60wt% exhibits 85% photocatalytic phenol degradation in 5 h under visible light, which is almost 7.5 and 5.2 folds greater than pure g-C_3N_4 and $LaCoO_3$, respectively. The performance improvement stems from enhanced visible-light absorption and superior charge-carrier separation efficiency enabled by Z-scheme heterostructure formation. The flake-like geometry of g-C_3N_4 provides a high surface area. So, the composite exhibited remarkable stability, with negligible performance decline across recycling cycles, establishing $LaCoO_3$/g-C_3N_4 as a promising, stable photocatalyst.

Further improvements on direct interface, this binary momentum, according to Liu et al., (2023), is the successful combination of direct Z-scheme BiOBr/CuI heterojunction that can effectively degrade phenol. This is based on the combination of the outstanding photoelectric properties of BiOBr with the efficient hole transfer of CuI to form an optimized photocatalysis. They exhibited a high efficiency of 98.4% phenol decomposition in 2.5 hours with an enhanced performance of 4.35- and 8.63-fold relative to pure BiOBr and CuI respectively. Such high photocatalytic performance is due to the charge-transfer effect under a Z-scheme scheme, leading to optimized band structures for redox reactions. The XPS spectrum analysis showed that the orbitals of Bi 4f, O 1s, Br 3d, Cu 2p, and I 3d in the nanocomposites are slightly different. This can indicate that the reaction of CuI with BiOBr results in the generation of heterojunctions. As regards the Bi 4f levels of BiOBr/CuI nanocomposites, it was noted that they shift toward the low energy levels as compared to pure BiOBr. Meanwhile, the Cu 2p states of BiOBr/CuI slightly move toward higher energy levels in comparison to those of bare CuI. The changes in the Bi 4f and Cu 2p states show that there is an electron transition from CuI to BiOBr, indicating the successful construction of the Z-scheme heterojunction.

Through the momentum thus created by Wang et al. (2023b), a quaternary and most advanced photocatalytic system using semimetallic bismuth (Bi^0) is doped into the quaternary Z-scheme heterojunctions. Using novel approaches such as sequential calcinations, iodide etching, and ultraviolet (UV) reduction techniques, they have introduced Bi^0, BiOI, Bi_2O_3, and C_3N_4. The use of SPR effects from semimetals, just like noble metals, is revolutionary in terms of reducing cost. This breakthrough exhibits an outstanding 93.2% degradation rate with a rate constant of 0.0161 min^{-1} in merely 75 minutes, thus creating a new standard for performance. The key advantage in this innovation is that it has a significantly lower concentration of the benzoquinone intermediate with the addition of C_3N_4, which implies its mineralization efficiency is higher.

This quaternary system excels because of the synergy between all four systems, and this was achieved by employing several methods. Bi^0 incorporation promotes SPR effects and production of electron–hole pairs. The XRD spectra of the synthesized NMs are shown in Figure 11.3(a). Thus, Figure 11.3 collectively proves that structural integrity, high surface area, and efficient interfacial contact synergistically contribute to improved phenol degradation performance. For the Bi/BiOI–Bi_2O_3 and Bi/BiOI–Bi_2O_3–C_3N_4 samples, the minor peaks at 27.1°, 37.9°, and 39.6° planes of metallic Bi indicate the successful introduction of Bi. Again, for the Bi/BiOI–Bi_2O_3–C_3N_4 sample, no diffraction peak of C_3N_4 was detected, which was due to the overlay of diffraction peaks as a consequence of the similar 2θ degree of Bi to C_3N_4, which confirms the successful in situ construction of multiple interfaces with minimal lattice defects contributed to the outstanding recycling stability (only a 3.8% activity decline after four cycles) and establishes this quaternary system as a transformative advance in phenolic wastewater remediation. Extending to niche pollutants, Omar et al. (2025) developed a novel Z-scheme MIL–53(Fe)/Mn–$SrTiO_3$ heterojunction photocatalyst that achieves 97.75% o-nitrophenol (ONP) degradation in just 90 minutes under Xe lamp irradiation, dramatically outperforming pure MIL–53Fe (71.94%) and Mn–$SrTiO_3$ (62.13%) components

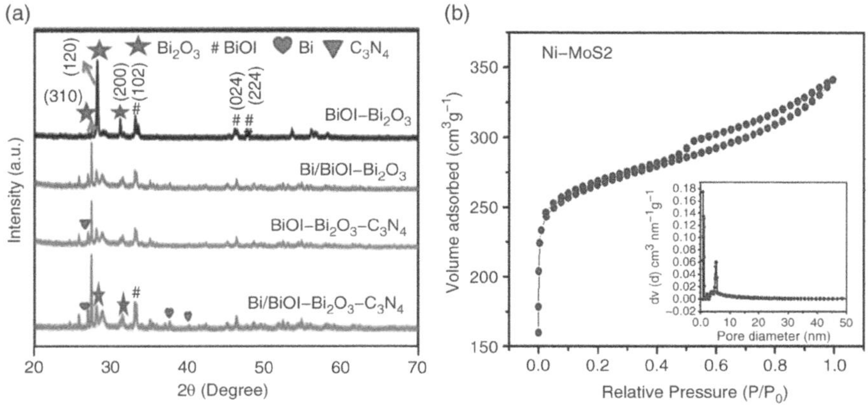

FIGURE 11.3 (a) XRD patterns (Wang et al., 2023b), License Number 6265880089568 (b) N_2 adsorption–desorption isotherms and distribution of pore size curves (inset) of Ni–MoS_2: License Number 6266510024856 (Mishra et al., 2025).

with a rate constant 3.4–4.4 times higher. Its innovation lies in combining MOF's high surface area with perovskite's enhanced charge mobility through strong interfacial interactions. Radical scavenging identifies $\cdot O_2^-$ and h^+ as dominant species in the direct Z-scheme mechanism, with excellent five-cycle stability making it superior for wastewater remediation. Culminating in ultra-high porosity, Mishra et al. (2025) developed a novel Ni–MoS_2/g-C_3N_5/CdS Z-scheme heterojunction photocatalyst via hydrothermal synthesis for efficient solar-driven phenol degradation from wastewater. This ternary nanocomposite achieves 60% degradation of 30 mg/L phenol in 180 min, which is 6.7-fold better than bare Ni–MoS_2 and 20 times superior to g-C_3N_5 due to enhanced charge separation (low PL intensity) of the Z-scheme heterojunction and ultra-high surface area (1,034.8 m^2/g from BET) (Table 11.2).

As shown in Figure 11.3(b), the specific surface area of Ni–MoS_2 was determined to be 1,034.8 m^2/g by the calculated BET analysis, which is substantially more than that of pure MoS_2. Again, excellent recyclability (57.5% retention after four cycles) positions it as a promising, stable environmental remediation solution.

11.2.4 Principle and Design of S-Scheme Heterojunction

Artificial S-scheme systems, inspired by the Z-scheme model, offer a promising approach to overcoming the long-standing challenges of charge recombination and light harvesting (Mansingh et al., 2023). The S-scheme heterojunction comprises two components: the RP and the OP (Wu et al., 2026). Such heterojunctions can preserve the light-harvesting capacity of the OP and RP while retaining the highest redox potential of SMs (Li et al., 2021). For the construction of S-scheme photocatalyst nanocomposite, the following specifications must be taken into account: (1) there should be a staggered band structure between the two semiconductors; (2) the RP should have a higher CB position and Fermi level than the OP; and (3) an intimate interface should exist, ensuring the effective transfer of charge carriers over the interface (Shawky & Mohamed, 2022). Figure 11.4 shows the S-scheme heterojunction system, which consists of an OP and an RP. Furthermore, the RP or OP might be an n-type or p-type semiconductor (Han et al., 2022). For example, in n-type semiconductors, free electrons are transferred from RP to the OP because of the different work functions of these two constituents, NMs (Li et al., 2021). This electron-transport mechanism continues until the E_FS of these two semiconductors match. As a result, the RP side is positively charged, while the OP side is negatively charged at the interfacial surface. In the meantime, an inherent electric field (IEF) is created at the contact, oriented from the RP toward the OP (Tao et al., 2021). In particular, on exposure to light, the existence of band bending and the IEF prevents photogenerated holes

TABLE 11.2 Comparative performance analysis of recent Z-scheme heterojunction photocatalysts for phenol degradation

PHOTOCATALYST/ REFERENCE	*YEAR*	*PHENOL CONC.*	*EFFICIENCY (%)*	*TIME (MIN)*	*RATE CONST. (MIN^{-1})*	*NOVELTY AND KEY CHARACTERIZATION*
Ni–MoS_2/gC_3N_5/CdS (Mishra et al., 2025)	2025	30 mg/L	60	180	0.007	Ni-doping boosts MoS_2 basal activity; BET 1,034 m^2/g; PL intensity drop confirms charge separation
Bi/BiOI–Bi_2O_3–gC_3N_4 (Wang et al., 2023b)	2023	20 mg/L	93.2	75	0.0161	Dual Z-scheme + Bi plasmonics; TEM shows intimate interfaces; $\cdot O_2^-$/h^+ dominant
Bi_2WO_6/gC_3N_4/TiO_2 (Sattari et al., 2022)	2022	10 mg/L	84.7	180	0.009	Ternary optimization via RSM; 2× single components; XRD verifies phase purity
MIL-53(Fe)/Mn–$SrTiO_3$ (Omar et al., 2025)	2025	50 mg/L	97.75	90	0.0374	MOF–MOF Z-scheme; fastest kinetics; retains 92% after five cycles (EIS)
In_2S_3/WO_3 (Qiu et al., 2021)	2021	10 mg/L	95	120	0.015	Direct Z-scheme; Mott–Schottky confirms band alignment; recyclable 5×
ZnO/CuO (2D/0D) (Akhtar et al., 2025)	2025	25 mg/L	91	150	0.012	Dimensional synergy; FESEM shows uniform dispersion; -OH key radical
MoS_2/CdS (Hamid et al., 2024)	2024	40 mg/L	88	120	0.011	Sunlight-driven; Raman confirms MoS_2 edges; 85% retention post-recycling

from transferring from the VB of OP to the VB of RP, and photoelectrons from moving from the CB of the reference material to the CB of the target material (Kumar et al., 2023).

Under visible-light illumination, photoexcited charge carriers with lower redox potentials undergo recombination, a process IEF facilitates. Ultimately, the accumulation of photogenerated electrons and holes with high redox potentials is separated into RP and OP regions, respectively (Bilal et al., 2025). As a result, a spatial distribution of charges is established at the oxidation and reduction sites. This process follows an 'S' shape, hence the name, and thus allows for efficient photocatalysis (Wang et al., 2022). Thus, the S-scheme design not only efficiently reduces charge rejoining but also facilitates significant redox capacity (Liang et al., 2021).

Compared with a single-junction S-scheme photocatalyst, the twin S-scheme heterojunction may generate two IEFs within the photocatalyst, leading to superior effective spatial charge separation and enhanced redox capability. Twin S-scheme photocatalysts are complex heterojunction structures that combine two different, well-matched photocatalysts with a secondary, usually two-dimensional, material to create a three-phase system. For instance, Ruan et al. (2023) created the twin S-scheme heterojunction by loading H-doped rutile TiO_2 nanorods and anatase TiO_2 NMs onto the surface of C_3N_4 nanosheets. Upon light irradiation, photoexcited electrons in the CB of rutile TiO_2 nanorods and anatase TiO_2 NMs reunite with photogenerated holes of C_3N_4. In the reduction and oxidation processes, the photoexcited electrons in the CB of C_3N_4 and the photogenerated holes in the VB of rutile TiO_2 nanorods and anatase TiO_2 NMs would be consumed, respectively (Figure 11.5).

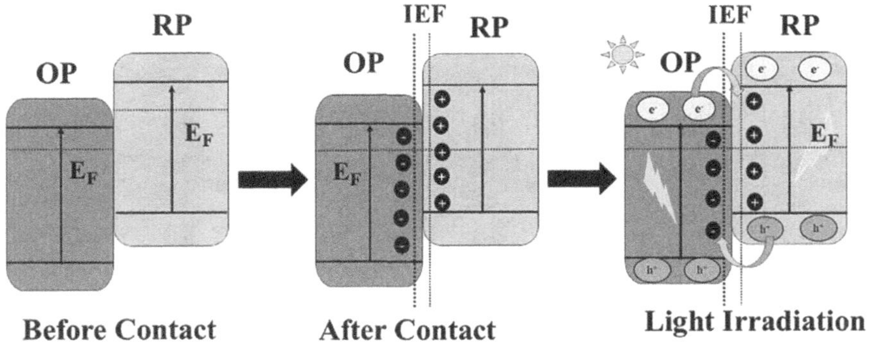

FIGURE 11.4 Mechanisms of charge transfer in S-scheme heterostructure.

S-scheme charge-transfer semiconductor heterojunction systems make a significant contribution to energy and environmental remediation. Compared with the Z-scheme photocatalyst, the S-scheme heterostructure is driven by the formation of an IEF, which minimizes charge-carrier transfer distance through synergistic interactions at the interface between the two photocatalysts (Cao et al., 2021). The formation of IEF is beneficial by removing the comparatively lower energetic/useless photoexcited electron in the CB and holes in the VB, which are unable to contribute to the production of ROS owing to the smaller potential of the originating band energy. The residual, valuable charge carriers have high redox potential, forming O_2 radicals in the reduction reaction and •OH radicals in the oxidation reaction, thereby boosting photocatalytic activity (Kumar et al., 2021). The major disadvantage of the Z-scheme photocatalyst is that a longer transmission distance results in greater charge losses. The type II heterostructure weakens the redox properties of photoexcited electrons and holes, thereby adversely affecting the overall photocatalytic activity. However, the S-scheme photocatalyst can retain photoexcited carriers with strong redox ability on two SCs, which is advantageous for photocatalytic performance (Kovalskiy et al., 2019; Zhang et al., 2019).

In recent times, the development of S-scheme heterojunctions has led to significant developments in the photodegradation of wastewater using photocatalysis, especially in terms of recalcitrant phenolic contaminants due to improved charge separation, redox capability, and robustness owing to novel combinations of materials. Extending from their basic two-element structure of $ZnTiO_3/Bi_2WO_6$ S-scheme heterojunction used for phenol removal, Zuo et al. (2023) developed 2D/2D structure of $ZnTiO_3/Bi_2WO_6$ S-scheme heterojunction for phenol degradation, which degraded phenol at a concentration of 10 mg/L at visible-light exposure (+0.5 V bias) in just a few hours at a rate threefold higher than that observed in Bi_2WO_6 with over 90% efficiency over five cycles against TiO_2 or g-C_3N_4 by boosting $\cdot O_2^-$/·OH radicals production. Hu et al. (2025) designed a CuO/TiO_2@C S-scheme heterojunction for phenol degradation in a photo-PMS system, obtaining 88% degradation of 40 mg/L phenol and 48% TOC in only 25 minutes of simulated sunlight irradiation. The CuO or TiO_2@C systems effectively overcome charge recombination and supercharge the Cu(II)/Cu(I) redox cycle, generating ·OH and $\cdot O_2^-$ radicals for efficient phenol oxidation. The superior degradation performance in this work stems from its seamless in situ calcination of MXene-derived TiO_2@C coated with CuO, which initiates spatial charge separation to activate PMS without harsh conditions, and from its outstanding recyclability for four cycles, making it ready to tackle volatile organic compounds such as phenol in tough industrial wastewater.

To enhance the synergy on ternary complexity, Liu et al. (2025b) proposed the $MoS_2/Bi_2O_3/CdS$ ternary S-scheme heterojunction photocatalyst. They achieved almost 99% degradation of 4-nitrophenol from a 100-ppm solution in 120 min under visible-light irradiation, much faster than the binary system Bi_2O_3/CdS at 86% or the simple CdS/Bi_2O_3 at 50–60%, owing to the efficient charge separation that produces more reactive oxygen radicals such as $\cdot O_2^-$ and ·OH for complete mineralization to non-toxic CO_2 and H_2O. The significance of this breakthrough is that this solvothermally mixed 2D nanosheet and

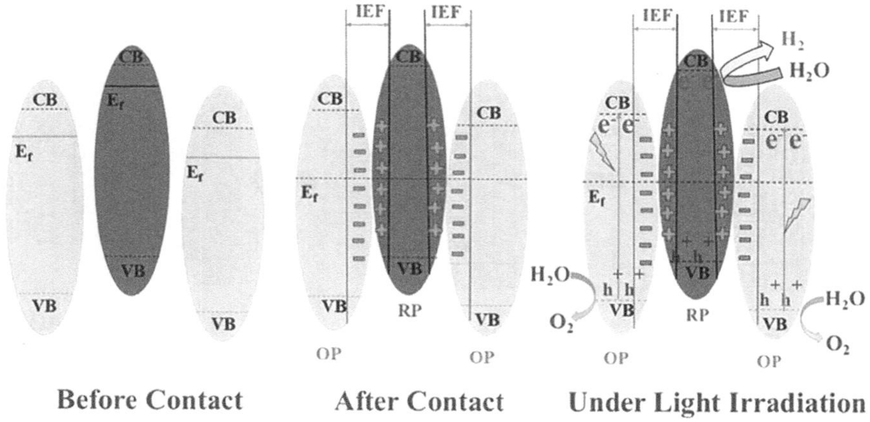

FIGURE 11.5 Twin S-scheme heterojunction for photocatalytic degradation of phenol.

nanosphere assembly with a high internal electric field for S-scheme operation effectively circumvents rapid electron–hole pair rejoining while preserving high redox capability and also boasts excellent durability, with less than 10% degradation even after four cycles. The culmination of this multifunctional innovation process, Liu et al. (2025a) proposed a MOF-derived ZnO/PANI S-scheme heterojunction photocatalyst. They reported complete mineralization of phenol (100% TOC removal) in less than 60 minutes of simulated sunlight irradiation while producing H_2O_2 at 0.75 $\mu mol \cdot L^{-1} \cdot h^{-1}$, which is far superior to the bare ZnO or ZIF-8-based photocatalysts due to the enhanced ·OH and $\cdot O_2^-$ radicals production. The novelty of this innovation, however, lies in this brilliant "two-for-one" strategy, where the toxic phenol, the pollutant itself, is combined with eco-friendly H_2O_2 production to circumvent the common limitations of ZnO, such as rapid charge recombination and poor visible-light response, all made possible by the porous structure of ZIF-8 and the seamless integration of PANI, which generates an electric field to facilitate efficient S-scheme charge transfer. Thus, these studies represent a clear-cut progression from binary photo-electrocatalytic milestones to PMS-activated binaries, ternary nanocomposites, and, finally, multifunctional hybrids, which have steadily increased degradation rates, depths of mineralization, and practicality while maintaining the fundamental superiority of the S-scheme in terms of recombination-resistant, high-potency charge dynamics. Table 11.3 represents the detailed potential of the S-scheme heterojunction reported by various researchers in the literature.

11.2.5 Scalability, Cost, and Practical Applicability of Z-Scheme vs S-Scheme Photocatalysts

Both Z-scheme and S-scheme heterostructures display great redox capability as well as enhanced charge separation properties, although their practical applicability differs significantly. For instance, Z-schemes usually consist of complicated structural designs that require the use of expensive electron mediators such as noble metals, making their large-scale application unfeasible (Chauhan et al., 2024; Ng et al., 2020; Tada et al., 2006). However, S-scheme photocatalysts operate using an internal electric field and band bending to drive selective charge recombination, eliminating the need for mediators and simplifying the system design (Bao et al., 2021; Wang et al., 2022). Therefore, S-scheme systems show better results with regard to low cost, stability, and scalability, and hence, they are more applicable in practical environmental cases like phenol degradation. However, Z-scheme systems still show impressive results in terms of having high redox activity, particularly in advanced multicomponent designs, highlighting that both systems have complementary advantages depending on the target application (Liao et al., 2022; Mohammed et al., 2025).

TABLE 11.3 Comparative analysis of phenol removal using S-scheme heterojunction as reported in the literature

PHOTOCATALYST/ REFERENCES	*YEAR*	*PHENOL CONC.*	*EFFICIENCY (%)*	*TIME (MIN)*	*RATE CONST. (MIN⁻¹)*	*NOVELTY*
ZnO/PANI (MOF-derived) (Liu et al., 2025a)	2025	Unspecified	100 (complete mineralization)	60	Not reported	Dual function (H_2O_2 + phenol); ROS synergy (·OH, ·O_2^-); high stability
$MoS_2/Bi_2O_3/CdS$ (ternary) (Liu et al., 2025b)	2025	Unspecified	99	120	Higher than binary (est. 0.02+)	Ternary S-scheme; 99% vs 86% binary; real wastewater applicable
h-$MoO_3/Bi_9O_7I_3$/ Cr–CdS (double S-scheme) (Swetha et al., 2024)	2024	Not specified (p-CP proxy)	94	300	Not reported	Ternary dual S-scheme; low-toxicity byproducts; 2.22 eV bandgap
g-C_3N_4/CdS (2D–2D) (Saad et al., 2024)	2024	Unspecified	99.47	50	10.4× > $ZnTiO_3$	Reduced recombination;anti-photocorrosion; h^+/·OH dominant
$ZnTiO_3/Bi_2WO_6$ (2D/2D) (Zuo et al., 2023)	2023	Wastewater	~90 (est. from kinetics)	Not specified	3.6× > Bi_2WO_6; 10.4× > $ZnTiO_3$	PEC-enhanced; S-scheme confirmed by carrier accumulation
CuO/TiO_2@C (Hu et al., 2025)	2025	40 mg/L	88	Not specified	Enhanced vs singles	Photo-PMS activation; ·OH/$^1O_2$ key; carbon interlayer

11.3 CONCLUSION AND FUTURE PROSPECTIVE

Photocatalysis is an inexpensive, eco-friendly, and sustainable technique for the selective and rapid degradation of phenolic pollutants in contaminated environments. Unlike adsorption, solvent extraction, and ion exchange, the photocatalytic degradation process occurs via molecular oxidation of phenol, controlled by interfacial charge transfer and the production of free-radical ROS. Z-scheme heterojunctions address the type II heterojunction limitation by restoring robust OP and RP via directed recombination pathways. However, their scalability and stability are often hindered by their reliance on mediators and the complexity of their interfaces. This chapter summarizes significant progress in Z-scheme and S-scheme heterojunctions and provides a comprehensive evaluation of the photocatalytic degradation mechanism of phenol. In recent times, the S-scheme photocatalytic nanocomposite has emerged as a more mechanistically sound and structurally straightforward substitute. The major advantage of the S-scheme is its fundamental nature, driven by the internal electric field, band-edge bending, and Coulombic interactions between opposite charges. The photogenerated electrons and holes, which have lower potentials, are consequently removed. In contrast, those with the desired potential remain in the heterojunction photocatalyst, thereby enhancing visible-light utilization, improving stability in complex aqueous environments, and improving pollutant degradation activity. In addition, the Fermi-level adjustment, thermodynamic barriers, and the optimal interfacial area between OP and RP photocatalysts in Z-scheme and S-scheme heterojunctions play major roles in overall photocatalytic activity. Further effort is required to tune the direct Z- and S-scheme heterostructure to improve the photoconversion efficiency of visible light, facilitate rapid charge transfer with minimal resistance, and enhance physicochemical stability by designing the structural, crystal-defect, and morphological properties of the photocatalyst. The selectivity and reusability of the NMs are also very crucial. Realizing practical, high-efficiency photocatalytic degradation of phenol will require interdisciplinary collaborative research across fields such as materials chemistry, charge-transfer physics, sustainability science, and reactor engineering, laying the foundation for effective, scalable solutions to the global challenge of phenolic pollution. Future research on Z- and S-scheme photocatalysts includes optimizing light-harvesting capacity, designing ternary or multicomponent structures, and applying non-toxic, long-stable materials, such as metal-free or stable oxide-based heterojunctions that prevent photocorrosion and formation of secondary pollution. Further research on solid-state Z- and S-scheme photocatalytic systems that resist fouling in complex industrial effluents or intimately coupled photocatalysis and biodegradation. Further research is needed on optimization of reactor geometry to enhance light-harvesting capacity and improve mixing and mass transfer profile.

REFERENCES

Ahuja, T., Brighu, U., & Saxena, K. (2023). Recent advances in photocatalytic materials and their applications for treatment of wastewater: A review. *Journal of Water Process Engineering*, 53, 103759. https://doi.org/10.1016/j.jwpe.2023.103759

Akhtar, T., Hill, A. J., Bukhari, S. A. B., Nasir, H., Ullah, S., & Schwank, J. W. (2025). Fabrication of 2D/0D ZnO/CuO Z-scheme heterojunctions for enhanced photocatalytic reduction of 2, 4-dinitrophenol and 2, 4, 6-trinitrophenol under visible light. *Solar Energy*, 289, 113330. https://doi.org/10.1016/j.solener.2025.113330

Al Miad, A., Saikat, S. P., Alam, M. K., Sahadat Hossain, M., Bahadur, N. M., & Ahmed, S. (2024). Metal oxide-based photocatalysts for the efficient degradation of organic pollutants for a sustainable environment: A review. *Nanoscale Advances,* 19, 4781–4803. https://doi.org/10.1039/d4na00517a

Alshabib, M., & Onaizi, S. A. (2019). A review on phenolic wastewater remediation using homogeneous and heterogeneous enzymatic processes: Current status and potential challenges. *Separation and Purification Technology*, 219, 186–207. https://doi.org/10.1016/j.seppur.2019.03.028

ATSDR. (2008). Toxicological profile for phenol. Agency for Toxic Substances and Disease Registry, Division of Toxicology and Environmental Medicine, Atlanta, Georgia, Available at https://www.ncbi.nlm.nih.gov/books/NBK599448/pdf/Bookshelf_NBK599448.pdf.

Bao, Y., Song, S., Yao, G., & Jiang, S. (2021). S-scheme photocatalytic systems. *Solar RRL*, 5(7), 2100118. https://doi.org/10.1002/solr.202100118

Bian, J., Zhang, Z., Feng, J., Thangamuthu, M., Yang, F., Sun, L., Li, Z., et al. (2021). Energy platform for directed charge transfer in the cascade Z-scheme heterojunction: CO_2 photoreduction without a cocatalyst. *Angewandte Chemie*, 60, 20906. https://doi.org/10.1002/anie.202106929

Bilal, M., Alfaifi, M. Q., Ben Ahmed, S., Abduljawad, M. M., Alrashed, Y. I., Aldurahim, E. S., & Alassmy, Y. A. (2025). A review of strategies to switch heterojunction system from type-II to S-scheme for photocatalytic applications. *Materials Science in Semiconductor Processing*, 186, 109051. https://doi.org/10.1016/j.mssp.2024.109051

Cao, W., Jiang, C., Chen, C., Zhou, H., & Wang, Y. (2021). A novel Z-scheme $CdS/Bi_4O_5Br_2$ heterostructure with mechanism analysis: Enhanced photocatalytic performance. *Journal of Alloys and Compounds*, 861, 158554. https://doi.org/10.1016/j.jallcom.2020.158554

Chauhan, A., Raizada, P., Singh, P., Ahamad, T., Nguyen, V.-H., Le, Q. V., Khan, A. A. P., et al. (2024). Recent developments in Ag_3VO_4-based photocatalysts towards environmental remediation: Properties, synthesis, strategies and applications. *Journal of Industrial and Engineering Chemistry*, 130, 25–53. https://doi.org/10.1016/j.jiec.2024.01.015

Che, Y., Lu, B., Qi, Q., Chang, H., Zhai, J., Wang, K., & Liu, Z. (2018). Bio-inspired Z-scheme g-C_3N_4/Ag_2CrO_4 for efficient visible-light photocatalytic hydrogen generation. *Scientific Reports*, 8(1), 13645. https://doi.org/10.1038/s41598-018-34287-w

Chowdhury, A. P., Anantharaju, K. S., Keshavamurthy, K., Swain, S., & Uma, B. (2025). A mini-review on Z and S scheme heterojunction bismuth-based photocatalysts towards H_2 production, CO_2 reduction, and pollutant degradation. *Next Materials*, 8, 100918. https://doi.org/10.1016/j.nxmate.2025.100918

Downs, J. W., & Wills, B. K. (2025). *Phenol toxicity*. StatPearls Publishing. https://europepmc.org/books/NBK542311

Grace Pavithra, G. K., Sundar Rajan, P., Arun, J., Brindhadevi, K., Le, Q. H., & Pugazhendhi, A. (2023). A review on recent advancements in extraction, removal and recovery of phenols from phenolic wastewater: Challenges and future outlook. *Environmental Research*, 236, 116747. https://doi.org/10.1016/j.envres.2023.117005

Hameed, B. H., & Rahman, A. A. (2008). Removal of phenol from aqueous solutions by adsorption onto activated carbon prepared from biomass material. *Journal of Hazardous Materials*, 160(2–3), 576–581. https://doi.org/10.1016/j.jhazmat.2008.03.028

Hamid, K., Mukhlish, M. Z. B., & Uddin, M. T. (2024). Sunlight-activated heterostructure MoS_2/CdS nanocomposite photocatalyst with enhanced photocatalytic activity: Band alignment and mechanism study. *RSC Advances*, 14(52), 38908–38923. https://doi.org/10.1039/d4ra06857b

Han, X., Lu, B., Huang, X., Liu, C., Chen, S., Chen, J., Zeng, Z., Deng, S., & Wang, J. (2022). Novel p- and n-type S-scheme heterojunction photocatalyst for boosted CO_2 photoreduction activity. *Applied Catalysis B: Environmental*, 316, 121587. https://doi.org/10.1016/j.apcatb.2022.121587

Hu, H., Hu, Y., Kong, W., Tao, Y., Jiang, Q., Wang, J., Li, C., et al. (2023). The photocatalytic mineralization of phenolic wastewater via self-generation and -activation of H_2O_2 technology. *Journal of Environmental Chemical Engineering*, 11(5), 111108. https://doi.org/10.1016/j.jece.2023.111108

Hu, Z., Zhang, W., Liu, Z., Zhang, X., & Wang, X. (2025). Construction of a CuO/TiO_2@C S-scheme heterojunction for phenol removal by activated peroxymonosulfate. *Environmental Research*, 277, 121564. https://doi.org/10.1016/j.envres.2025.121564

Jin, Z., Hu, R., Wang, H., Hu, J., & Ren, T. (2019). One-step impregnation method to prepare direct Z-scheme $LaCoO_3$/g-C_3N_4 heterojunction photocatalysts for phenol degradation under visible light. *Applied Surface Science*, 491, 432–442. https://doi.org/10.1016/j.apsusc.2019.06.036

Kovalskiy, V. A., Eremenko, V. G., Vergeles, P. S., Soltanovich, O. A., Khodos, I. I., Dorokhin, M. V., & Danilov, Y. A. (2019). On the mechanism of cross-hatch pattern formation in heterostructures with a small lattice mismatch. *Applied Surface Science*, 479, 930–941. https://doi.org/10.1016/j.apsusc.2019.02.032

Kumar, A., Rana, S., Wang, T., Dhiman, P., Sharma, G., Du, B., & Stadler, F. J. (2023). Advances in S-scheme heterojunction semiconductor photocatalysts for CO_2 reduction, nitrogen fixation and NOx degradation. *Materials Science in Semiconductor Processing*, 163, 107869. https://doi.org/10.1016/j.mssp.2023.107869

Kumar, G. J., Sarathbavan, M., Sivalingam, Y., Navaneethan, M, & Bharathi, K. K. (2021). Mechanism of analog bipolar resistive switching and work function in Au/Na0.5Bi0.5TiO_3/Pt heterostructure thin films. *Materials Chemistry and Physics*, 257, 123765. https://doi.org/10.1016/j.matchemphys.2020.123765

Ladeia Ramos, R., Moreira, V. R., & Amaral, M. C. S. (2024). Phenolic compounds in water: Review of occurrence, risk, and retention by membrane technology. *Journal of Environmental Management*, 351, 119772. https://doi.org/10.1016/j.jenvman.2023.119772

Li, J., Ding, Z., Li, J., Wang, C., Pan, L., & Wang, G. (2021). Synergistic coupling of NiS1.03 nanoparticle with S-doped reduced graphene oxide for enhanced lithium and sodium storage. *Chemical Engineering Journal*, 407, 127199. https://doi.org/10.1016/j.cej.2020.127199

Liang, J., Jing, C., Wang, J., & Men, Y. (2021). Photocatalytic reduction of Cr(VI) over g-C_3N_4 photocatalysts synthesized by different precursors. *Molecules*, 26(22), 7054. https://doi.org/10.3390/molecules26227054

Liao, G., Li, C., Liu, S.-Y., Fang, B., & Yang, H. (2022). Z-scheme systems: From fundamental principles to characterization, synthesis, and photocatalytic fuel-conversion applications. *Physics Reports*, 983, 1–41. https://doi.org/10.1016/j.physrep.2022.09.001

Liu, B., Hu, X., Yang, J., Yang, C., & Huang, Y. (2023). Construction of direct Z-scheme BiOBr/CuI heterojunction for boosting photocatalytic degradation of phenol. *Catalysis Science & Technology*, 13(2), 504–515. https://doi.org/10.1039/d2cy01840c

Liu, B., Zhang, J., Li, H., Cheng, B., & Bie, C. (2025a). MOF-derived ZnO/PANI S-scheme heterojunction for efficient photocatalytic phenol mineralization coupled with H_2O_2 generation. *Acta Physico-Chimica Sinica*, 41(10), 100121. https://doi.org/10.3866/PKU.WHXB202402016

Liu, X., Zhang, Q., & Ma, D. (2021). Advances in 2D/2D Z-scheme heterojunctions for photocatalytic applications. *Solar RRL*, 5(2), 2000397. https://doi.org/10.1002/solr.202000397

Liu, Z., Roosh, M., Lu, M., Arshad, A., Xian, W., Shen, Y., Liu, G. et al. (2025b). Empowering wastewater treatment with step scheme heterojunction ternary nanocomposites for photocatalytic degradation of nitrophenol. *Scientific Reports*, 15(1), 3299. https://doi.org/10.1038/s41598-025-86975-z

Low, J., Jiang, C., Cheng, B., Wageh, S., Al-Ghamdi, A. A., & Yu, J. (2017). A review of direct Z-scheme photocatalysts. *Small Methods*, 1(5), 1700080. https://doi.org/10.1002/smtd.201700080

Low, J., Yu, J., & Jiang, C. (2020). Design and fabrication of direct Z-scheme photocatalysts. *In Interface Science and Technology*, 31, 193–229. https://doi.org/10.1016/B978-0-08-102890-2.00006-3

Ma, G., Chen, S., Kuang, Y., Akiyama, S., Hisatomi, T., Nakabayashi, M., Shibata, N., Katayama, M., Minegishi, T., & Domen, K. (2016). Visible light-driven Z-scheme water splitting using oxysulfide H_2 evolution photocatalysts. *The Journal of Physical Chemistry Letters*, 7(19), 3892–3896. https://doi.org/10.1021/acs.jpclett.6b01802

Mafa, P. J., Malefane, M. E., Idris, A. O., Mamba, B. B., Liu, D., Gui, J., & Kuvarega, A. T. (2021). Cobalt oxide/copper bismuth oxide/samarium vanadate (Co_3O_4/$CuBi_2O_4$/$SmVO_4$) dual Z-scheme heterostructured photocatalyst with high charge-transfer efficiency: Enhanced carbamazepine degradation under visible light irradiation. *Journal of Colloid and Interface Science*, 603, 666–684. https://doi.org/10.1016/j.jcis.2021.06.024

Malefane, M. E., Mafa, P. J., Managa, M., Nkambule, T. T. I., & Kuvarega, A. T. (2023). Understanding the principles and applications of dual Z-scheme heterojunctions: How far can we go? *The Journal of Physical Chemistry Letters*, 14(4), 1029–1045. https://doi.org/10.1021/acs.jpclett.2c03387

Mansingh, S., Das, K. K., Priyadarshini, N., Sahoo, D. P., Prusty, D., Sahu, J., Mohanty, U. A., & Parida, K. (2023). Minireview elaborating S-scheme charge dynamic photocatalysts: Journey from Z to S, mechanism of charge flow, characterization proof, and H_2O_2 evolution. *Energy & Fuels*, 37(14), 9873–9894. https://doi.org/10.1021/acs.energyfuels.3c00717

Menzikov, S. A., & Morozov, S. G. (2019). Involvement of brain GABAA R-coupled Cl^-/HCO_3^--ATPase in phenol-induced the head-twitching and tremor responses in rats. *NeuroToxicology*, 71, 122–131. https://doi.org/10.1016/j.neuro.2018.12.007

Mhlongo, N. L., Akharame, M. O., Pereao, O., Human, I. S., & Opeolu, B. O. (2023). Phenolic compounds occurrence and human health risk assessment in potable and treated waters in Western Cape, South Africa. *Frontiers in Toxicology*, 5, 1269601. https://doi.org/10.3389/ftox.2023.1269601

Mishra, S., Sahoo, N. K., Sahoo, P. K., Panda, P., Sahoo, S., Sahoo, G. S., & Rout, P. R. (2025). Construction of a novel Ni-MoS_2/gC_3N_5/CdS heterojunction for efficient photocatalytic degradation of phenol. *Inorganic Chemistry Communications*, 174, 114033. https://doi.org/10.1016/j.inoche.2025.114033

Mishra, S., Sahoo, N. K., Sahoo, P. K., Sahoo, S., Rout, P. R., & Rath, G. (2024). Synergistic effect of Zn-$AgIn_5S_8$/CdS Z-scheme heterojunction and S-doped RGO for efficient removal of chromium from contaminated water. *Nanoscale Advances*, 6(23), 5925–5938. https://doi.org/10.1039/D4NA00350K

Mohammed, J., Elsharkawy, E. R., El-Bahy, S. M., Hafeez, H. Y., Musa, R. I., Idris, S. A., Maikudi, S., & El-Bahy, Z. M. (2025). Z scheme and S scheme heterostructured photocatalyst for photocatalytic water splitting: A comprehensive review towards efficient H_2 generation. *Materials Today Sustainability*, 32, 101227. https://doi.org/10.1016/j.mtsust.2025.101227

Molaei, M. J. (2024). Recent advances in hydrogen production through photocatalytic water splitting: A review. *Fuel*, 365, 131159. https://doi.org/10.1016/j.fuel.2024.131159

Mumtaz, F., Li, B., Al Shehhia, M. R., Feng, X., & Wang, K. (2024). Treatment of phenolic-wastewater by hybrid technologies: A review. *Journal of Water Process Engineering*, 62, 104695. https://doi.org/10.1016/j.jwpe.2023.104695

Narayanan, C. M., & Narayan, V. (2019). Biological wastewater treatment and bioreactor design: A review. *Sustainable Environment Research*, 29(1), 33. https://doi.org/10.1186/s42834-019-0036-1

Ng, B.-J., Putri, L. K., Kong, X. Y., Teh, Y. W., Pasbakhsh, P., & Chai, S.-P. (2020). Z-scheme photocatalytic systems for solar water splitting. *Advanced Science*, 7(7), 1903171. https://doi.org/10.1002/advs.201903171

Omar, M., Omar, S., El-Subruiti, G. M., Attia, N. F., & Eltaweil, A. (2025). Z-scheme MIL-53(Fe)/Mn-$SrTiO_3$ heterostructure for efficient photocatalytic degradation of o-nitrophenol. *Royal Society Open Science*, 12(11), 241073. https://doi.org/10.1098/rsos.241073

Panigrahy, N., Priyadarshini, A., Sahoo, M. M., Verma, A. K., Daverey, A., & Sahoo, N. K. (2022). A comprehensive review on eco-toxicity and biodegradation of phenolics: Recent progress and future outlook. *Environmental Technology & Innovation*, 28, 102423. https://doi.org/10.1016/j.eti.2022.102423

Pet, I., Sanad, M. N., Farouz, M., ElFaham, M. M., El-Hussein, A., El-sadek, M. S. A., Althobiti, R. A., & Ioanid, A. (2024). Review: Recent developments in the implementation of activated carbon as heavy metal removal management. *Water Conservation Science and Engineering*, 9(2), 62. https://doi.org/10.1007/s41101-024-00287-3

Qiu, Q., Zhu, P., Liu, Y., Liang, T., Xie, T., & Lin, Y. (2021). Highly efficient In_2S_3/WO_3 photocatalysts: Z-scheme photocatalytic mechanism for enhanced photocatalytic water pollutant degradation under visible light irradiation. *RSC Advances*, 11(6), 3333–3341. https://doi.org/10.1039/D0RA09315G

Rong, X., Chen, H., Rong, J., Zhang, X., Wei, J., Liu, S., Zhou, X., Xu, J., Qiu, F., & Wu, Z. (2019). An all-solid-state Z-scheme $TiO_2/ZnFe_2O_4$ photocatalytic system for the N_2 photofixation enhancement. *Chemical Engineering Journal*, 371, 286–293. https://doi.org/10.1016/j.cej.2019.04.071

Ruan, X., Huang, C., Cheng, H., Zhang, Z., Cui, Y., Li, Z., Xie, T., et al. (2023). A twin S-scheme artificial photosynthetic system with self-assembled heterojunctions yields superior photocatalytic hydrogen evolution rate. *Advanced Materials*, 35(6), 2209141. https://doi.org/10.1002/adma.202209141

Saad, E. M., Abd-Elhafiz, M. F., Ahmed, E. M., & Abo Markeb, A. (2024). Hexavalent chromium ion removal from wastewater using novel nanocomposite based on the impregnation of zero-valent iron nanoparticles into polyurethane foam. *Scientific Reports*, 14(1), 55803. https://doi.org/10.1038/s41598-024-55803-1

Sahoo, S., Sahoo, N. K., Sahoo, P. K., Mishra, S., Kumar, A., Naik, B., & Rout, P. R. (2024). Treatment of chromium-contaminated water using a highly efficient, novel ternary synergistic S–RGO–BiOBr–In_2S_3 heterojunction. *Materials Advances*, 5(24), 9626–9640. https://doi.org/10.1039/D4MA00431K

Sahoo, S. K., Das, A. A., Deka, D., Naik, B., & Sahoo, N. K. (2021). Organic-inorganic hybrid hydroquinone bridged V-CdS/HAP/Pd-TCPP: A novel visible light active photocatalyst for phenol degradation. *Journal of Molecular Liquids*, 339, 116721. https://doi.org/10.1016/j.molliq.2021.116721

Samsudin, M. F. R., Bacho, N., Sufian, S., & Ng, Y. H. (2019). Photocatalytic degradation of phenol wastewater over Z-scheme g-C_3N_4/CNT/$BiVO_4$ heterostructure photocatalyst under solar light irradiation. *Journal of Molecular Liquids*, 277, 977–988. https://doi.org/10.1016/j.molliq.2019.01.083

Sattari, M., Farhadian, M., Solaimany Nazar, A. R., & Moghadam, M. (2022). Enhancement of phenol degradation using novel Z-scheme $Bi_2WO_6/C_3N_4/TiO_2$ composite: Catalyst and operational parameters optimisation. *Journal of Photochemistry and Photobiology A: Chemistry*, 431, 114065. https://doi.org/10.1016/j.jphotochem.2022.114065

Shawky, A., & Mohamed, R. M. (2022). S-scheme heterojunctions: Emerging designed photocatalysts toward green energy and environmental remediation redox reactions. *Journal of Environmental Chemical Engineering*, 10(5), 108249. https://doi.org/10.1016/j.jece.2022.108249

Swetha, S., Alahmadi, T. A., Ansari, M. J., & Khan, S. S. (2024). Strategically tailored double S-scheme heterojunction in h-MoO_3 doped $Bi_7O_9I_3$ decorated with Cr–CdS quantum dots for efficient photocatalytic degradation of phenolics. *Journal of Cleaner Production*, 449, 141656. https://doi.org/10.1016/j.jclepro.2024.141656

Tada, H., Mitsui, T., Kiyonaga, T., Akita, T., & Tanaka, K. (2006). All-solid-state Z-scheme in CdS–Au–TiO_2 three-component nanojunction system. *Nature Materials*, 5(10), 782–786. https://doi.org/10.1038/nmat1734

Tao, J., Yu, X., Liu, Q., Liu, G., & Tang, H. (2021). Internal electric field induced S–scheme heterojunction MoS_2/CoAl LDH for enhanced photocatalytic hydrogen evolution. *Journal of Colloid and Interface Science*, 585, 470–479. https://doi.org/10.1016/j.jcis.2020.11.049

Tu, W., Zhou, Y., Feng, S., Xu, Q., Li, P., Wang, X., Xiao, M., & Zou, Z. (2015). Hollow spheres consisting of $Ti_{0.91}O_2$/CdS nanohybrids for CO_2 photofixation. *Chemical Communications*, 51(69), 13354–13357. https://doi.org/10.1039/c5cc03905c

Wang, C., Zhao, Y., Cheng, C., Li, Q., Guo, C., & Hu, Y. (2024). S-scheme heterojunction photocatalysts: Mechanism, challenges and opportunities. *Coordination Chemistry Reviews*, 521, 216177. https://doi.org/10.1016/j.ccr.2024.216177

Wang, L., Sun, J., Cheng, B., He, R., & Yu, J. (2023a). S-scheme heterojunction photocatalysts for H_2O_2 production. *The Journal of Physical Chemistry Letters*, 14(20), 4803–4814. https://doi.org/10.1021/acs.jpclett.3c00811

Wang, L., Zhu, B., Zhang, J., Ghasemi, J. B. G., Mousavi, M., & Yu, J. (2022). S-scheme heterojunction photocatalysts for CO_2 reduction. *Matter*, 5(12), 4187–4211. https://doi.org/10.1016/j.matt.2022.09.017

Wang, Q., Li, N., Tan, M., Deng, M., Yang, G., Li, Q., & Du, H. (2023b). Novel dual Z-scheme Bi/BiOI-Bi_2O_3-C_3N_4 heterojunctions with synergistic boosted photocatalytic degradation of phenol. *Separation and Purification Technology*, 307, 122733. https://doi.org/10.1016/j.seppur.2022.122733

Wu, X., Sayed, M., Wang, G., Yu, W., & Zhu, B. (2026). COF-based S-scheme heterojunction photocatalyst. *Advanced Materials*, 38(2), e11322. https://doi.org/10.1002/adma.202311322

Xu, M., Wu, C., & Zhou, Y. (2020). Advancements in the Fenton process for wastewater treatment. In *Advanced oxidation processes-applications, trends, and prospects*. IntechOpen. https://doi.org/10.5772/intechopen.90256

Xu, Q., & Yu, J. (2022). Graphene oxide-based heterojunction photocatalysts. In *Graphene oxide–metal oxide and other graphene oxide-based composites in photocatalysis and electrocatalysis*. Elsevier, 173–188. https://doi.org/10.1016/B978-0-12-824526-2.00006-4

Xu, Y., Guo, L., Zhang, H., Zhai, H., & Ren, H. (2019). Research status, industrial application demand and prospects of phenolic resin. *RSC Advances*, 9, 1–15. https://doi.org/10.1039/C9RA06487G

Yang, N., He, T., Chen, X., He, Y., Zhou, T., Zhang, G., & Liu, Q. (2024). TiO_2-based heterojunctions for photocatalytic hydrogen evolution reaction. *Microstructures*, 7, 1–20. https://doi.org/10.20517/microstructures.2024.06

Yu, J., Wang, S., Low, J., & Xiao, W. (2013). Enhanced photocatalytic performance of direct Z-scheme g-C_3N_4–TiO_2 photocatalysts for the decomposition of formaldehyde in air. *Physical Chemistry Chemical Physics*, 15(39), 16883–16890. https://doi.org/10.1039/c3cp53131g

Yu, L., Gamliel, D. P., Markunas, B., & Valla, J. A. (2021). A promising solution for food waste: Preparing activated carbons for phenol removal from water streams. *ACS Omega*, 6(13), 8870–8883. https://doi.org/10.1021/acsomega.0c06029

Zhang, G., Wang, Z., & Wu, J. (2021). Construction of a Z-scheme heterojunction for high-efficiency visible-light-driven photocatalytic CO_2 reduction. *Nanoscale*, 13(8), 4359–4389. https://doi.org/10.1039/d0nr08442e

Zhang, R., Sun, M., Zhao, G., Yin, G., & Liu, B. (2019). Hierarchical Fe_2O_3 nanorods/TiO_2 nanosheets heterostructure: Growth mechanism, enhanced visible-light photocatalytic and photoelectrochemical performances. *Applied Surface Science*, 475, 380–388. https://doi.org/10.1016/j.apsusc.2019.01.150

Zhou, D., Li, D., & Chen, Z. (2024). Recent advances in ternary Z-scheme photocatalysis on graphitic carbon nitride based photocatalysts. *Frontiers in Chemistry*, 12, 1359895. https://doi.org/10.3389/fchem.2024.1359895

Zhou, P., Yu, J., & Jaroniec, M. (2014). All-solid-state Z-scheme photocatalytic systems. *Advanced Materials*, 26(29), 4920–4935. https://doi.org/10.1002/adma.201400288

Zuo, C., Tai, X., Jiang, Z., Liu, M., Jiang, J., Su, Q., & Yan, X. (2023). S-scheme 2D/2D heterojunction of $ZnTiO_3$ nanosheets/Bi_2WO_6 nanosheets with enhanced photoelectrocatalytic activity for phenol wastewater under visible light. *Molecules*, 28(8), 3495. https://doi.org/10.3390/molecules28083495

Nano-Photocatalysis for the Degradation of Persistent Organic Pollutants in Water

12

Ajitha E., Sushma M., Kavanya Srinivasan, Vidhya Bhojan, and Raju Nandhakumar

12.1 INTRODUCTION

Water scarcity and pollution are two of the greatest environmental crises of this century, increasingly viewed as global threats to human health, ecosystem integrity, and long-term sustainability. Rapid population growth has increased the competition for limited resources and intensified agricultural, industrial, and urban demands on freshwater systems, thus reducing per capita availability.[1] The situation is compounded by untreated industrial effluents, agricultural runoff carrying pesticides, and municipal wastewater discharged into natural water bodies all of which hasten the deterioration of water quality.[2] Spatial and temporal variability in water distribution further complicates effective management in developing regions with infrastructure and governance limitations.[3]

Persistent organic pollutants (POPs) are some of the most alarming contaminants. They belong to a group of synthetic compounds known for their extreme stability, resistance to breakdown, and ability to accumulate in living organisms across different levels of the food chain. Halogenated aromatic structures make them highly stable, as the carbon–halogen bonds resist both hydrolysis and microbial degradation. This allows these pollutants to remain in aquatic and terrestrial environments for decades.[4] Many POPs are lipophilic and semi-volatile; hence, they accumulate in fatty tissues and magnify through food chains, leading to higher exposure risks for apex organisms, including humans. The global awareness regarding ecological as well as health risks posed by POPs brought regulatory actions under frameworks like the Stockholm Convention. However, even with regulatory controls, legacy contamination and ongoing emissions continue to create severe environmental challenges, mostly in water systems where removal is technically complex. Adsorption methods are widely used but create secondary waste streams that need management and may be released again under different environmental conditions. Membrane systems like reverse osmosis provide physical separation but are associated with high energy consumption and membrane fouling problems and generate concentrated brines with high pollutant loads.[5] Figure 12.1

DOI: 10.1201/9781003741626-12

FIGURE 12.1 Example structures and sources of the main POPs are pesticides, combustion byproducts, and industrial chemicals.[57]

presents representative molecular structures for major classes of POPs, including organochlorine pesticides, industrial chemicals, and combustion-derived byproducts. The prevalence of halogenated aromatic rings and multiple carbon–halogen bonds is indicative of structural factors contributing toward their high thermodynamic stability against degradation as well as long-term persistence in the environment.

12.2 FUNDAMENTALS OF NANO-PHOTOCATALYSIS

12.2.1 Basic Mechanism of Nano-Photocatalysis

The basic principle of photocatalysis is based on light interaction with semiconductor materials. The electronic structure, especially the band-gap energy (E_g), plays a key role in determining the ability of a catalyst to absorb photons and effect redox reactions. The band-gap energy and thermodynamic alignment of band edges with respect to redox potentials necessary for reactive oxygen species (ROS) formation dictates the ability of a photocatalyst to produce ROS, as shown in Figure 12.2, where the photon energy (hν) excites electrons from the valence band (VB) to the conduction band (CB) to produce electron–hole pairs. These charge carriers take part in redox reactions that create reactive species and eventually degrade pollutants into benign products like CO_2 and H_2O. The CB of the catalyst must be positioned sufficiently negatively relative to $O_2/{\bullet}O_2^-$, which means it should facilitate superoxide generation; at the same time, the VB must be sufficiently positive in order to ensure hydroxyl radical formation. Therefore, appropriate band structure design of photocatalysts is very important for optimizing redox activity.[6]

Even though degradation based on ROS has a bright future, recombination between electron and hole is still an important problem since it directly competes with redox reactions. There are many strategies that have been used to increase the lifetime of excited charges such as band-gap engineering, defect modulation, and heterojunction formation.[7] Good performance in photocatalysis depends on light absorption, mobility of charge carriers, and prevention against recombination all at once. The basic mechanism starts from absorption by

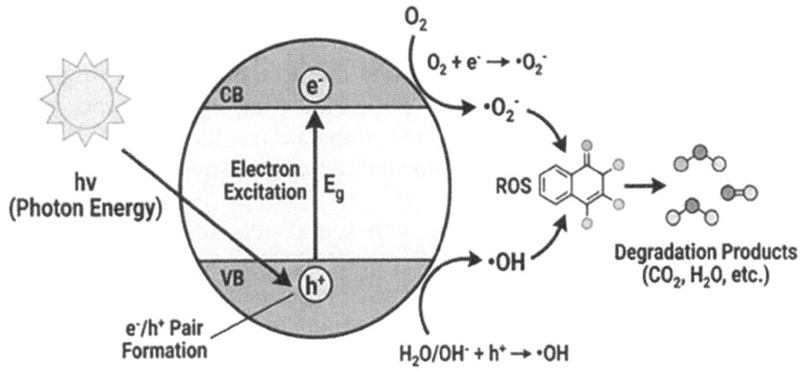

FIGURE 12.2 Formation of reactive oxygen species (ROS) (•OH and $\bullet O_2^-$) from photoexcited electrons and holes during the photocatalytic process.

photons going through steps until ROS is formed; this knowledge can help in creating better catalysts for future generations that will respond more readily to solar irradiation and degrade stubborn pollutants effectively.

12.3 CLASSIFICATION OF NANOCATALYSTS

12.3.1 Metal Oxide Nanocatalysts: TiO_2 and ZnO

Metal oxides are still the most studied and used photocatalysts for environmental remediation, with titanium dioxide (TiO_2) and zinc oxide (ZnO) leading the field due to their suitable electronic structures, chemical stability, and low-cost synthesis methods, with TiO_2 (anatase) and ZnO (wurtzite) nanocatalysts regarding crystal structure, band positions, and photocatalytic pathways. In addition, TiO_2 is chemically stable over a wide pH range, resistant to photocorrosion, and non-toxic; these factors make it popular in laboratory-scale and pilot-scale photocatalytic applications.[8,9] These inherent properties have made TiO_2 an industry standard photocatalyst against which new materials are usually compared.

Nanoscale engineering of TiO_2 and ZnO greatly enhances their surface-area-to-volume ratios and optimizes charge separation dynamics. Nanosized TiO_2 helps in reducing the electron–hole recombination by providing shorter transport pathways plus more surface exposure, whereas doped or defect-engineered ZnO nanoparticles can be improved with regard to visible-light absorption, which is the common limitation among UV-dependent photocatalysts. Moreover, tuning band structures through metal and non-metal doping enables these oxides to harness solar light even more efficiently, which is an essential criterion for environmental applications demanding cost-effective sustainable water treatment.[10] Hence, the compatibility of TiO_2 and ZnO with heterojunctions as well as carbon-based materials increases even more the potential for new generations of photocatalyst designs involving integrated systems meant for efficient degradation of POPs.

12.3.2 Carbon-Based Nanostructures: Graphene, CNTs, and $g\text{-}C_3N_4$

Carbon-based nanostructures like graphene, CNTs, and $g\text{-}C_3N_4$ are regarded as some of the most versatile and highly performing materials in photocatalytic systems due to their excellent conductivity, tunable surface chemistry, chemical stability, and compatibility with different semiconductor photocatalysts. Within this group of compounds, graphene, carbon nanotubes (CNTs), and graphitic carbon nitride (g-C_3N_4) have attracted considerable attention due to their peculiar properties that enhance charge mobility, reducing recombination losses while extending light absorption ranges. The combination of metal oxides such as TiO_2 or ZnO and graphene results in the formation of conductive networks which enhance the efficiency of photocatalytic degradation through improved interfacial charge transport.[11]

CNTs have been effectively used together with semiconductors to reinforce π–π interactions between pollutant molecules adsorbed on the surface of CNTs, improving adsorption as well as enhancing separation efficiency for photogenerated charges. These mechanisms greatly boost degradation performance in water-treatment applications.[12] They are also suitable for surface functionalization, such as carboxylation and nitrogen doping. This provides adjustable pathways for optimizing charge transfer and extending photocatalytic response into the visible-light region. Among carbon-based materials, g-C_3N_4 is particularly interesting because it is a metal-free photocatalyst that can absorb visible light; thus, it has become an important material in the new era of green, sustainable photocatalysis. The moderate band gap of ~2.7 eV allows g-C_3N_4 to produce photogenerated charge carriers under visible illumination unlike classical metal oxides. Also, due to its layered structure and high nitrogen content, there are many lone-pair electrons that can help with surface activation as well as pollutant adsorption.[13]

12.3.3 Nanocomposites and Heterojunctions

Nanocomposites and designed heterojunctions have become effective strategies to solve the problems of short charge-carrier lifetimes and low visible-light absorption in single-component photocatalysts. By combining semiconductors with complementary band structures, these systems promote directional charge separation, enhance redox capability, and extend optical response into the visible region. The resulting interfacial interactions generate synergistic effects that significantly improve photocatalytic performance. Figure 12.3 compares Z-scheme and S-scheme heterojunction mechanisms, showing different charge-transfer pathways between two photocatalysts under light irradiation for both strategies that improve charge separation and redox capability while suppressing recombination to enhance overall performance in photocatalysis.

Traditional type II heterojunctions are characterized by the migration of electrons to the semiconductor with lower conduction-band energy, while holes accumulate on the higher valence-band counterpart. Although this arrangement aims at promoting spatial charge separation, it often reduces redox potential because the electrons and holes are now found in bands with less reducing and oxidizing power. To handle this shortcoming, Z-scheme heterojunctions were formed, mimicking natural photosynthesis by letting low-energy electrons and holes recombine while preserving highly oxidative holes and strongly reductive electrons in their own semiconductors. This method retained powerful redox capacity while at the same time suppressing recombination efficiently, making Z-scheme architectures highly effective in degrading POPs under solar irradiation.[14]

S-scheme heterojunctions have become a recent development of Z-scheme systems. Built-in electric fields and bending and selective Coulombic interactions at the S-scheme interface render highly directional charge transport with reduced recombination, without having to resort to external electron mediators. Such systems allow selective transfer of electrons in the semiconductor with a low Fermi level to the catalyst with a higher Fermi level while retaining high-energy charged particles possessing strong oxidation and reduction potential. Due to improved charge separation and stronger redox reaction pathways under visible-light irradiation, S-scheme heterojunctions thereby exhibit enhanced photocatalytic performance.[15] By enhancing charge separation, redox potential, and light absorption, Z-scheme and S-scheme systems bring about effective degradation of pollutants that resist conventional oxidative treatments.

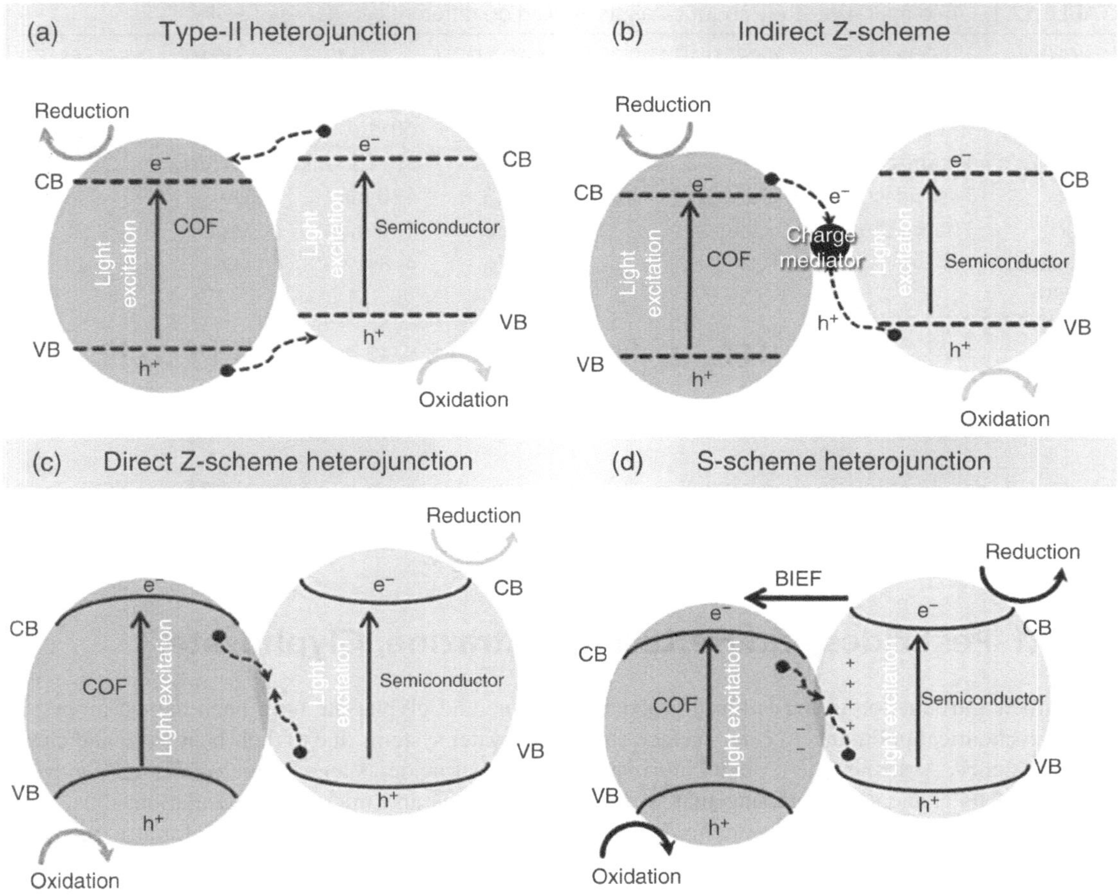

FIGURE 12.3 Schematic representation of Type-II, Z-scheme, and S-scheme heterojunction charge-transfer pathways (reproduced with permission from Ref. 58).

12.4 DEGRADATION OF SPECIFIC POPS

The photocatalytic degradation of POPs has gained widespread attention due to its ability to mineralize highly stable contaminants that resist conventional water-treatment methods and the photocatalytic degradation mechanism of POPs under light irradiation, which generates electron–hole pairs that drive surface redox reactions to form reactive species that degrade pollutants into intermediate products and mineralize them eventually to CO_2 and H_2O. Unlike physical or biological processes, photocatalysis generates highly ROS, including hydroxyl and superoxide radicals, which can non-selectively oxidize C–Cl and C–N bonds and aromatic rings typical of POP structures. This makes it particularly effective for the breakdown of pesticides, pharmaceuticals, dyes, and per- and polyfluoroalkyl substances (PFAS), many of which persist in aquatic systems due to their chemical stability, low biodegradability, and tendency for bioaccumulation.[16] Table 12.1 summarizes the major classes of POPs and their relevant structural features affecting photocatalytic degradation. It describes how halogenation, aromaticity, and functional groups influence reactivity and mineralization efficiency.

Table 12.1 presents a comparative study of catalysts applied on different categories of POPs; from this table it can be deduced that degradation efficiency is highly dependent on the structure of the pollutant. In general, removal efficiencies for dyes and pharmaceuticals are high (>80%) because they possess

TABLE 12.1 A comparative study about catalysts applied on different categories of POPs

CATALYST SYSTEM	*POP TYPE*	*REPRESENTATIVE POLLUTANT*	*DEGRADATION EFFICIENCY (%)*	*TIME*	*LIGHT SOURCE*	*REFERENCE*
TiO_2 (P25)	Industrial dyes	Methylene Blue	90–98	60–120 min	UV	17
TiO_2	Pharmaceuticals	Ciprofloxacin	75–92	90–150 min	UV/Vis	18
ZnO	Pesticides	Atrazine	65–85	120 min	UV	19
g-C_3N_4	Pharmaceuticals	Diclofenac	70–88	120 min	Visible	20
TiO_2-based hetero-junction	PFAS	PFOA	40–75	4–8 h	UV	21
Modified composite catalyst	PFAS	PFOS	55–80	6–10 h	UV/Vis	22

chromophoric and aromatic rings, which make them susceptible to oxidative attack. Among POP classes, PFAS are the most recalcitrant due to their very high bond dissociation energies. However, dyes and pharmaceuticals are more easily oxidized by photocatalysts.

12.4.1 Pesticides and Herbicides (Atrazine, Glyphosate)

Since their introduction in agricultural practices, atrazine and glyphosate have been named among the first agrochemical pollutants to enter surface and groundwater systems due to their heavy use and chemical persistence. Atrazine, one type of chlorinated triazine herbicide, is an extremely resistant compound because of its stable s-triazine ring that withstands hydrolysis and microbial degradation; thus, it can persist for a long time in aquatic systems.[23] Nano-TiO_2- and ZnO-based photocatalysts for atrazine degradation are highly effective due to their excellent oxidative potential and the ability to generate substantial amounts of hydroxyl radicals, which are required to cleave C–N bonds.

Glyphosate poses a different threat because of its phosphonate group, which is known to be highly resistant to degradation and strongly chelates metal ions. Active ROS-driven photocatalysis would be directed toward oxidative attacks on the C–P bond and the carboxylate group, leading to mineralization through intermediates like aminomethylphosphonic acid (AMPA). Modified photocatalysts with dopants or carbonaceous supports are more effective at glyphosate degradation because of improved adsorption interactions and visible activation.[24] The development of heterojunctions requires the formation of TiO_2/g-C_3N_4 or ZnO/graphene, which has been confirmed to significantly improve electron–hole separation and increase degradation rates.

Hence, photocatalytic treatments of pesticides and herbicides are one of the key remedies to an environmental catastrophe, providing a means of degrading highly resistant chemical structures. Nanostructured photocatalysts, especially doped or composite systems, always perform better than their bulk versions because they have better charge mobility and interaction with pesticide molecules at the catalyst surface.[25]

12.4.2 Pharmaceuticals (Antibiotics and NSAIDs)

Pharmaceutical contaminants, especially antibiotics and non-steroidal anti-inflammatory drugs (NSAIDs), have emerged as potential pollutants in aquatic environments due to high medical usage, poor metabolism in both animals and humans, and persistent effects during conventional wastewater treatment. Among antibiotics such as ciprofloxacin, tetracycline, and sulfamethoxazole, little removal generally occurs at municipal treatment plants because of their stable heterocyclic structures and resistance to biodegradation. In that perspective, photocatalysis indeed proved itself to be quite effective in degrading these compounds because it cleaves the important functional groups obtained within structures such as amine, carboxyl, and fluorinated moieties via ROS-driven cleavage, followed by rapid breakdown into smaller intermediates and eventually mineralization.[26] Along with ibuprofen, naproxen, and diclofenac, NSAIDs concern not only surface waters but also groundwater and even drinking water due to the incomplete removal from secondary and tertiary treatment. Among other compounds, diclofenac is particularly known to be quite recalcitrant because of its dichlorinated aromatic structure and poor biodegradability. Photocatalytic degradation effectively cleaves these aromatic rings with the addition of a hydroxyl radical, ring hydroxylation, dechlorination, and decarboxylation to become increasingly simpler intermediates more prone to mineralization.[27] Rapid radical generation is facilitated by these materials compared with the traditional oxidation or biological processes, leading to faster and more complete NSAID breakdown.

Figure 12.4 shows the pathways for photocatalytic degradation of ciprofloxacin under UV/Vis light by semiconductor catalysts like ZnO@NiO. It depicts radical-driven oxidation processes leading through structural transformation, formation of intermediates, and finally mineralization into CO_2, H_2O, and inorganic ions.

12.4.3 Industrial Dyes (Methylene Blue, Rhodamine B)

Methylene blue and rhodamine B constitute one of the most common classes of industrial pollutants discharged during textile dyeing, printing, leather processing, and the cosmetics industry. These dyes remain in the aqueous environment because of their strong chromophores, high water solubility, and resistance to biodegradation. These dyes limit light penetration in water and thus hinder aquatic photosynthesis, while simultaneously posing toxic, mutagenic, or carcinogenic threats to aquatic organisms and humans. Conventional treatment methods have a much more limited degree of success against these dyes due to intrinsic stability imparted by the aromatic ring system and conjugated structure. Photocatalytic degradation has therefore emerged as one of the most efficient oxidation-based technologies for dye removal that allows the breakdown of the complex chromophores into colorless and non-toxic mineralized products.[28]

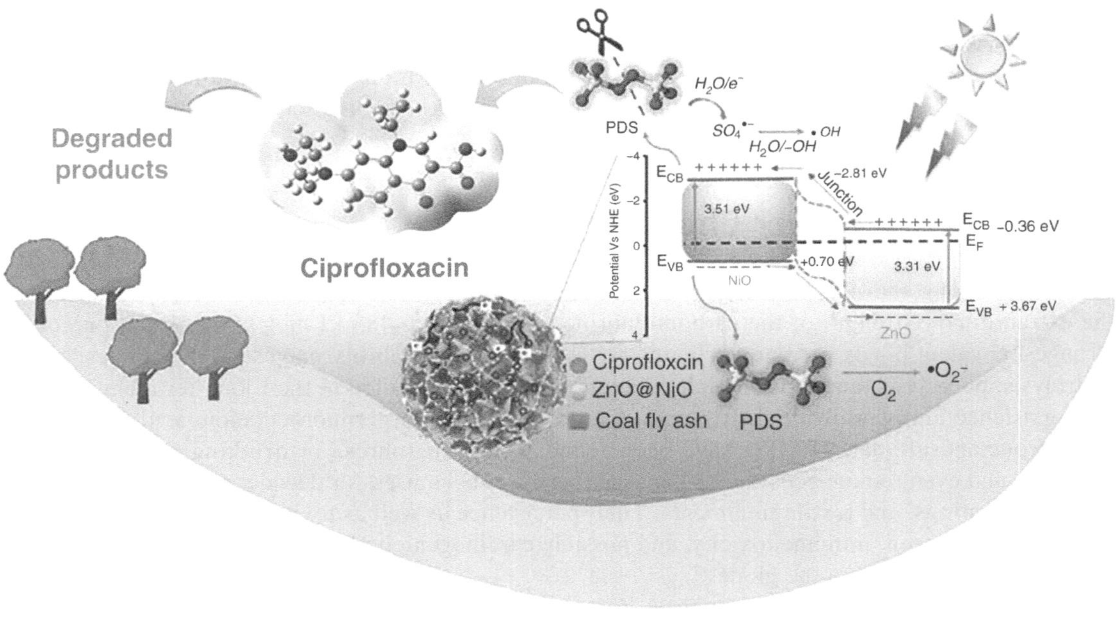

FIGURE 12.4 Representative photocatalytic degradation pathways of antibiotics (e.g., ciprofloxacin) (reproduced with permission from Ref. 59).

TABLE 12.2 Typical photocatalytic degradation pathways and main reactive species involved for different POP categories

POP CATEGORY	*PRIMARY ROS INVOLVED*	*MAIN DEGRADATION PATHWAYS*	*FINAL PRODUCTS*
Pesticides/ Herbicides	•OH, $•O_2^-$	Dealkylation, triazine-ring opening, C-P cleavage	Smaller organics, CO_2, H_2O
Pharmaceuticals	•OH, h^+	Aromatic hydroxylation, dechlorination, ring cleavage	Low-molecular-weight acids, CO_2, H_2O
Industrial Dyes	•OH, $•O_2^-$	N-deethylation, chromophore cleavage, oxidation of intermediates	Colorless species, CO_2, H_2O
PFAS	e^-, $SO_4•^-$ (in advanced systems)	Reductive/oxidative defluorination, chain-shortening	Fluoride ions, CO_2

The photocatalytic degradation of MB normally starts with an oxidative attack on its heterocyclic aromatic structure followed by demethylation, cleavage of chromophores, and then oxidation of intermediate fragments generated during the process. Hydroxyl radicals (•OH) are especially significant because they disrupt the delocalized π-electron system responsible for the dye's deep blue color. Modified TiO_2 and ZnO nanocatalysts, especially doped with metal or non-metal ions, were found to exhibit a commendable ability to degrade MB under UV and visible light, owing to enhanced charge separation and active-site availability. Carbon-based nanocomposites assimilated with TiO_2 also showcase better MB degradation owing to pronounced electron transport and pollutant adsorption. Table 12.2 describes typical pathways for photocatalytic degradation applicable to various categories of POPs. It also indicates major reactive species involved and transformation mechanisms that are primarily responsible for the breakdown and mineralization of pollutants.[29]

Dye degradation has great environmental significance that stretches beyond color removal; total photocatalytic mineralization circumvents the formation of toxic aromatic byproducts that might result from the partial oxidation process. Hence, the new types of nanocomposite photocatalysts would be most suitable for real wastewater treatment applications because they possess a combination of high efficiency, strong oxidative power, and good operational stability under solar or visible-light conditions. The photocatalytic processes seem to offer a promising route toward sustainable and efficient purification of dye-laden wastewater as the industrial effluents keep on discharging dyes into the water systems.

12.4.4 PFAS (Forever Chemicals)

PFAS are among the most chemically stable and environmentally persistent contaminants due to the extraordinary strength of the carbon–fluorine (C–F) bond (~485 kJ mol^{-1}); making it one of the strongest covalent bonds in organic chemistry. This structural stability makes PFAS very resistant to hydrolysis, photolysis, and microbial degradation, which has resulted in their long-term environmental persistence and worldwide distribution. Compounds such as perfluorooctanoic acid (PFOA) and perfluorooctanesulfonate (PFOS) have been found widely in sources of drinking water, industrial effluents, and even remote ecosystems because they have been used for decades in firefighting foams, non-stick coatings, and textile treatments. Their persistence as well as toxicological risks that include endocrine disruption, immune toxicity, and possible carcinogenicity have raised regulatory flags with remediation efforts across the globe.[30]

Quantitative degradation performance: Reported photocatalytic degradation efficiencies for PFAS are significantly lower than those observed for dyes and pharmaceuticals. For example: TiO_2-based systems under UV irradiation typically achieve 40–70% PFOA degradation within 4–8 h with apparent

TABLE 12.3 Flowchart-style representation of photocatalytic degradation

S.NO	*STAGE*	*FLOWCHART REPRESENTATION*	*DESCRIPTION*
1	Reaction kinetics initiation	Pollutant adsorption → catalyst surface (Langmuir–Hinshelwood adsorption step)	POP molecules first adsorb onto the photocatalyst surface; this is typically the rate-limiting step in the Langmuir–Hinshelwood kinetic model
2	Photocatalyst activation	Light irradiation → e^-/h^+ pair generation	Photons excite electrons from the valence band to the conduction band, producing electron–hole pairs necessary for ROS formation
3	ROS formation	e^- → $\bullet O_2^-$* *h^+ → •OH / h^+ oxidation	Electrons reduce dissolved oxygen to superoxide radicals; holes oxidize water or OH^- to form hydroxyl radicals—the primary oxidants in photocatalysis
4	Primary degradation pathways	*•OH / $\bullet O_2^-$ attack → bond cleavage* (aromatic ring opening, dechlorination, demethylation, C–P bond cleavage)	ROS initiate structural breakdown of POPs by oxidizing aromatic rings, cleaving side chains, removing halogens, and opening stable ring systems
5	Intermediate formation	Fragmented organics → smaller molecules	Highly persistent molecules form short-chain acids, aldehydes, amines, or other intermediates before full mineralization
6	Mineralization	Intermediates → CO_2 + H_2O + inorganic ions	Final oxidative steps convert intermediates into harmless mineralized products such as CO_2, water, fluoride ions (PFAS), nitrate, or sulfate

pseudo-first-order rate constants (k_app) in the range of 0.003–0.010 min^{-1}.[31,32] Modified heterojunction systems and carbon-assisted composites have demonstrated improved efficiencies of 55%–80% for PFOS under optimized conditions with rate constants up to 0.015 min^{-1}. Complete mineralization is, however, seldom attained at short reaction times, with fluoride release efficiencies often trailing behind apparent degradation percentages.[33] Table 12.3 provides an overview in flowchart style about the photocatalytic degradation process by presenting sequential steps from light activation to mineralization of pollutants. This visual presentation summarizes effectively all major stages involved in charge generation, formation of reactive species, transformation of intermediates, and final product formations. Hybrid systems integrating photocatalysis with electrochemical oxidation persulfate activation or plasma processes have further improved degradation efficiencies, reporting mineralization improvements of 20–30% compared to standalone photocatalysis. These multiphase strategies partially overcome kinetic barriers imposed by strong C–F bonds.

12.5 FACTORS INFLUENCING PHOTOCATALYTIC EFFICIENCY

The degradation of POPs through photocatalysis is significantly influenced by operational parameters such as pH, catalyst dosage, light intensity, initial concentration of the pollutant, and temperature. These parameters control the dynamics of charge carriers, utilization of light, and interaction between the pollutant and catalyst that finally defines the degradation efficiency. Therefore, it is important to optimize these parameters for any laboratory performance to be applicable in a scalable and reliable treatment system.

12.5.1 Operational Parameters Influencing Photocatalytic Efficiency

Photocatalytic degradation efficiency is highly dependent on operational conditions under which pollutants can interact with catalyst surfaces, charge carriers can behave, and how ROS are generated. Among them, pH, catalyst dosage, light intensity, initial pollutant concentration, and temperature are the most influential parameters controlling surface chemistry, adsorption equilibria, and reaction kinetics in heterogeneous photocatalytic systems.

12.5.1.1 pH

The pH of the reaction medium plays a pivotal role in the photocatalytic performance of a catalyst as it affects both the surface charge of the catalyst and the ionization state of the pollutants. As a result, one can predict that the changes will decide whether electrostatic interactions favor adsorption or hinder it to initiate a photocatalytic reaction. It is clear that pH undoubtedly affects the degradation efficiency by shifting the speciation of pollutants and causing effects in the control of hydroxyl radical generation. In addition, oxidation by VB holes and the formation of ROS differed under acidic and alkaline conditions due to pH.[34]

12.5.1.2 Catalyst Dosage

The dosage of catalyst will determine the number of active sites available toward pollutant adsorption as well as radical formation. Catalytic loadings typically enhance degradation up to an optimum beyond which higher dosages cause scattering of light and particle agglomeration, thus reducing efficiency. Therefore, it is necessary to optimize the amount of catalyst since too much can block photon penetration and lower overall photocatalytic efficiency.[35] Also, application of the same research on TiO_2–ZnO–g-C_3N_4 systems indicated that catalyst loading has direct influence on adsorption behavior and reaction rates from within Langmuir–Hinshelwood kinetics.[36]

12.5.1.3 Light Intensity

Degradation, at low intensities, increases linearly with light intensity, while too-high light intensities favor recombination and, therefore, reduce the productive yield of ROS. It has been shown that UV-active semiconductors require irradiation at lower wavelengths, while catalysts that effectively utilize visible light are doped or heterojunction catalysts.[37] Deviations in solar irradiance proved to greatly impact dye, pharmaceutical, and PFAS precursor degradation, signifying an important factor in reliable performance; stable irradiation conditions.[38]

12.5.1.4 Initial Pollutant Concentration

The superimposed concentration profiles of pollutants will affect adsorption equilibria and the availability of ROS. At low concentrations, degradation generally follows pseudo-first-order kinetics due to a surplus of active sites; however, higher concentrations cause exhaustion on the catalyst surface, making the kinetics tend toward zero-order behavior. This saturation effect has been greatly reported in the degradation of herbicides and dyes, where increasing concentrations slow down the reaction rates.[39] Pollutants with intense colors, such as methylene blue or rhodamine dyes, also absorb the incident light, thereby inhibiting the penetration of photons and further reducing their photocatalytic efficiency.[40]

12.5.1.5 Temperature

Photocatalysis depends on temperature in such a way that it controls mobility of molecules, adsorption–desorption equilibrium, and recombination of charge carriers. Generally, small temperature increases bring about higher reaction rates, while big temperature increases create situations where adsorption becomes unstable and recombination accelerates and, in the long run, efficiency during degradation decreases. Reports show that, on most occasions, near ambient temperature gave the best performance in semiconductor photocatalysis.[41] Therefore, temperature management will be crucial for consistent photocatalytic results.

12.5.2 Reaction Kinetics: Langmuir–Hinshelwood Model

The kinetics of photocatalytic degradation must be understood in order to predict pollutant removal efficiency and develop scaled-up treatment systems. The Langmuir–Hinshelwood (L–H) model is perhaps the most commonly used for describing heterogeneous photocatalysis, combining surface adsorption and reaction rate dynamics into a single kinetic expression. According to the Langmuir–Hinshelwood model, degradation routes undergo different steps: (i) pollutant adsorption onto the surface of the catalyst; (ii) either oxidative or reductive reaction with photogenerated species; and finally, (iii) desorption of intermediates and final products. It thus assumes that the adsorption mechanism is the first step in the reaction pathway, and for most cases, surface reactions become the rate-limiting step. The Langmuir–Hinshelwood kinetic model for photocatalysis is shown with three main steps—adsorption of pollutants on the catalyst surface, a reaction on the surface induced by light, and desorption of final products. This emphasizes how reactive species generated under irradiation degrade adsorbed contaminants into mineralized products such as CO_2 and H_2O.

In terms of mathematical modeling, the Langmuir–Hinshelwood model assumes pseudo-first-order kinetics at low concentrations, as most active sites remain unsaturated, and reaction rate becomes directly proportional to pollutant concentration. At higher concentrations, these active sites become occupied, and the system shifts toward zero-order kinetics, where reaction rate is limited by available oxidative species, rather than pollutant concentration. This concentration-dependent shift is widely observed in dye and pharmaceutical, pesticide, and PFAS precursor degradation, with the distinct feature of different adsorption coefficients due to differing molecular structures and charges[42] is followed by an expression given below,

$$\ln(C_0/C) = k\_app\ t \tag{12.1}$$

where

- C_0 = initial concentration
- C = concentration at time t
- k_{app} = apparent first-order rate constant
- t = irradiation time

In conclusion, the Langmuir–Hinshelwood model provides a strong basis to understand photocatalytic reaction mechanisms and optimize operational conditions. A central and powerful tool to evaluate catalyst performance in general, its applicability to many different classes of pollutants makes the Langmuir–Hinshelwood model highly useful for guiding next-generation photocatalytic system design.

12.5.3 Influence of Real Wastewater Matrix Effects on Photocatalytic Performance

Photocatalytic degradation efficiencies are frequently reported under ideal laboratory conditions with deionized water and single-pollutant systems; however, actual performance in real wastewater matrices is usually drastically lowered. Real wastewater contains natural organic matter (NOM), bicarbonate

(HCO_3^-), carbonate (CO_3^{2-}), chloride, sulfate ions, dissolved metals, and suspended solids which may interfere with photocatalytic processes. Bicarbonate and carbonate ions serve as radical scavengers that decrease oxidative species availability while NOM competes for active sites on the catalyst surface and reduces photon penetration by scattering light and absorbing it.[43]

The matrix effect is even more pronounced for PFAS treatment. The presence of NOM and co-existing anions reduce defluorination efficiency by competing for adsorption sites and scavenging reactive species where the degradation rates are often lowered by more than 30% compared to lab conditions. In addition, turbidity plus suspended solids reduce light penetration depth; thus, photon utilization efficiency in slurry-based systems is directly limited. This indicates that laboratory-scale photocatalytic performance may be an overestimate of real-world treatment efficiency. This necessitates pilot-scale validation using authentic wastewater matrices to assess degradation kinetics accurately as well as catalyst stability and mineralization potential under environmentally relevant conditions.[44] Addressing matrix-induced limitations through catalyst surface engineering, immobilization strategies, and hybrid treatment integration represents a critical step toward practical implementation.

12.6 CHALLENGES AND TECHNICAL BARRIERS

The process of nano-photocatalysis has achieved significant laboratory results but still faces important technical and operational challenges which prevent its application in actual wastewater treatment facilities. The main restriction exists because conventional photocatalysts like TiO_2 and ZnO have low efficiency in using visible light because their broad band gaps permit only UV light activation, which reduces their capacity to harness solar energy.[45] The development of solar activation systems through heterojunction engineering and visible-light-responsive materials has succeeded, but their ability to remain stable while preventing charge recombination shows difficulties during actual operational testing.[46] The need for energy-intensive separation in slurry systems and the risk of mass-transfer problems in immobilized systems both prevent continuous usage because catalyst recovery and long-term reusability situations need to address these two requirements.[47] The use of NOM and suspended solids together with biofilm growth in real-time wastewater treatment leads to fouling, which decreases both photon penetration and catalytic activity, showing the necessity for antifouling reactor designs and hybrid treatment configurations.[48] The production of engineered nanomaterials creates environmental safety issues because they can cause oxidative stress and toxicity while releasing ions into aquatic environments; this situation requires the development of materials which incorporate safety features from their design and through evaluation of their entire life cycle. The high energy consumption together with operational costs of UV-dependent systems creates economic viability problems, thus making solar-driven reactors and energy-efficient process integration vital for UV-dependent systems to operate economically. The process of pollutant degradation needs to reach higher efficiency levels because its current state shows toxic effects from transformation intermediates; therefore, standardized ecotoxicological testing must happen before the government will approve products for actual use in the field.[49]

12.7 FUTURE PERSPECTIVES

12.7.1 Scale-Up and Real-Time Applications

The laboratory results of nano-photocatalysis show excellent performance, but the technology encounters difficulties when it needs to move from laboratory testing to pilot and full-scale water-treatment operations. The majority of research studies work with controlled batch reactors, which do not accurately simulate actual wastewater conditions that feature changing pollutant levels and varying turbidity and complicated water movement patterns. The successful process of scale-up depends on three main elements, which include optimized reactor design and proper light distribution and catalyst material performance over extended periods.[50]

The evidence found that solar-driven reactors which use compound parabolic collectors and immobilized catalysts can successfully function in continuous-flow mode during daylight hours. The requirements for real-world applications include three important elements, which are catalyst stability and methods to recover and regenerate the catalyst and assessments of economic viability. The pilot studies show that the improved systems can achieve sustainable pollutant removal results while using fewer chemicals and energy than their laboratory counterparts, although their operational efficiency falls slightly below laboratory standards.[51,52] The process of taking laboratory achievements into field testing stands as the fundamental requirement for product commercialization (Figure 12.5).

12.7.2 Hybrid Systems and Fouling Control

The practical wastewater treatment solution employs hybrid photocatalysis–membrane systems for its operation. Photocatalytic processes enable the breakdown of persistent pollutants through oxidation, while membrane filtration systems provide both physical pollutant removal and better treatment results.[53] The operation of photonic membrane reactors (PMRs) functions through their ability to in situ degrade organic materials that accumulate on membranes, which enables them to sustain consistent water flow while extending their usage time. Organic matter and suspended solids together with catalyst buildup create major operational issues that hinder real-world operations. The latest nanocomposite and nanofibrous membranes show enhanced antifouling capabilities together with their capacity to endure irradiation, which enables water recycling and decentralized treatment systems.[54]

12.7.3 Green Synthesis and Sustainable Materials

The green synthesis method uses plant extracts to produce photocatalysts, which serve as an environmentally friendly alternative to traditional chemical production methods. The phytochemicals serve as natural agents which reduce and stabilize metal oxide nanomaterials to achieve environmentally

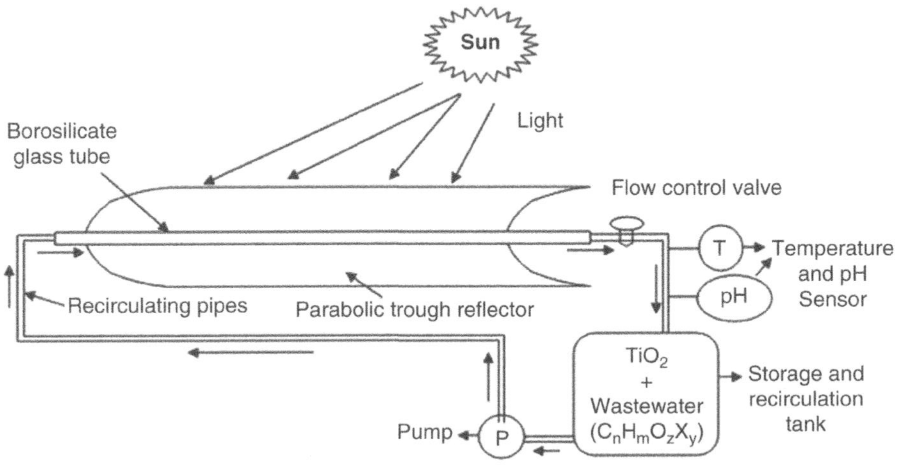

FIGURE 12.5 Schematic of a pilot-scale solar photocatalytic reactor.[60]

friendly synthesis and improved surface reactivity and visible-light activity. The plant-based TiO_2 and ZnO nanoparticles proved effective for dye and pharmaceutical pollutant degradation while producing less secondary toxic effects.[55] The standardization of synthesis procedures combined with mechanistic knowledge about bio-nanointeractions holds the key to solving reproducibility issues and particle size regulation challenges and biological precursor variability problems, which currently affect their practical applications.[56] The combination of green synthesis with solar energy use and hybrid treatment methods establishes an effective route for developing photocatalytic water-treatment systems which achieve both large-scale operation and environmental sustainability.

12.8 CONCLUSION

The use of nano-photocatalysis technology provides an innovative solution to combat POPs, which endanger aquatic ecosystems, water quality, and human health. This chapter demonstrates that standard treatment methods show limitations because POPs can remain in the environment while they resist biodegradation and their toxic effects move through food chains, but nano-photocatalytic systems achieve complete mineralization using their advanced oxidation processes. Pollutant degradation efficiency depends on three essential mechanisms, which include band-gap excitation, charge-carrier generation, and ROS formation. The transition from bulk materials to nanoscale materials together with heterojunction engineering and carbon-based photocatalysts improves the ability of materials to use visible light and separate electric charge. The case studies demonstrated that these systems can effectively degrade pesticides and pharmaceuticals and dyes and PFAS. The operational parameters, which include pH, catalyst dosage, light intensity, temperature, and pollutant concentration, determine the reaction rate, which scientists frequently model using the Langmuir–Hinshelwood equation. The combination of green synthesis methods with hybrid treatment technologies creates a sustainable solution for water purification which is both environmentally friendly and capable of widespread application, despite facing challenges from nanotoxicity and catalyst recovery and recombination losses and large-scale system implementation.

REFERENCES

1. Musie, W., & Gonfa, G. (2023). Fresh water resource, scarcity, water salinity challenges and possible remedies: A review. *Heliyon, 9*(8), e18685. https://doi.org/10.1016/j.heliyon.2023.e18685
2. Tzanakakis, V. A., Paranychianakis, N. V., & Angelakis, A. N. (2020). Water supply and water scarcity. *Water, 12*(9), Article 2347. MDPI. https://doi.org/10.3390/w12092347
3. Liu, J., Yang, H., Gosling, S. N., Kummu, M., Flörke, M., Pfister, S., Hanasaki, N., Wada, Y., Zhang, X., Zheng, C., Alcamo, J., & Oki, T. (2017). Water scarcity assessments in the past, present, and future. *Earth's Future, 5*(6), 545–559. Wiley. https://doi.org/10.1002/2016EF000518
4. Negrete-Bolagay, D., Sierra, M. A., Tarras-Wahlberg, N., & Armijos, E. (2021). Persistent organic pollutants (POPs) in aquatic environments: Sources, distribution, and ecological risks. *Journal of Environmental Management, 299*, 113659. https://doi.org/10.1016/j.jenvman.2021.113659
5. Ren, J., Wang, X., Wang, C., Gong, P., Wang, X., & Yao, T. (2017). Biomagnification of persistent organic pollutants along a high-altitude aquatic food chain in the Tibetan Plateau: Processes and mechanisms. *Environmental Pollution, 220*(Part A), 636–643. https://doi.org/10.1016/j.envpol.2016.10.019
6. Wang, L., Cai, M., Sun, W., He, L., & Zhang, X. (2018). Promoting charge separation in semiconductor nanocrystal superstructures for enhanced photocatalytic activity. *Advanced Materials Interfaces, 5*(10), 1701694. https://doi.org/10.1002/admi.201701694
7. Muiruri, J. K., Ye, E., Zhu, Q., Loh, X. J., & Li, Z. (2022). Recent advance in nanostructured materials innovation towards photocatalytic CO_2 reduction. *Applied Catalysis A: General, 648*, 118927. https://doi.org/10.1016/j.apcata.2022.118927

8. Tinoco Navarro, L. K., & Cíhalová, J. (2023). Enhancing photocatalytic properties of TiO_2 photocatalyst and heterojunctions: A comprehensive review of the impact of biphasic systems in aerogels and xerogels synthesis, methods, and mechanisms for environmental applications. *Gels, 9*(12), 976. https://doi.org/10.3390/gels9120976
9. Xing, Z., Zhang, J., Cui, J., Yin, J., Zhao, T., Kuang, J., Xiu, Z., Wan, N., & Zhou, W. (2018). Recent advances in floating TiO_2-based photocatalysts for environmental application. *Applied Catalysis B: Environmental, 225*, 452–467. https://doi.org/10.1016/j.apcatb.2017.12.005
10. Li, J., & Chu, D. (2018). Energy band engineering of metal oxide for enhanced visible light absorption. In Z. Lin, M. Ye, & M. Wang (Eds.), *Semiconductor photocatalysis: Materials, mechanisms and applications* (pp. 49–78). Elsevier. https://doi.org/10.1016/B978-0-08-101977-1.00005-3
11. Karthik, V., Selvakumar, P., Senthil Kumar, P., Vo, D. V. N., & Nguyen, V. H. (2021). Graphene-based materials for environmental applications: A review. *Environmental Chemistry Letters, 19*, 3631–3644. https://doi.org/10.1007/s10311-021-01262-3
12. Aseena, S., Abraham, N., & Suresh Babu, V. (2020). Synthesis of CNT-based nanocomposites and their application as photoanode material for improved efficiency in DSSC. *Ceramics International, 46*(18, Part A), 28355–28362. https://doi.org/10.1016/j.ceramint.2020.07.339
13. Saraswat, V., Kohli, S., Sinha, N. K., Purohit, G., Sandhu, N., Swami, S., Rathee, G., Rawat, M., & Kumari, N. (2025). Metal-free graphitic carbon nitride as a promising nano-adsorbent for the efficient photocatalytic degradation of toxic dyes: Recent trends and future perspectives. *Chemical Engineering Research and Design, 221*, 240–275. https://doi.org/10.1016/j.cherd.2025.08.016
14. Low, J., Yu, J., Jaroniec, M., Wageh, S., & Al-Ghamdi, A. A. (2017). Heterojunction photocatalysts. *Advanced Materials, 29*(20), 1601694. https://doi.org/10.1002/adma.201601694
15. Zhang, Y., Zhang, M., Luo, C., Li, Y., Zhang, X., & Zhang, L. (2024). S-scheme heterojunction efficient extraction of hot carriers in $CsPbBr_3/Bi_4O_5I_2$ for enhanced photocatalytic H_2 evolution and CO_2 reduction. *ACS Applied Energy Materials*. https://doi.org/10.1021/acsaem.4c02419
16. Rasool, B. S., Abbas, A. K., & Haddad, R. (2025). Photocatalytic innovations in environmental remediation: Mechanisms, materials, and challenges for persistent organic pollutant removal. *Environmental Monitoring and Assessment, 197*, 1086. https://doi.org/10.1007/s10661-025-14531-3
17. Zhang, F., Li, X., Duan, L., Zhang, H., Gu, W., Yang, X., Li, J., He, S., Yu, J., & Ren, M. (2021). Effect of different DOM components on arsenate complexation in natural water. *Environmental Pollution, 270*, 116221. https://doi.org/10.1016/j.envpol.2020.116221
18. Hadjer, F., Nacera, Y. M., Chafia, B., Meriem, K., & Zohra, B. (2025). Optimization of solar-driven photocatalytic degradation of metformin in wastewater using TiO_2/Ca-alginate composites: A response surface methodology for study of parameter interactions and ROS generation. *Ozone: Science & Engineering, 47*(6), 617–636. https://doi.org/10.1080/01919512.2025.2538564
19. Ahmad, F., Tahir, S., Wali, A., Khan, M. I., & Shanableh, A. (2025). A review article on the photocatalytic degradation of atrazine by potential catalysts. *Next Materials, 8*, 100534. https://doi.org/10.1016/j.nxmate.2025.100534
20. Liu, W., Li, Y., Liu, F., Jiang, W., Zhang, D., & Liang, J. (2019). Visible-light-driven photocatalytic degradation of diclofenac by carbon quantum dots modified porous g-C_3N_4: Mechanisms, degradation pathway and DFT calculation. *Water Research, 151*, 8–19. https://doi.org/10.1016/j.watres.2018.11.084
21. Kasalica, K., Stojadinović, S., Lješević, M., Ivanov, P., Yamamoto, A., Nakano, T., Radulović, J., Slavković Beškoski, L., Radić, N., & Beškoski, V. (2025). Photocatalytic degradation of PFOA over rGO-doped TiO_2 coatings formed by plasma electrolytic oxidation. *Journal of Environmental Chemical Engineering, 13*(5), 117452. https://doi.org/10.1016/j.jece.2025.117452
22. Leonello, D., Fendrich, M. A., Parrino, F., Patel, N., Orlandi, M., & Miotello, A. (2021). Light-induced advanced oxidation processes as PFAS remediation methods: A review. *Applied Sciences, 11*(18), 8458. https://doi.org/10.3390/app11188458
23. Daramola, I. O., Ojemaye, M. O., Okoh, A. I. et al. (2025). α-Fe_2O_3 co-precipitation synthesis: Evaluating the role of hydrogen peroxide as oxidizing agent in the photocatalytic degradation of atrazine. *Discover Chemistry, 2*, 193. https://doi.org/10.1007/s44371-025-00269-x
24. Cao, L., Ma, D., Zhou, Z., Xu, C., Cao, C., Zhao, P., & Huang, Q. (2019). Efficient photocatalytic degradation of herbicide glyphosate in water by magnetically separable and recyclable $BiOBr/Fe_3O_4$ nanocomposites under visible light irradiation. *Chemical Engineering Journal, 368*, 212–222. https://doi.org/10.1016/j.cej.2019.02.100
25. Sun, Z., Ou, H., Tan, B., Song, Z., Yan, J., Yan, X., & Xu, Y. (2025). Enhanced photocatalytic degradation of imidacloprid using a peroxymonosulfate-activated all-day-responsive $BaFe_{12}O_{19}$/NiS II-scheme heterojunction. *Journal of Environmental Chemical Engineering, 13*(6), 119863. https://doi.org/10.1016/j.jece.2025.119863

26. He, J., Zhang, Y., Guo, Y., Rhodes, G., Yeom, J., Li, H., & Zhang, W. (2019). Photocatalytic degradation of cephalexin by ZnO nanowires under simulated sunlight: Kinetics, influencing factors, and mechanisms. *Environment International, 132*, 105105. https://doi.org/10.1016/j.envint.2019.105105
27. Méndez-Arriaga, F., Esplugas, S., & Giménez, J. (2008). Photocatalytic degradation of non-steroidal anti-inflammatory drugs with TiO_2 and simulated solar irradiation. *Water Research, 42*(3), 585–594. https://doi.org/10.1016/j.watres.2007.08.002
28. Nayna, O. K., & Tareq, S. M. (2019). Application of semiconductor nanoparticles for removal of organic pollutants or dyes from wastewater. In A. Ahsan & A. F. Ismail (Eds.), *Nanotechnology in water and wastewater treatment* (pp. 267–290). Elsevier. https://doi.org/10.1016/B978-0-12-813902-8.00013-7
29. Khan, I., Saeed, K., Zekker, I., Zhang, B., Hendi, A. H., Ahmad, A., Ahmad, S., Zada, N., Ahmad, H., Shah, L. A., Shah, T., & Khan, I. (2022). Review on methylene blue: Its properties, uses, toxicity and photodegradation. *Water, 14*(2), 242. https://doi.org/10.3390/w14020242
30. Sunderland, E. M., Hu, X. C., Dassuncao, C., Tokranov, A. K., Wagner, C. C., & Allen, J. G. (2019). A review of the pathways of human exposure to poly- and perfluoroalkyl substances (PFASs) and present understanding of health effects. *Journal of Exposure Science & Environmental Epidemiology, 29*(2), 131–147. https://doi.org/10.1038/s41370-018-0094-1
31. Cardoso, I. M. F., Pinto da Silva, L., & Esteves da Silva, J. C. G. (2023). Nanomaterial-based advanced oxidation/reduction processes for the degradation of PFAS. *Nanomaterials, 13*(10), 1668. https://doi.org/10.3390/nano13101668
32. Ali, Z. A., Yamijala, S. R. K. C., & Wong, B. M. (2023). A review of emerging photoinduced degradation methods for per- and polyfluoroalkyl substances in water. *Current Opinion in Chemical Engineering, 41*, 100947. https://doi.org/10.1016/j.coche.2023.100947
33. Guo, Y., Siretanu, I., Mugele, F., Mul, G., & Mei, B. (2024). pH-dependent photocatalytic performance of faceted BiOBr semiconductor particles in degradation of dyes. *Molecular Catalysis, 553*, 113753. https://doi.org/10.1016/j.mcat.2023.113753
34. Shibuya, S., Sekine, Y., & Mikami, I. (2015). Influence of pH and pH adjustment conditions on photocatalytic oxidation of aqueous ammonia under airflow over Pt-loaded TiO_2. *Applied Catalysis A: General, 496*, 73–78. https://doi.org/10.1016/j.apcata.2015.02.024
35. Bazrafshan, E., Mohammadi, L., & Zarei, A. A. (2023). Optimization of the photocatalytic degradation of phenol using superparamagnetic iron oxide (Fe_3O_4) nanoparticles in aqueous solutions. *RSC Advances, 13*(36), 25408–25424. https://doi.org/10.1039/d3ra03612j
36. Molla, M. A. I., Furukawa, M., Tateishi, I., Katsumata, H., & Kaneco, S. (2019). Optimization of alachlor photocatalytic degradation with nano-TiO_2 in water under solar illumination: Reaction pathway and mineralization. *Clean Technologies, 1*(1), 141–153. https://doi.org/10.3390/cleantechnol1010010
37. Podborska, A., Suchecki, M., Mech, K., et al. (2020). Light intensity-induced photocurrent switching effect. *Nature Communications, 11*, 854. https://doi.org/10.1038/s41467-020-14675-5
38. Enesca, A., & Isac, L. (2020). The influence of light irradiation on the photocatalytic degradation of organic pollutants. *Materials, 13*(11), 2494. https://doi.org/10.3390/ma13112494
39. Pirilä, M., Saouabe, M., Ojala, S. et al. (2015). Photocatalytic degradation of organic pollutants in wastewater. *Topics in Catalysis, 58*, 1085–1099. https://doi.org/10.1007/s11244-015-0477-7
40. Wang, H., Cui, H., Lan, Y., Yang, L., Zhao, J., Ye, Y., Zhang, W., Deng, R., Liu, H., Lv, Y., & Yang, H. (2025). Enhanced visible-light photocatalytic degradation of antibiotics using aminoguanidine-functionalized iron-based metal–organic frameworks. *Journal of Colloid and Interface Science, 695*, 137797. https://doi.org/10.1016/j.jcis.2025.137797
41. Koe, W. S., Lee, J. W., Chong, W. C., Pang, Y. L., & Sim, L. C. (2020). An overview of photocatalytic degradation: Photocatalysts, mechanisms, and development of photocatalytic membrane. *Environmental Science and Pollution Research, 27*, 2522–2565. https://doi.org/10.1007/s11356-019-07193-5
42. Purkait, M. K., Gupta, B., & Das, P. P. (2025). Kinetic modeling and rate of reactions for the photocatalytic degradation of pharmaceuticals. In M. K. Purkait, B. Gupta, & P. P. Das (Eds.), *Advances in Chemical Engineering* (Vol. 66, pp. 35–52). Academic Press. https://doi.org/10.1016/bs.ache.2024.12.007
43. Hübner, U., Spahr, S., Lutze, H., Wieland, A., Rüting, S., Gernjak, W., & Wenk, J. (2024). Advanced oxidation processes for water and wastewater treatment – Guidance for systematic future research. *Heliyon, 10*(9), e30402. https://doi.org/10.1016/j.heliyon.2024.e30402
44. Mohamed, M. S., & Reddy, K. R. (2026). Sustainability assessment of alternative methods for destruction of per- and polyfluoroalkyl substances (PFAS) in water. *Remediation Journal, 36*(2), 1–12. https://doi.org/10.1002/rem.70054
45. Chen, X., Shen, S., Guo, L., & Mao, S. S. (2010). Semiconductor-based photocatalytic hydrogen generation. *Chemical Reviews, 110*(11), 6503–6570. https://doi.org/10.1021/cr1001645

46. Han, R., Melo, M. A., Jr., Zhao, Z., Wu, Z., & Osterloh, F. E. (2020) Light intensity dependence of photochemical charge separation in the $BiVO_4$/Ru-$SrTiO_3$:Rh direct contact tandem photocatalyst for overall water splitting (2020). *The Journal of Physical Chemistry C, 124*(18), 9724–9733. https://doi.org/10.1021/acs.jpcc.0c00772
47. Fisher, I., Kohpaei, A. J., & Sathasivan, A. (2010). Comment on "Using Bayesian statistics to estimate the coefficients of a two-component second-order chlorine bulk decay model for a water distribution system" by Huang, J. J., McBean, E. A. *Water Research, 44*(10), 3309–3310. https://doi.org/10.1016/j.watres.2010.01.039
48. Mamba, G., & Mishra, A. (2016). Advances in magnetically separable photocatalysts: Smart, recyclable materials for water pollution mitigation. *Catalysts, 6*(6), 79. https://doi.org/10.3390/catal6060079
49. Backhaus, T., & Faust, M. (2012). Predictive environmental risk assessment of chemical mixtures: A conceptual framework. *Environmental Science & Technology, 46*(5), 2564–2573. https://doi.org/10.1021/es2034125
50. Nagaya, S., Nishikiori, H., Mizusaki, H., Sato, K., Wagata, H., & Teshima, K. (2016). Crystal structure and photoelectric conversion properties of eosin Y-adsorbing ZnO films prepared by electroless deposition. *Applied Catalysis B: Environmental, 189*, 51–55. https://doi.org/10.1016/j.apcatb.2016.02.013
51. Bahadur, N., & Bhargava, N. (2019). Novel pilot scale photocatalytic treatment of textile & dyeing industry wastewater to achieve process water quality and enabling zero liquid discharge. *Journal of Water Process Engineering, 32*, 100934. https://doi.org/10.1016/j.jwpe.2019.100934
52. Mohamadpour, F., & Amani, A. M. (2024). Photocatalytic systems: Reactions, mechanism, and applications. *RSC Advances, 14*, 20609–20645. https://doi.org/10.1039/D4RA03259D
53. Zheng, X., Shen, Z.-P., Shi, L., Cheng, R., & Yuan, D.-H. (2017). Photocatalytic membrane reactors (PMRs) in water treatment: Configurations and influencing factors. *Catalysts, 7*(8), 224. https://doi.org/10.3390/catal7080224
54. Ahuja, T., Brighu, U., & Saxena, K. (2023). Recent advances in photocatalytic materials and their applications for treatment of wastewater: A review. *Journal of Water Process Engineering, 53*, 103759. https://doi.org/10.1016/j.jwpe.2023.103759
55. Mousa, S. A., Wissa, D. A., Hassan, H. H., et al. (2024). Enhanced photocatalytic activity of green synthesized zinc oxide nanoparticles using low-cost plant extracts. *Scientific Reports, 14*, 16713. https://doi.org/10.1038/s41598-024-66975-1
56. Negi, P., Chandra, V Kumar, A., Kumar, P., Priya, Yashmin, K., Kumar, N., Joshi, N. C., Saxena, R., Kumar, A., Dhiman, N., & Rawat, B. S. (2025). Plant-mediated synthesis of metal oxide nanocomposites: Emerging prospects in electrochemical corrosion inhibition and biosensing. *Sustainable Chemistry for Climate Action, 7*, 100107. https://doi.org/10.1016/j.scca.2025.100107
57. Boulkhessaim, S., Gacem, A., Khan, S. H., Amari, A., Yadav, V. K., Harharah, H. N., Elkhaleefa, A. M., Yadav, K. K., Rather, S.-u., Ahn, H.-J., & Jeon, B.-H. (2022). Emerging trends in the remediation of persistent organic pollutants using nanomaterials and related processes: A review. *Nanomaterials, 12*(13), 2148. https://doi.org/10.3390/nano12132148
58. Ahmad, I., Babar, Z. U. D., Zhang, Y., Al-Qattan, A., Ben Ahmed, S., & Li, G. (2025). Unlocking sunlight driven photocatalysis: Synthesis, diversity, and applications of COF-based S-scheme heterojunctions. *Journal of Materials Chemistry A, 13*, 15517–15553. https://doi.org/10.1039/D5TA01232E
59. Labidi, A., Ren, H., Liang, X., Dong, Q., Li, X., Tian, Q., Sial, A., Cui, Y., Kang, H., Liang, J., Zhao, K., Lichtfouse, E., Padervand, M., & Wang, C. (2024). Visible–light–driven photocatalytic degradation of ciprofloxacin antibiotic by novel heterostructured coal fly ash waste: Mechanism insight, toxicity pathway and DFT calculation. *Materials Today Chemistry, 42*, 102388. https://doi.org/10.1016/j.mtchem.2024.102388
60. Singh, C., Chaudhary, R., & Gandhi, K. (2013). Preliminary study on optimization of pH, oxidant and catalyst dose for high COD content: Solar parabolic trough collector. *Iranian Journal of Environmental Health Science & Engineering, 10*, 13. https://doi.org/10.1186/1735-2746-10-13

Artificial Intelligence and Machine Learning in Water Treatment and Quality Assessment

13

Nageswar Sahu, Shelly Verma,
Achlesh Daverey, and Biswanath Mahanty

13.1 INTRODUCTION

Rapid urbanization, intensified agriculture, and climate change have affected the entire biosphere, disrupting the ecosystem services and the environment (Kurniawan et al., 2024). Overexploitation of groundwater, discharge of untreated or improperly treated wastewater, and pollution of surface water have degraded the quality of water resources (Akhtar et al., 2021). Protecting available water, implementing effective wastewater treatment, and monitoring water quality have become more critical than ever.

National and international organizations have recommended different water quality standards based on their suitability for intended applications, such as drinking water, recreational use, or wastewater reuse for agriculture (WHO, 2017). However, mapping numerous water quality measurements poses significant logistical and operational challenges. The water quality index (WQI) has been adopted to convert the raw data into a dimensionless subindex, aggregated (i.e., importance or expert opinion) into a single, interpretable number, facilitating spatial–temporal comparisons of water quality status (Chidiac et al., 2023; Uddin et al., 2021). In recent times, WQI has been used to monitor (Manna and Biswas, 2023) water quality for surface water, groundwater, lakes, and rivers.

In addition to monitoring water quality, the recycling and reuse of treated wastewater are critical for sustainable water resources and for advancing the UN Sustainable Development Goals (SDGs), thereby increasing water security and resilience (Federigi et al., 2024). To protect the receiving environment and maintain acceptable quality for reuse (Srivastava and Singh, 2023), a myriad of physical and chemical methods, i.e., coagulation (Damaraju et al., 2017) or advanced oxidation processes (AOPs) (Zou et al., 2022), combined with biological treatment (anaerobic digestion or activated sludge), are adopted to

DOI: 10.1201/9781003741626-13

improve the removal of recalcitrant compounds. The selection of an industrial wastewater treatment configuration and the performance of an individual treatment unit, as well as overall operation of industrial and municipal wastewater treatment plants (WWTPs), depend on influent characteristics and process complexity (Trianni et al., 2021), which can easily affect operational parameters.

Mechanistic modeling for biological wastewater treatment, e.g., the activated sludge model variants (ASM1, ASM2, and ASM3) from the International Water Association (IWA), enables predictions of water treatment and nutrient removal efficiencies (Gujer et al., 1999). Extension and modification of ASMs considering the removal of xenobiotics (Polesel et al., 2016) or Anaerobic Digestion Model No. 1 (ADM1) have been integrated into WWTP. However, the need for domain-specific expertise, a strict parametric framework, limited flexibility, and estimation of biokinetic parameters could pose challenges (Rossi et al., 2025).

With the advancement of rapid analytical methods, real-time multiparametric measurements, soft sensors, and digitalization, the quality and quantity of data available for process modeling have greatly improved. In this data-intensive environment, the adoption of time-consuming mechanistic models may not always be warranted. Machine learning (ML) models have emerged as a viable alternative for mapping the relationships between predictors and response variables hidden in the dataset, without requiring mechanistic insight or process descriptions.

13.2 WATER AND WASTEWATER TREATMENT SYSTEMS – NEED FOR MACHINE LEARNING

Complex environmental engineering infrastructure in modern water treatment plants and WWTPs is essential to public health, environmental protection, and water sustainability (UN-Habitat and WHO, 2021). These WTP/WWTP are integrated systems that can eliminate a wide range of organic/inorganic contaminants, suspended solids, nutrients, pathogens, heavy metals, and emerging contaminants (Sharma et al., 2022), thereby ensuring reliable access to clean and safe water amid rising wastewater complexity and stricter regulations (Babu Ponnusami et al., 2023). Multiple treatment technologies can function simultaneously within a single treatment train in a WWTP under dynamic environmental and operational conditions (Ghazal et al., 2022). These technologies can be broadly categorized into physical, chemical, biological, or hybrid, each governed by distinct mechanisms and process dynamics (Table 13.1).

13.2.1 Physical Methods

Two widely used physical technologies for treating wastewater are adsorption and membrane separation (Sak et al., 2025). Physical or chemical interactions of adsorbate molecules with the adsorbent surface result in a higher surface concentration than in the bulk (Sukmana et al., 2021). In WWTPs, adsorption enables the efficient removal of contaminants, including heavy metals, organic pollutants, and emerging contaminants, through their attachment onto various adsorbents such as activated carbon, zeolites, biochar, metal–organic frameworks, ion-exchange resins, graphene oxide, and mesoporous nanomaterials (Rathi and Kumar, 2021). Despite its cost-effectiveness and versatility, large-scale application is constrained by limited adsorption capacity and selectivity, material instability, regeneration challenges, economic and regulatory concerns, and the transfer of contaminants from wastewaters to the adsorbent rather than their degradation (Satyam and Patra, 2024). Membrane separation employs semi-permeable barriers to selectively retain contaminants based on properties such as molecular size, charge, and solubility (Sak et al., 2025). The microfiltration, ultrafiltration, and nanofiltration technologies offer high treatment efficiency, operational simplicity, and strong adaptability to fluctuating influent characteristics (Tang et al., 2025). Membrane-based technologies are widely adopted to treat discharge from dyeing and printing,

TABLE 13.1 Comprehensive overview of different physical, chemical, and biological wastewater treatment technologies

CATEGORY	*TREATMENT SYSTEM*	*MODE OF OPERATION*	*DOMINANT MECHANISM*	*KEY PERFORMANCE INDICATORS*	*MAJOR INFLUENTIAL FACTORS*	*REFERENCES*
Physical	Adsorption	Batch/ Continuous	Surface adsorption, ion exchange	Removal efficiency (%), adsorption capacity (mg g^{-1}), and breakthrough time	pH, contact time, adsorbent dose, influent concentration, temperature	(Chinedu and Chukwudum, 2025; Sukmana et al., 2021; Tran, 2023)
	Membrane separation	Continuous	Physical sieving (size exclusion), electrostatic repulsion	Permeate flux, salt/TSS rejection, fouling rate	Trans-membrane pressure, MLSS, temperature, feed quality	(Tang et al., 2025; Zhou et al., 2023)
Chemical	Electrochemical treatment	Batch/ Continuous	Electrooxidation, electroreduction, or electrocoagulation	% Removal of COD, TOC, TSS, color, metals, nutrients, pathogens	Current density, electrode type, electrolyte type, reaction time, pH	(Adeoye et al., 2026; AlJaberi et al., 2023)
	Photo-electro chemical oxidation	Continuous	Light-enhanced electrooxidation	COD/TOC removal, pollutant degradation rate, energy efficiency, byproduct formation	Light intensity, current density, catalyst and electrode type, pH, pollutant concentration, reaction time	(Divyapriya et al., 2021; Mukherjee et al., 2023)
	Advanced oxidation processes (AOPs)	Batch/ Continuous	Hydroxyl-radical oxidation	% Removal of COD, TOC, color, toxicity, pollutant degradation rate	pH, oxidant dose, UV intensity, reaction time	(Sak et al., 2025; Yuan et al., 2025; Zou et al., 2022)
Biological	Activated sludge process	Batch/ Continuous	Aerobic biodegradation, nitrification	BOD, COD, NH_4^+–N removal, sludge volume index	DO, food-to-microbes (F/M) ratio, sludge retention time, temperature	(Chen et al., 2026; Singh et al., 2024)
	Anaerobic digestion (AD)	Continuous	Anaerobic biodegradation, methanogenesis	BOD, COD removal, biogas yield	Substrate composition and nutrient content, Temperature, pH, organic loading rate, hydraulic retention time	(Ahmad et al., 2025; Alengebawy et al., 2025; Kegl et al., 2025)
	Nitrification–denitrification	Batch/ Continuous	Ammonia oxidation → nitrate → N_2	NH_4^+–N, NO_3^-–N, TN removal	DO, carbon source, C/N ratio, sludge retention time, temperature	(Farazaki and Gikas, 2019; Singh et al., 2024)
	Anammox	Continuous	Anaerobic ammonium oxidation	Nitrogen removal rate, NH_4^+ and NO_2^- conversion	Temperature, biomass retention, hydraulic retention time NO_2^-/NH_4^+ ratio, DO, pH	(Hu et al., 2025; Mu et al., 2025; Verma and Daverey, 2024)

Abbreviation: TSS – total suspended solids, MLSS – mixed liquor suspended solids, COD – chemical oxygen demand, BOD – biological oxygen demand, TOC – total organic carbon, DO – dissolved oxygen.

the coking process, or wastewater laden with pharmaceutical residues and emerging contaminants (Zhou et al., 2023). However, high cost for membrane fabrication, energy consumption in operation, propensity for fouling, and limited chemical and thermal resistance are major deterrents in large-scale application of membrane technologies (Nure and Nkambule, 2023).

13.2.2 Chemical Methods

Chemical treatment methods, including electrochemical and photoelectrochemical oxidation and AOPs, are widely referred to as treatment options as they transform contaminants into possibly less harmful or mineralized products (Ghazal et al., 2022; Sak et al., 2025). Electrochemical methods, including electrocoagulation, electrooxidation, electroflotation, electrodialysis, electroflocculation, and electroreduction (Lee et al., 2024), are selective and energy efficient, offering removal or conversion of contaminants with low sludge generation (Adeoye et al., 2026). However, short electrode lifetimes and low current efficiency often restrict their practical implementation (AlJaberi et al., 2023). The selection of a specific treatment method depends on the target pollutants, effluent characteristics, and the electrode material. While electrocoagulation, electrooxidation, and electro-Fenton are selected for emerging pollutants, electroflocculation and electroflotation are applied to remove colloids, oils, or grease (Yakamercan et al., 2023). Photoelectrochemical methods, such as photo-electrocatalysis, photo/solar electron-Fenton, and photoanodic oxidation, offer higher treatment efficiency and mineralization of persistent organics through synergistic photochemical and electrolysis reactions (Mukherjee et al., 2023).

Inter alia, AOPs are highly effective for degrading recalcitrant organic pollutants and inorganic contaminants due to their high oxidation efficiency and minimal secondary pollution (Saravanan et al., 2022; Yuan et al., 2025). AOPs are based on the generation of highly reactive oxidizing species, predominantly hydroxyl radicals (·OH), which exhibit high oxidation potential (2.80 V) and can effectively degrade diverse pollutants through non-selective oxidation pathways (Zheng et al., 2025). They mineralize complex organic molecules, including pesticides, pharmaceuticals, dyes, and endocrine-disrupting compounds, into environmentally safe products such as H_2O, CO_2, and inorganic ions (Sak et al., 2025). The most common AOPs include ozonation, photocatalytic oxidation, Fenton-based processes, and electrochemical oxidation. Recently, ozonation has become increasingly popular due to its demonstrated effectiveness in reducing sludge and removing new and emerging contaminants, including synthetic dyes, phenolics, amoxicillin, and other medications (Mukherjee et al., 2023). Ozonation occurs via direct oxidation by molecular ozone, targeting electron-rich species (e.g., phenols and olefins), and indirect oxidation via hydroxyl radicals generated from ozone decomposition, thereby enabling non-selective pollutant oxidation. While effective, it is constrained by rapid ozone decomposition, which results in low utilization efficiency, increased energy demand, and inconsistent performance in complex wastewater matrices (Sak et al., 2025).

13.2.3 Biological Methods

Biological treatment adopting the activated sludge process (ASP), anaerobic digestion (AD), and biological nitrogen removal is the backbone of conventional and advanced WWTPs (Singh et al., 2024). ASP is the most widely used biological treatment integrated into municipal or industrial WWTPs, offering operational simplicity and high treatment efficiency (Chen et al., 2026). Aeration of wastewater improves microbial activity and removal of organic and inorganic contaminants, while continuous biomass recycling ensures long-term operation (Daverey et al., 2019). However, high energy demand, operational costs, and fluctuations in influent characteristics warrant process optimization and the development of advanced reactor designs.

In AD, the microorganism degrades complex organic matter under anoxic conditions, producing biogas (primarily CH_4 and CO_2) and nutrient-rich digestate (Kegl et al., 2025). This mature yet rapidly developing technology offers multiple benefits, including renewable energy generation, organic waste reduction, and greenhouse gas mitigation, and is extensively used in the treatment of food waste, sewage sludge, agricultural residues, municipal solid waste, and industrial wastewaters (Ahmad et al., 2025; Alengebawy et al., 2025). Nevertheless, the large-scale deployment of AD remains a challenge (Gupta et al., 2025), as anaerobic effluents contain substantial ammonium nitrogen, warranting further treatment (Sheng et al., 2020).

Conventionally, nitrogen removal in wastewater is achieved through combined nitrification–denitrification (Singh et al., 2024). However, despite its high nitrogen removal efficiency, this process is often techno-economically unfeasible, as it requires intensive aeration for nitrification and the addition of external carbon to sustain denitrification (Verma and Daverey, 2021). Alternatively, anaerobic ammonium oxidation (anammox) has emerged as an excellent, sustainable nitrogen removal technique, involving direct anaerobic/anoxic conversion of ammonium and nitrite into dinitrogen gas (Verma and Daverey, 2023). The eco-friendly and economical process has been extensively applied for the treatment of ammonium-rich industrial and municipal wastewater (Negi et al., 2025; Verma and Daverey, 2024). Nevertheless, the extended application of the anammox process is limited by delayed start-up periods and high vulnerability toward environmental conditions (Verma and Daverey, 2026), driving research into process optimization, including the integration of ML (Hu et al., 2025).

13.2.4 Need for ML and AI in Wastewater Treatment

Most treatment processes in a WWTP are intrinsically complex and exhibit strongly nonlinear dynamics. Maintaining rapid and effective responses to fluctuations in the dynamics of complex nonlinear processes is highly challenging for effluent quality, especially when operational cost is a key concern (Zhou et al., 2023). Conventional mechanistic models often struggle to capture time-varying, highly nonlinear process behavior (influenced by both known and unknown factors), limiting their capacity to fully represent complex systems and leading to high implementation costs and poor generalization performance (Ahmad et al., 2025; D. Wang et al., 2021). In light of this, artificial intelligence (AI), specifically ML, can be a robust choice to improve the performance of WTP/WWTP, particularly when the underlying mechanisms are not yet fully understood. Being strictly 'data-driven' and solely relying on extracting relationships between input–output data, ML models circumvent many significant limitations of conventional mechanistic models (Vanrolleghem et al., 2025). Some recent applications include predicting treatment efficiency of WWTPs (Zamfir et al., 2025) and/or optimizing system performance (Ismail et al., 2025), investigating relationships between operational variables and effluent quality (D. Wang et al., 2021), estimating the removal efficiency of adsorption processes, calculating the adsorption capacity of the adsorbent and selection of adsorbent (Tran, 2023; Zhao et al., 2022), early detection of process failures (e.g., foaming, bulking, inhibition) (Kegl et al., 2025), optimizing antibiotic removal efficiency in ASP (Chen et al., 2026), estimating biogas yield in anaerobic digesters (Gupta et al., 2025), and optimization, real-time control, and predicting nitrogen removal efficiency in anammox reactors (Mu et al., 2025).

13.3 MAJOR ML MODEL VARIANTS FOR WATER QUALITY AND/OR WATER TREATMENT

A variety of ML models can be used in the context of wastewater treatment or water quality measurement. Unsupervised learning requires no labeled dataset; instead, it uncovers the patterns inherent to the data (Ghahramani, 2004). Clustering, as an unsupervised learning algorithm, groups similar unlabeled data points into clusters based on their similarity measures (Kim et al., 2024). Samples in a given cluster are close to one another, indicating an interplay of possibly common geochemical processes or a source of contamination (Chin, 2025). A supervised ML model, however, takes labeled information or response data to "learn" the relationship between input and output variables in training for a classification or regression task (Figure 13.1). While parameters of ML models need to be identified through training, the hyperparameters (HPs) must be known even before training is initiated (Rizkallah, 2025). Ultimately, a well-trained model can be used to generate reasonable predictions for new test data.

The choice of ML model depends on data quality (variance, observations/features ratio), the response type, and the purpose (Arena et al., 2024). No ML algorithm can efficiently handle every dataset or all use cases; instead, the choice is often made based on predictive performance, generalization capability, computational efficiency, and the trade-off between accuracy and interpretability (Table 13.2).

A wide variety of ML models, with different architectures and complexities, can be used (Figure 13.2) to model water quality or water treatment processes, depending on the size of the dataset, the number and types of predictors, and the anticipated complexity of the process. A brief description of the most widely used model variants is presented in this section.

Multiple linear regression (MLR): This algorithm is suitable for predicting the values of a continuous response with a limited set of predictors, assuming a linear or moderate non-linearity $Y = \beta_0 + \sum_{i=1}^{p} \beta_i x_i + \sum_{i=1}^{p} \sum_{j=i+1}^{p} \beta_{ij} x_i x_j + \varepsilon$. The regression coefficients are estimated while minimizing the sum of squared errors (SSE), $\sum_{k=1}^{n} \varepsilon_k^2$ for the dataset. This algorithm is not suitable for inherently nonlinear processes and is sensitive to outliers. A regularization version with a tuning parameter (γ) can be used in parameter estimation for ill-conditioned data and to control overfitting, i.e., $\hat{\beta} = \arg\min_{\beta} \sum_{k=1}^{n} \varepsilon_k^2 + \gamma \|\beta|$ (Pillonetto et al., 2022).

Support vector machines (SVM), as supervised classification and regression models, find an optimal separation hyperplane between classes, maximizing distance to the majority of all elements in the respective class (Guido et al., 2024). The creation of the optimal hyperplane is influenced by a small subset of training samples, known as support vectors (SVs). Although defined in a linear, two-dimensional

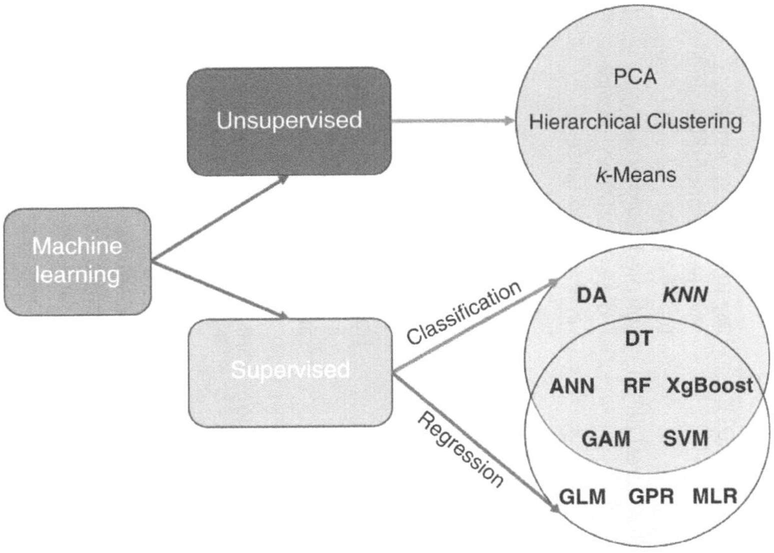

FIGURE 13.1 ML model variants used in classification, regression, and clustering.

TABLE 13.2 The advantages and limitations of ML models in water treatment and quality

ML ALGORITHM	*ADVANTAGE*	*LIMITATION*	*USE*	*REFERENCES*
Multiple linear regression (MLR)	Simple, interpretable, and computationally efficient	Inability to capture nonlinear or complex relationships	Baseline model for WQI estimation and initial feature significance analysis; screening of treatment-process parameters (e.g., removal efficiencies)	(Suphawan and Chaisee, 2021; M. Zhu et al., 2022)
Decision tree (DT)	Easy to interpret, handles nonlinear relationships, and missing data	Prone to overfitting; performance instability with small datasets	Classification of water quality categories and identification of key determinants	(Nishat et al., 2025)
Random forest (RF)	High predictive accuracy; reduces overfitting; provides variable importance	Limited interpretability; requires more computational resources	WQI prediction; modeling effluent quality from treatment plants; variable importance in treatment-process optimization	(Galiyal et al., 2025; Nagpal et al., 2024; F. Wang et al., 2021)
Support vector machine (SVM)	Effective in high-dimensional feature spaces; robust generalization	Computationally expensive for large data; needs careful kernel tuning	Regression and classification of WQI for moderate-sized datasets; modeling membrane treatment-process performance (permeability/fouling) from operational settings	(Duarte et al., 2024; Leong et al., 2021; Waqas et al., 2022)
Artificial neural network (ANN)	Handle complex nonlinear interactions; adaptable architecture	Requires large datasets; less interpretable; risk of overfitting	Regression of WQI and individual parameters (DO, turbidity); modeling kinetics in treatment processes (e.g., bioreactor performance)	(Chen et al., 2020; Gyparakis et al., 2024; Lin et al., 2022)
Gradient boosting (XGBoost)	High predictive performance; handles mixed features; robust to outliers	Reduced interpretability; sensitive to tuning of learning parameters	Accurate regression- and classification-based WQI prediction for influent raw-water quality parameters	(Ortiz-Lopez et al., 2024)
Convolutional neural network (CNN)	Efficient in capturing spatial and structural patterns; suitable for image or remote sensing data	Requires structured or image input; computationally intensive	Image-based estimation of turbidity, TSS, or spatial WQI mapping	(Lopez-Betancur et al., 2022)
Principal component analysis (PCA)	Reduces data dimensionality, retaining most variance; identifies key parameters, aids preprocessing for ML models	Captures only linear correlations; less effective for nonlinear datasets	Extraction of key variables influencing WQI; identification of pollution sources; monitoring and fault detection in wastewater treatment plants, variable selection	(Cui et al., 2021; Kini et al., 2025; Naik et al., 2022)

Long short-term memory (LSTM)	Captures temporal dependencies and sequential patterns in time series data	Requires large time series datasets and HP optimization	Forecasting dynamic variations in water quality parameters	(Mahesh et al., 2024)
Gaussian process regression (GPR)	Provides probabilistic predictions and uncertainty quantification	Computationally intensive; limited scalability to large datasets	WQI estimation where prediction uncertainty is essential	(Wan et al., 2022)
Bayesian belief network (BBN)	Explicitly models uncertainty and causal relationships	Complex structure learning; requires expert knowledge for variable dependencies	Probabilistic water quality forecasting and decision-support systems	(Daniel et al., 2020)
K-nearest neighbors (KNN)	Nonparametric; simple implementation; adaptable to various data types	Sensitive to noise and feature scaling; poor scalability with large data	Localized or short-term water quality classification and anomaly detection	(Huang et al., 2023)

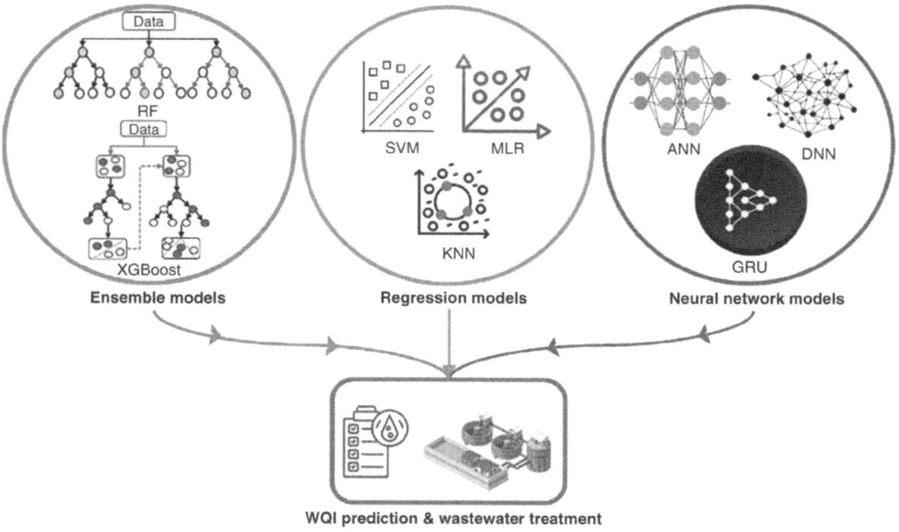

FIGURE 13.2 Common ML models used in water quality and in wastewater treatment modeling.

space, SVM can handle nonlinear problems by applying kernel functions to map data into higher-dimensional spaces.

A decision tree (DT) is a nonparametric algorithm used for regression and classification, organized hierarchically in a tree-like structure comprising a root node, branches, internal nodes, and leaf nodes. Each internal node evaluates a test criterion/condition for a predictor, and each leaf represents a categorical or numeric response (Qi et al., 2024). The choice of predictor in a decision split is made based on split purity, i.e., into distinct and non-overlapping regions, guided by information gain, information gain ratio, entropy, and Gini impurity. A leaf may group multiple observations, retaining the mean of each response. These models can process large, multi-feature datasets without preprocessing. However, excessive decision splits without pruning results in an overfitted model.

Random forest is an ensemble learning algorithm that combines multiple independent decision trees to make the final prediction. Random forest adopts a bootstrapped aggregation (bagging) approach, where bootstrapped training data on a random set of selected predictors is used to create an independent DT (Breiman, 2001). The average predictions or voted outcome (in classification) from the constructed trees is used for prediction.

Boosted trees, such as extreme gradient boosting (XGBoost), are an ensemble learning algorithm similar to bagging. However, it does not develop multiple trees; rather, it reiterates (improves) the tree-growing exercise using information from the previous cycle (Elith et al., 2008). In this sequential learning approach, the residuals of the prior model are used to improve predictive performance in previously underperforming areas.

Artificial neural networks (ANNs) are considered excellent nonlinear ML models for both classification and regression tasks. In an ANN, information from predictor data at the input layer is processed through one or more hidden-layer neurons before projecting to the output layer. Hidden-layer neurons combine input from preceding-layer neurons with their weights, add a bias term, and apply a nonlinear activation function before passing the result to the next layer. The output from an intermediate i-th layer can be represented as $y_i = \sigma(w_i \bullet X_i + b_i)$, where X_i and w_i are the input vector and the weight matrix to the i-th layer, b_i is the bias vector to the i-th layer, and σ is the nonlinear activation function. The number of hidden layers determines the model's complexity and the type of neural network it employs. The training aims to optimize the network's weights and biases to minimize the loss function.

13.4 ML MODEL DEVELOPMENT STAGES – TRAINING, HP OPTIMIZATION

13.4.1 Data Collection – Sources of Water Quality Data

To integrate AI and ML into water systems, reliable and comprehensive physicochemical and biological water quality parameters (WQPs) are required. Although in situ sampling and laboratory analyses still set the benchmark (Gholizadeh et al., 2016), recent advances in sensor-based field monitoring networks, in-stream probes, and automated stations have increased the availability of real-time or high-frequency measurements of key WQPs. The use of remote sensing platforms (satellite imagery, airborne, and drone-based) provides large-scale dynamics across lake, river, reservoir, and wetland systems. Reflectance-based algorithms can retrieve optically active indicators, such as chlorophyll a, suspended sediments, and dissolved organic matter, complementing ground-based monitoring by providing spatial depth and temporal continuity (Batina and Krtalić, 2023; Yang et al., 2022). Despite its spatial coverage, remote sensing cannot fully substitute for in situ observations, as many critical parameters, such as biochemical oxygen demand (BOD), chemical oxygen demand (COD), and nutrient concentrations, lack distinct optical signatures and thus require ground-truth measurements and rigorous calibration (Gholizadeh et al., 2016). Integration of multimodal measurements expands the depth, diversity, and reliability of datasets, thereby improving the accuracy and generalizability of ML models for water quality prediction (Casey et al., 2020).

13.4.2 Data Preprocessing in ML

Preprocessing involves an array of tools and strategies to address missing data and outliers and rescale the heterogeneous multivariate dataset (Xu et al., 2015). The data imputation approach accounts for the pattern of missing values within the dataset. While features with high missing frequency and of nominal importance can be dropped, an appropriate imputation strategy is required for features with low frequency of missingness and noticeable influence on the prediction results (Sun et al., 2023). In the univariate technique, missing values are imputed using summary statistics, such as the mean, median, and mode. The multivariate technique, however, takes into account the correlation among predictors and the labeled response (Joel et al., 2025). A regression-based imputation builds a localized regression model that uses all features in the dataset, excluding the feature being imputed, as predictors to predict the missing value.

The outlier in a data set can be defined as a value that deviates three times the standard deviation from the mean (J. Zhu et al., 2022). Outliers can be removed from the dataset before the modeling exercise. Feature engineering, though not widely used, involves selecting and extracting meaningful features from raw data, combining multiple features, or generating new features.

Data normalization is essential to keep predictor variables on the same scale and maintain consistency across indicators, thereby reducing dimensionality. Normalization can be min–max normalization $x' = (x - x_{\min})/(x_{\max} - x_{\min})$, mean-centering $x' = (x - \overline{x})$, autoscaling, or z-score standardization $x' = (x - \mu_x)/\sigma_x$.

13.4.3 ML Model Training and Hyperparameter Optimization

For a given HP, such as the architecture, the ML model parameters are estimated through an iterative training process. For example, a backpropagation ANN is trained by minimizing the difference between

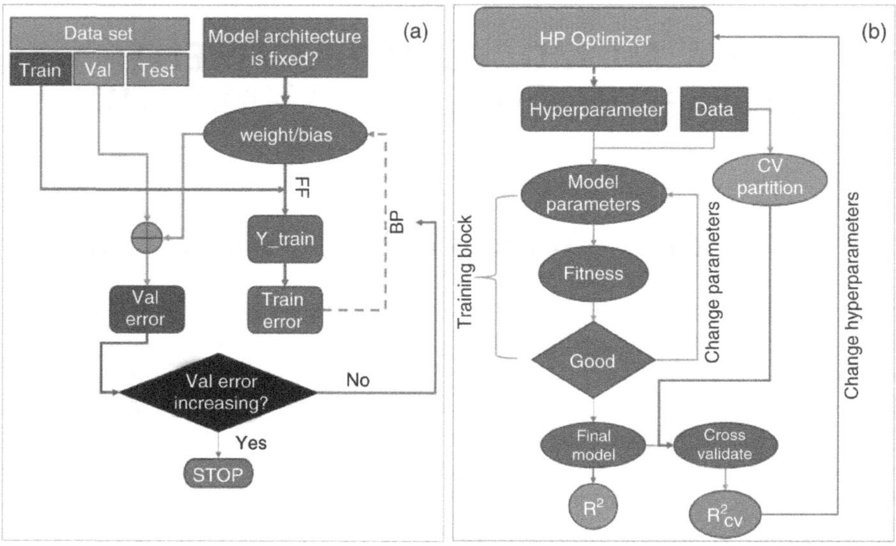

FIGURE 13.3 ML model training (a) using an independent training, test, and validation partition (no HP optimization) and (b) combined with automatic HP optimization based on cross-validation performance.

predicted and actual outputs on the training dataset, while propagating errors backward through the network and iteratively updating the weights and biases. Generally, the dataset is randomly divided into three partitions: training, test, and validation sets (Korjus et al., 2016). While the model is trained on the training set, intermittent validation checks provide early stopping and help prevent overtraining (Figure 13.3a). The test data partition remains untouched until the completion of the training exercise and final verification. For a limited-size dataset, reserving test and validation subsets might not be practical, and the entire dataset is used for training. In this situation, the model's generalization capability is assessed through cross-validation.

The model HPs (e.g., number of hidden layers and number of neurons) are part of the ML model configuration and must be fixed before any model training (Yang and Shami, 2020). An ML model with fewer HPs can be finalized by iteratively evaluating a set of values based on intuition and experience. However, systematic HP selection involves either an exhaustive grid search or the use of an optimization algorithm (meta-heuristic or Bayesian approach) (Ali et al., 2023). In each iteration of the HP optimization algorithm (Figure 13.3b), an ML model is trained on sampled HPs from the design space using the dataset. A cross-validated model is built using a fixed cross-validation partition, and the cross-validated R^2 (R^2_{CV}) is used as the objective to be minimized.

13.5 MACHINE LEARNING IN WATER QUALITY PREDICTION

Water quality assessment captures the key physicochemical parameters, such as pH, dissolved oxygen (DO), BOD, COD, total dissolved solids (TDS), electrical conductivity (EC), turbidity, nutrients (nitrate and phosphate), and alkalinity, to assess the degree of contamination, potability, treatment efficiency, and collective impact on the aquatic ecosystem. However, collecting, analyzing, and, more importantly, interpreting large volumes of water quality data can be challenging. WQI combines that information into a single, interpretable number. Although variations exist (Chidiac et al., 2023), there are four key steps: selection of WQPs;

conversion of measurements to unitless sub-indices; assignment of parameter weights based on their significance; and use of an aggregation function (Uddin et al., 2021). Simplistically, the subindex could be $q_i = \left(C_i / S_i\right)$, the ratio of measured concentration (C_i) to the permissible limit (S_i). The parameter weight (w_i) can be based on expert opinion, on variance information from principal component analysis (PCA), or inversely proportional to the permissible limit (Chidiac et al., 2023).

Finally, WQI is calculated as $WQI = \sum_{i=1}^{p} w_i q_i \Big/ \sum_{i=1}^{p} w_i$.

Each parameter contributes to the overall quality of water. These statistical methods rely on linearity, independence, and normal distribution of data, conditions rarely satisfied by non-stationary, multicollinear, and spatiotemporally separated measurements (Hussein et al., 2023). ML algorithms capture nonlinear dependencies by uncovering hidden patterns within large, heterogeneous datasets, leading to improved estimation of WQI (Islam et al., 2024; M. Zhu et al., 2022). Although a parametric mathematical expression is still required to generate labeled WQI, the ML model can significantly reduce the number of parameters necessary for future WQI prediction, saving cost and time (Lap et al., 2023).

Various ML models have been adopted in WQI prediction (Table 13.3), each offering particular advantages over others in terms of complexity, generalization capability, and resilience to overfitting (Alqahtani et al., 2022). Linear regression and decision tree models serve as baseline learners, while ensemble methods such as random forest (RF) (Mohseni et al., 2024) and gradient boosting (e.g., XGBoost and LightGBM) aggregate multiple weak learners for training and prediction (Islam Khan et al., 2022; Zheng et al., 2024). SVM effectively captures nonlinear interactions among parameters, improving WQI estimation (Leong et al., 2021). The long short-term memory (LSTM) networks map temporal dependencies within sequential data, enabling reliable forecasting of parameters (Liu et al., 2025). Integrated feature selection, retaining the influential subset of parameters, enhances computational efficiency and interpretability of ML models (Lap et al., 2023).

Beyond traditional measurements of WQPs, the availability of high-resolution remote sensing data has become an essential tool for water resource surveys and monitoring. Spectral band water indexes, including normalized difference water index (NDWI), normalized water index (NWI), enhanced water index (EWI), automated water extraction index-non-shadow (AWEI-nsh), and 2D modeling spectral index, such as difference index (DI), ratio index (RI), and normalized difference index (NDI), can be correlated with experimentally measured WQPs, leading to WQI prediction by adopting appropriate ML models (Li et al., 2021). Mapping WQP measurements with time-synchronized spectral bands and indices from Landsat 8 images can be used as a cost-effective alternative to frequent monitoring (Ahmed et al., 2023). Although the use of satellite imagery and remote sensing measurements is typically restricted to large open water bodies, a well-calibrated chemometric tool can be beneficial in water treatment facilities.

13.6 ML-DRIVEN OPTIMIZATION IN WASTEWATER TREATMENT

Wastewater treatment primarily depends on physical, chemical, and biological processes to remove contaminants before discharge or reuse (Sharma et al., 2022). In wastewater treatment, ML applications primarily focus on identifying important features (process conditions and operating parameters) to predict contaminant and nutrient removal and optimize process control. Nonlinear and dynamic interactions among parameters such as COD, BOD, TOC, nitrogen levels, and microbial activity highlight the potential of a data-driven approach.

Multiple ML models have been used either in WWTPs or in lab-scale reactors to optimize the process and treatment conditions (Table 13.4). Ensemble models, such as XGBoost, CatBoost, and LightGBM, are

TABLE 13.3 Machine learning applications for water quality prediction, incorporating performance and additional attributes

WATER QUALITY RESOURCE	*ML MODELS FOR WQI*	*PERFORMANCE*	*ADDITIONAL ATTRIBUTES*	*REFERENCE*
Water quality for 1,987 samples of Indian rivers on seven parameters	XGBoost, CatBoost, RF, GB, Extra Trees, and AdaBoost	CatBoost and GB, R^2>0.98	SHAP analysis revealed that DO, BOD, conductivity, and pH	(Choudhary et al., 2025)
Water sample (262) of four major rivers – Buriganga, Balu, Tongi Khal, and Turag from Dhaka, on six parameters	DT, LR, RR, SGD, XGBoost, LightGBM, ENR, SVM, RF, BR, ANN, CatBoost, Extra Trees, and AdaBoost	ANN: RMSE –2.34, MAE –1.24, NSE –0.97, R^2 > 0.97, Adjusted R^2 > 0.965	Using SHAP and LIME to identify key factors (Fe, DO, pH, TDS, EC) influencing WQI revealed Fe and DO as major negative contributors	(Nishat et al., 2025)
Groundwater samples (40) on seven physicochemical parameters from six locations in Bangladesh	RPR, SVM, BPANN, RF, XGBoost, BR	XGBoost: R^2 > 0.97, RMSE – 1.72, MAE – 1.38	Gaussian noise-based data augmentation improved generalizability	(Sarker et al., 2025)
A total of 33 groundwater samples from 6 districts of Yemen, on 14 parameters	Three clustering and one ML model, RF	R^2 > 0.95	Feature selection revealed the EC, Na, and Cl are major features	(Eid et al., 2025)
Surface water sample (21) from different locations of the Mahanadi River, Odisha, on 20 parameters	RF, ARAS	RF, R^2>0.97	Feature importance analysis suggests Cl^-, EC, SO_4^{2-}, and NO_3^- were the major contributors	(Das, 2025)
Groundwater samples (62) from Saudi Arabia, on 16 parameters	CNN	R^2>0.95	SHAP analysis highlighted Cl^-, SO_4^{2-}, and TSS as key contributors to the WQI	(Mallick et al., 2025)
Groundwater sample (772) from Rajasthan, India, on 16 parameters	XGBoost	R^2 – 0.99, RMSE – 0.57, MSE – 0.32	HP optimization improved the ML model; SHAP-based analysis revealed that NO_3^-, K^+, and Cl^- are the most essential features	(Khan et al., 2025)
Total 110 parameters from 176 WWTP effluents, China	Four clustering and five classifiers (XGBoost, CatBoost, LightGBM, RF, and CART)	F1-score > 0.9; a total of 20 critical parameters selected	Healthy Effluent Quality Index model (HEQi)	(Chen et al., 2025)
Water sample (98) from the river and coastal area of Ireland, on 11 parameters	XGBoost, LR, RF, Ridge Regression, DT, SVM, KNN, Lasso Regression	LR, RF, XGBoost, DT, R^2 – 0.99	The utilization of CCME, Brown, and SRDD base models provides complementary strengths in comprehensiveness, simplicity, and regional sensitivity	(Rahman et al., 2025)

Abbreviation: AdaBoost – adaptive boosting, ANN – artificial neural network, ARAS – additive ratio assessment, BPANN – backpropagation artificial neural network, BR – Bayesian ridge, BSA – beetle swarm algorithm, CART – classification and regression tree, CatBoost – categorical boosting, CNN – convolutional neural network, DT – decision tree, ENR – elastic net regression, Extra Trees – extremely randomized trees, GB – gradient boosting, KNN – K-nearest neighbors, LightGBM – light gradient boosting machine, LR – linear regression, ML – machine learning, MLP – multilayer perceptron, NB – Naïve Bayes, RF – random forest, RPR – reduced polynomial regression, RR – ridge regression, SGD – stochastic gradient descent, SVM – support vector machine, XGBoost – extreme gradient boosting, WQI – water quality index.

TABLE 13.4 Machine learning model used for predicting wastewater treatment plant (WWTP) performance parameters based on operational and physicochemical predictors

PROCESS SYSTEM	*PREDICTORS*	*RESPONSE(S)*	*MODEL USED*	*PERFORMANCE AND FINDINGS*	*REFERENCE*
Daily operational data (574) from WWTP, Guangzhou, China, MBR-enhanced anaerobic–anoxic–oxic configuration	SS_{inf}, FR_{inf}, COD_{inf}, TN_{inf}, $NH_4^+–N_{inf}$, TP_{inf}, aeration	COD_{eff}	XGBoost, CatBoost, RF, AdaBoost	XGBoost, R^2> 0.91; SHAP analysis revealed that aeration rate, FR_{inf}, and SS_{inf} have a significant impact on COD_{eff} prediction	(Wei et al., 2026)
WWTP samples (1095) from, Zhejiang, China	T_{inf}, pH_{inf}, COD_{inf}, $NH_4^+–N_{inf}$, TN_{inf}, DO_{inf}, ECS, EC, FR_{inf}	TN_{eff}	MLR, SVM, LightGBM, RF, XGBoost, CatBoost	XGBoost, R^2 >0.91; TN_{inf}, EC, DO_{inf}, and FR_{inf} are the most influential features in determining TN_{eff} concentration	(Yun et al., 2025)
North West WWTP, Illinois, USA	Aeration, FR_{inf}, T_{inf}, SS_{inf}, pH_{inf}, alum dose	TP_{eff}	LightGBM, XGBoost, RF, GBDT, KNN, SVC	LightGBM, R^2 = 0.92; SHAP analysis revealed that alum dose, SS_{inf}, and aeration are the most essential features	(Sun et al., 2024)
WWTP samples from Saudi Arabia	FR_{inf}, TDS_{inf}	SS_{eff}	FT, ILR, RLR, LSVM, MGSVM	ILR, R^2 >0.90; in-between randomization technique used to increase the size	(Maddah, 2025)
Daily operational data (365) and hourly data (8761) from Umbilo WWTP, in South Africa	FR_{inf}, COD_{inf}, $NH_4^+–N_{inf}$, TKN_{inf}, TN_{inf}, SS_{inf}, BOD_{inf}, TP_{inf}, PO_{4inf}	COD_{eff}	SVM, MLP, RF, and LSTM	MLP: daily data (R^2 0.87), LSTM: hourly data (R^2 0.90); PDP and permutation importance-based study revealed SS_{inf} is a significant predictor for COD_{eff}	(Cechinel et al., 2024)
Data (127) from a laboratory-scale anaerobic digester at the Shiraz WWTP	Time, TS_{inf}, TVS_{inf}, and the ratio of excess biological sludge to total sludge	Biogas yield, TVS destruction, and TS_{eff}	MLP, SVM, DT, RF, GB, BMA	BMA, R^2– 0.981, 0.927, 0.983 for biogas yield, TVS destruction, and TS_{eff}, respectively	(Karami et al., 2025)
Data (570) from a lab-scale membrane bioreactor used synthetic wastewater	Salinity, DO_{inf}, HRT, pH, T_{inf}, COD_{inf}, COD_{eff}, $NH_4^+–N_{inf}$, $NO_2^-–N_{eff}$, $NO_3^-–N_{eff}$, C/N ratio, COD removal efficiency, TN removal efficiency	$NH_4^+–N_{eff}$, TN_{eff}	GBDT, LightGBM, RF, XGBoost, AdaBoost, CatBoost	CatBoost, R^2– 0.88 and 0.91 for $NH_4^+–N_{eff}$ and TN_{eff}, respectively; SHAP analysis revealed DO_{inf}, pH, and COD removal efficiency as key regulators	(Wei et al., 2025)

(Continued)

TABLE 13.4 (Continued)

PROCESS SYSTEM	*PREDICTORS*	*RESPONSE(S)*	*MODEL USED*	*PERFORMANCE AND FINDINGS*	*REFERENCE*
Data from the membrane bioreactor of the WWTP located in Saudi Arabia	pH_{inf}, EC_{inf}, TDS_{inf}, TSS_{inf}, COD_{inf}, $NH_4^+–N_{inf}$, $NO_3^-–N_{inf}$, FOG_{inf}, CaH_{inf}, MgH_{inf}, PO_{4inf}, Cl^-_{inf}, alkalinity	COD removal efficiency	GPR, SWLR, RF	RF, R^2 –0.91; PSO, GA, and SA techniques were used for HP optimization, where the GA-based technique outperformed the others	(Ismail et al., 2025)
Data (1131) of pilot-scale membrane bioreactor treating real oil refinery wastewater	SRT, sludge filterability, MLVSS, pH, T_{inf}, membrane permeability, COD_{inf}, PO_4, NH_{4inf}, sulfides, OG, COD_{eff}, COD removal efficiency	NH_4 removal	ANN, PCA, MSPC	ANN, R^2 –0.87; sensitivity analysis conducted with the ANN model suggested that SRT, Tinf, and NH_{4inf} are the most critical variables	(Santos et al., 2024)
Removal of 80 pharmaceuticals across 17 WWTP, China, over 4 seasons, operating parameters	HRT, SRT, T, pH, COD_{inf}, NH_4^+-N_{inf}, TN_{inf}, $TP_{inf,}$ molecular descriptor	Removal efficiency	GBC, LightGBM, XGBoost, RF, ET, DT, KNN	GBC: classification accuracies of 0.68	(Zhang et al., 2025)
Four-phase data (46) from a gas-liquid-solid circulating fluidized bed bioreactor at the Powder Technology Research Center (PTRC), London	Empty bed contact times, riser-riser recirculation ratio, downer-riser recirculation ratio, HRT, VSS, SRT, run time, C:N:P ratio for feed stream to the riser, TKN, FR_{inf}, COD_{inf}, $NO_3^-–N_{inf}$	Anoxic NH_4–N, Anoxic NO_3–N	MLR, BOA–SVM	SVM, R^2 –0.99 for both responses; feature selection revealed FR_{inf}, TKN, and $NO_3^-–N_{inf}$ were important features for nitrogen removal	(Razzak et al., 2024)

Abbreviation: AdaBoost – adaptive boosting, ANN – artificial neural network, BOA–SVM – Butterfly optimization algorithm–support vector machine, BMA – Bayesian model averaging, BOD – biochemical oxygen demand , CaH – calcium hardness, CatBoost – categorical boosting, COD – chemical oxygen demand, DO– dissolved oxygen, DT – decision tree, EC – electrical conductivity, ECS – electrical conductivity sensor, ET – Extra Trees, FOG – fats, oils, and grease, FR– flow rate, FT – Fourier transform, GB – gradient boosting, GBC – gradient boosting classifier, GBDT – gradient boosting decision tree, GPR – Gaussian process regression, HRT – hydraulic retention time, ILR – improved linear regression, KNN – K-nearest neighbors, LightGBM – light gradient boosting machine, LSVM – least squares support vector machine, LR – linear regression, LSTM – long short-term memory, MBR – membrane bioreactor, MGSVM – modified Gaussian support vector machine, ML – machine learning, MLP – multilayer perceptron, MLR – multiple linear regression, MLVSS – mixed liquor volatile suspended solids, MSPC – multivariate statistical process control, NB – Naïve Bayes, OG – oil and grease, PCA – principal component analysis, PTRC – Powder Technology Research Center, RF – random forest, RLR – ridge linear regression, SRT – solids retention time, SS – suspended solids, SWLR – stepwise linear regression, SVC – support vector classifier, SVM – support vector machine, SVR – support vector regression, T – temperature, Ta – air temperature, TDS – total dissolved solids, TK-N – total Kjeldahl nitrogen, TN – total nitrogen, TP – total phosphorus, TS – total solids , TVS – total volatile solids , TVS destruction – total volatile solids destruction, VSS – volatile suspended solids, WWTP – wastewater treatment plant, XGBoost – extreme gradient boosting. Suffix: *inf* – influent, *eff* – effluent.

primarily used in WWTPs due to their high accuracy, ability to handle multicollinearity, prevention of overfitting, and resilience to missing data scenarios. Hybrid and deep learning models, such as LSTM, Bayesian model averaging (BMA), and ANN, when integrated into optimization frameworks, like genetic algorithm (GA) and particle swarm optimization (PSO), capture temporal variations and nonlinear dependencies in WWTP (Cechinel et al., 2024; Ismail et al., 2025). ML models trained on continuous sensor data can be used to characterize biofilm dynamics in WWTP, enabling biologically interpretable process control (Fisher et al., 2025). Explainable ML tools, such as Shapley additive explanation (SHAP) and permutation predictor importance, suggest the significance of operational variables over the chemical factors – indicating the possibility of dynamic control with process parameters. Aeration rate, flow rate, suspended solids, and nutrient concentrations are identified as the dominant predictors influencing COD, total nitrogen (TN), and total phosphate (TP) removal in most of the WWTP samples (Sun et al., 2024; Wei et al., 2026; Yun et al., 2025). These insights show a shift from traditional fixed designs to real-time, data-driven optimization, where ML improves understanding, allowing proactive and sustainable WWTP management intervention (Razzak et al., 2024).

13.7 CHALLENGES AND FUTURE PERSPECTIVES

The use of predictive modeling in water quality analysis and wastewater treatment is increasingly common. However, a statistical comparison of their predictive performance with simpler alternatives has not been widely explored. Unexpected variation in environmental conditions can introduce measurement uncertainties (Ghiasi et al., 2022), compromising the reliability and reproducibility of ML models. The complexity of ML models makes it impossible to grasp the rationale behind predictions, particularly in time series modeling, where WQPs in WWTPs are often correlated with their previous values (Toivonen and Räsänen, 2024). A complex deep learning algorithm may not be justified when autoregressive integrated moving average (ARIMA), exponential smoothing, or multivariate linear regression could be equally effective. The use of a deep neural network (DNN) introduces unnecessary complexity, obscuring the interpretation of results. HP tuning is similarly crucial as a complex ML architecture (Bischl et al., 2023). Increasing the number of neurons or hidden layers in an ANN, or the number of boosted trees, may not be reasonable. ML continues to revolutionize wastewater treatment, prediction, and process optimization. ML-integrated water quality monitoring and anomaly detection in treatment plants (Prabu et al., 2025) offer a sustainable future for water resource management.

13.8 CONCLUSION

AI and ML have emerged as transformative tools in water resource management, revolutionizing both water quality assessment and wastewater treatment. These data-driven approaches enable the interpretation of complex interactions that conventional models cannot capture. In water quality studies, the integration of remote sensing, sensor-based monitoring, and predictive algorithms has improved the spatial and temporal understanding of aquatic ecosystems, allowing forecasting of contamination levels. Similarly, in wastewater treatment, ML-based optimization models enhance process efficiency by identifying key operational variables and accurately predicting treatment performance. Despite these advances, the limited availability of data, model interpretability, and subjective ML workflows remain challenges. Future directions lie in developing explainable, hybrid models that integrate mechanistic understanding with AI-driven intelligence, supported by IoT-enabled real-time monitoring and automated control. Such integrated intelligent frameworks will enable sustainable, adaptive, and resilient water systems in the face of growing environmental and climatic challenges.

REFERENCES

Adeoye, A.E., Ukoba, K.O., Oloore, L.E., Aborisade, C.A., Jen, T., 2026. Electrochemical technologies for wastewater treatment, in: *Water Remediation Methods and Wastewater Treatment*. Elsevier, pp. 743–758. https://doi.org/10.1016/B978-0-443-33038-4.00027-1

Ahmad, A., Yadav, A.K., Pal, A., Hasan, S., 2025. A novel intensified anaerobic co-digestion process for high-quality biomethane production using ultrasound energy: Optimization and robustness tests with ML-RSM-TLBO algorithm. *Energy Convers. Manag.* 325, 119408. https://doi.org/10.1016/j.enconman.2024.119408

Ahmed, W., Mohammed, S., El-Shazly, A., Morsy, S., 2023. Tigris River water surface quality monitoring using remote sensing data and GIS techniques. *Egypt. J. Remote Sens. Sp. Sci.* 26, 816–825. https://doi.org/10.1016/j.ejrs.2023.09.001

Akhtar, N., Syakir Ishak, M.I., Bhawani, S.A., Umar, K., 2021. Various natural and anthropogenic factors responsible for water quality degradation: A review. *Water* 13, 2660. https://doi.org/10.3390/w13192660

Alengebawy, A., Eraky, M., Mahrous, T., Touil, K., Elmahsoub, N.M., Ibrahim, A., Elgarhy, A.H., 2025. Advances in anaerobic digestion for municipal wastewater treatment, in: *Municipal Wastewater Treatment*. Elsevier, pp. 309–352. https://doi.org/10.1016/B978-0-443-24826-9.00010-3

Ali, Y., Awwad, E., Al-Razgan, M., Maarouf, A., 2023. Hyperparameter search for machine learning algorithms for optimizing the computational complexity. *Processes* 11, 349. https://doi.org/10.3390/pr11020349

AlJaberi, F.Y., Ahmed, S.A., Makki, H.F., Naje, A.S., Zwain, H.M., Salman, A.D., Juzsakova, T., Viktor, S., Van, B., Le, P.-C., La, D.D., Chang, S.W., Um, M., Ngo, H.H., Nguyen, D.D., 2023. Recent advances and applicable flexibility potential of electrochemical processes for wastewater treatment. *Sci. Total Environ.* 867, 161361. https://doi.org/10.1016/j.scitotenv.2022.161361

Alqahtani, A., Shah, M.I., Aldrees, A., Javed, M.F., 2022. Comparative assessment of individual and ensemble machine learning models for efficient analysis of river water quality. *Sustainability* 14, 1183. https://doi.org/10.3390/su14031183

Arena, S., Florian, E., Sgarbossa, F., Sølvsberg, E., Zennaro, I., 2024. A conceptual framework for machine learning algorithm selection for predictive maintenance. *Eng. Appl. Artif. Intell.* 133, 108340. https://doi.org/10.1016/j.engappai.2024.108340

Babu Ponnusami, A., Sinha, S., Ashokan, H., Paul, M.V., Hariharan, S.P., Arun, J., Gopinath, K.P., Hoang Le, Q., Pugazhendhi, A., 2023. Advanced oxidation process (AOP) combined biological process for wastewater treatment: A review on advancements, feasibility and practicability of combined techniques. *Environ. Res.* 237, 116944. https://doi.org/10.1016/j.envres.2023.116944

Batina, A., Krtalić, A., 2023. A review of remote sensing applications for determining lake water quality. https://doi.org/10.20944/preprints202309.0489.v1

Bischl, B., Binder, M., Lang, M., Pielok, T., Richter, J., Coors, S., Thomas, J., Ullmann, T., Becker, M., Boulesteix, A., Deng, D., Lindauer, M., 2023. Hyperparameter optimization: Foundations, algorithms, best practices, and open challenges. *WIREs Data Min. Knowl. Discov.* 13. https://doi.org/10.1002/widm.1484

Breiman, L., 2001. Random Forests. *Mach. Learn.* 45, 5–32. https://doi.org/10.1023/A:1010933404324

Casey, K.A., Rousseaux, C.S., Gregg, W.W., Boss, E., Chase, A.P., Craig, S.E., Mouw, C.B., Reynolds, R.A., Stramski, D., Ackleson, S.G., Bricaud, A., Schaeffer, B., Lewis, M.R., Maritorena, S., 2020. A global compilation of in situ aquatic high spectral resolution inherent and apparent optical property data for remote sensing applications. *Earth Syst. Sci. Data* 12, 1123–1139. https://doi.org/10.5194/essd-12-1123-2020

Cechinel, M.A.P., Neves, J., Fuck, J.V.R., de Andrade, R.C., Spogis, N., Riella, H.G., Padoin, N., Soares, C., 2024. Enhancing wastewater treatment efficiency through machine learning-driven effluent quality prediction: A plant-level analysis. *J. Water Process Eng.* 58, 104758. https://doi.org/10.1016/j.jwpe.2023.104758

Chen, J., Li, L., Lv, X., Liu, T., Zhang, J., Luo, L., Tian, Y., 2026. Mechanistic insights into sulfamethoxazole removal in activated sludge systems through machine learning and microcosm experiments. *Water Res.* 292, 125331. https://doi.org/10.1016/j.watres.2026.125331

Chen, L., Wang, J., Zhu, M., He, R., Mu, H., Ren, H., Wu, B., 2025. Quality evaluation parameter and classification model for effluents of wastewater treatment plant based on machine learning. *Water Res.* 268, 122696. https://doi.org/10.1016/j.watres.2024.122696

Chen, Y., Song, L., Liu, Y., Yang, L., Li, D., 2020. A review of the artificial neural network models for water quality prediction. *Appl. Sci.* 10, 5776. https://doi.org/10.3390/app10175776

Chidiac, S., El Najjar, P., Ouaini, N., El Rayess, Y., El Azzi, D., 2023. A comprehensive review of water quality indices (WQIs): History, models, attempts and perspectives. *Rev. Environ. Sci. Bio/Technology* 22, 349–395. https://doi.org/10.1007/s11157-023-09650-7

Chin, D.A., 2025. Identifying associations of water quality parameters using cluster analysis: A case study of Biscayne Bay, USA. *Mar. Pollut. Bull.* 211, 117465. https://doi.org/10.1016/j.marpolbul.2024.117465

Chinedu, N., Chukwudum, Q.C., 2025. Addressing wastewater treatment challenges in developing nations: A standardized framework for sustainable adsorption techniques in small and medium industries. *Pollut. Study* 6, 3106. https://doi.org/10.54517/ps3106

Choudhary, R., Kumar, A., C., P., Naik, M.M., Choudhury, M., Khan, N.A., 2025. Predicting water quality index using stacked ensemble regression and SHAP based explainable artificial intelligence. *Sci. Rep.* 15, 31139. https://doi.org/10.1038/s41598-025-09463-4

Cui, F., Kim, M.M., Park, C., Kim, D., Mo, K., Kim, M.M., 2021. Application of principal component analysis (PCA) to the assessment of parameter correlations in the partial-nitrification process using aerobic granular sludge. *J. Environ. Manage.* 288, 112408. https://doi.org/10.1016/j.jenvman.2021.112408

Damaraju, M., Bhattacharyya, D., Kurilla, K.K., 2017. Removal of recalcitrant carbon from an industrial wastewater using electrocoagulation. *Int. J. Civ. Eng.* 15, 697–703. https://doi.org/10.1007/s40999-017-0187-6

Daniel, D., Iswarani, W.P., Pande, S., Rietveld, L., 2020. A Bayesian Belief Network model to link sanitary inspection data to drinking water quality in a medium resource setting in rural Indonesia. *Sci. Rep.* 10, 18867. https://doi.org/10.1038/s41598-020-75827-7

Das, A., 2025. Spatiotemporal evaluation and impact of superficial factors on surface water quality for drinking using innovative techniques in Mahanadi River Basin, Odisha, India. *J. Hydrol. Reg. Stud.* 59, 102366. https://doi.org/10.1016/j.ejrh.2025.102366

Daverey, A., Pandey, D., Verma, P., Verma, S., Shah, V., Dutta, K., Arunachalam, K., 2019. Recent advances in energy efficient biological treatment of municipal wastewater. *Bioresour. Technol. Reports* 7, 100252. https://doi.org/10.1016/j.biteb.2019.100252

Divyapriya, G., Singh, S., Martínez-Huitle, C.A., Scaria, J., Karim, A. V., Nidheesh, P.V., 2021. Treatment of real wastewater by photoelectrochemical methods: An overview. *Chemosphere* 276, 130188. https://doi.org/10.1016/j.chemosphere.2021.130188

Duarte, M.S., Martins, G., Oliveira, P., Fernandes, B., Ferreira, E.C., Alves, M.M., Lopes, F., Pereira, M.A., Novais, P., 2024. A review of computational modeling in wastewater treatment processes. *ACS ES&T Water* 4, 784–804. https://doi.org/10.1021/acsestwater.3c00117

Eid, M.H., Saeed, O., Szűcs, P., Kovács, A., Székács, A., Mörtl, M., Alrakhami, M.S., Al-Mashreki, M.H., Elsherbiny, O., Elsayed, S., Gaagai, A., Bence, C., Yaseen, Z.M., Gad, M., 2025. Impacts and sources of potential toxic elements on water quality and optimizing machine learning models for sustainable management. *Model. Earth Syst. Environ.* 11, 375. https://doi.org/10.1007/s40808-025-02548-z

Elith, J., Leathwick, J.R., Hastie, T., 2008. A working guide to boosted regression trees. *J. Anim. Ecol.* 77, 802–813. https://doi.org/10.1111/j.1365-2656.2008.01390.x

Farazaki, M., Gikas, P., 2019. Nitrification-denitrification of municipal wastewater without recirculation, using encapsulated microorganisms. *J. Environ. Manage.* 242, 258–265. https://doi.org/10.1016/j.jenvman.2019.04.054

Federigi, I., Salvadori, R., Lauretani, G., Leone, A., Lippi, S., Marvulli, F., Pagani, A., Verani, M., Carducci, A., 2024. Wastewater treatment plants performance for reuse: Evaluation of bacterial and viral risks. *Water* 16, 1399. https://doi.org/10.3390/w16101399

Fisher, O.J., Wang, Y., Ahmed, A., 2025. Making waves: Transforming biofilm-based wastewater treatment using machine learning-driven real-time monitoring. *Water Res.* 287, 124491. https://doi.org/10.1016/j.watres.2025.124491

Galiyal, A., Sayyed, U., Kusalkar, A., Kolhe, P., Borkar, A., Darade, M.M., 2025. "A Machine Learning-Based Approach for Troubleshooting Sewage Treatment Plant Process." *Int. J. Res. Innov. Appl. Sci.* X, 417–428. https://doi.org/10.51584/IJRIAS.2025.10060028

Ghahramani, Z., 2004. Unsupervised Learning. pp. 72–112. https://doi.org/10.1007/978-3-540-28650-9_5

Ghazal, H., Koumaki, E., Hoslett, J., Malamis, S., Katsou, E., Barcelo, D., Jouhara, H., 2022. Insights into current physical, chemical and hybrid technologies used for the treatment of wastewater contaminated with pharmaceuticals. *J. Clean. Prod.* 361, 132079. https://doi.org/10.1016/j.jclepro.2022.132079

Ghiasi, B., Noori, R., Sheikhian, H., Zeynolabedin, A., Sun, Y., Jun, C., Hamouda, M., Bateni, S.M., Abolfathi, S., 2022. Uncertainty quantification of granular computing-neural network model for prediction of pollutant longitudinal dispersion coefficient in aquatic streams. *Sci. Rep.* 12, 4610. https://doi.org/10.1038/s41598-022-08417-4

Gholizadeh, M., Melesse, A., Reddi, L., 2016. A comprehensive review on water quality parameters estimation using remote sensing techniques. *Sensors* 16, 1298. https://doi.org/10.3390/s16081298

Guido, R., Ferrisi, S., Lofaro, D., Conforti, D., 2024. An overview on the advancements of support vector machine models in healthcare applications: A review. *Information* 15, 235. https://doi.org/10.3390/info15040235

Gujer, W., Henze, M., Mino, T., Loosdrecht, M. van, 1999. Activated sludge model no. 3. *Water Sci. Technol.* 39, 183–193. https://doi.org/10.1016/S0273-1223(98)00785-9

Gupta, R., Murray, C., Sloan, W.T., You, S., 2025. Predicting the methane production of microwave-pretreated anaerobic digestion of food waste : A machine learning approach. *Energy* 328, 136613. https://doi.org/10.1016/j.energy.2025.136613

Gyparakis, S., Trichakis, I., Diamadopoulos, E., 2024. Using artificial neural networks to predict operational parameters of a Drinking Water Treatment Plant (DWTP). *Water* 16, 2863. https://doi.org/10.3390/w16192863

Hu, Y., Xu, D., Zhang, M., Ge, S., Shi, D., Ruan, Y., 2025. Research on the denitrification efficiency of anammox sludge based on machine vision and machine learning. *Water* 17, 2084. https://doi.org/10.3390/w17142084

Huang, M., Hu, B.Q., Jiang, H., Fang, B.W., 2023. A water quality prediction method based on k-nearest-neighbor probability rough sets and PSO-LSTM. *Appl. Intell.* 53, 31106–31128. https://doi.org/10.1007/s10489-023-05024-2

Hussein, E.E., Jat Baloch, M.Y., Nigar, A., Abualkhair, H.F., Aldawood, F.K., Tageldin, E., 2023. Machine Learning algorithms for predicting the water quality index. *Water* 15, 3540. https://doi.org/10.3390/w15203540

Islam Khan, M.S., Islam, N., Uddin, J., Islam, S., Nasir, M.K., 2022. Water quality prediction and classification based on principal component regression and gradient boosting classifier approach. *J. King Saud Univ. - Comput. Inf. Sci.* 34, 4773–4781. https://doi.org/10.1016/j.jksuci.2021.06.003

Islam, M.J., Salekin, S.U., Asif, A., Nafis, Z., Abdullah, A.A.K., Dilip, S., Md. Liton, R., Md. Tarek, H., 2024. Machine learning-driven water quality index prediction: Enhancing accuracy with gradient boosting and explainable AI for sustainable water monitoring. *Appl. Agric. Sci.* 2, 1–7. https://doi.org/10.25163/agriculture.2110031

Ismail, U.M., Bani-Melhem, K., Khan, M.F., Elnakar, H., 2025. Optimizing membrane bioreactor performance in wastewater treatment using machine learning and meta-heuristic techniques. *Results Eng.* 25, 104626. https://doi.org/10.1016/j.rineng.2025.104626

Joel, L.O., Doorsamy, W., Paul, B.S., 2025. A comparative study of imputation techniques for missing values in healthcare diagnostic datasets. *Int. J. Data Sci. Anal.* 20, 6357–6373. https://doi.org/10.1007/s41060-025-00825-9

Karami, A., Karimi-Jashni, A., Nikoo, M.R., Zare, B., 2025. Simulation of anaerobic digestion process under variable feeding sludge using a hybrid machine learning model. *Chemosphere* 387, 144672. https://doi.org/10.1016/j.chemosphere.2025.144672

Kegl, T., Torres Jiménez, E., Kegl, B., Kovač Kralj, A., Kegl, M., 2025. Modeling and optimization of anaerobic digestion technology: Current status and future outlook. *Prog. Energy Combust. Sci.* 106, 101199. https://doi.org/10.1016/j.pecs.2024.101199

Khan, I., Nizam, S., Bamal, A., Sajib, A.M., Mahammad Diganta, M.T., Shaida, M.A., Ashekuzzaman, S.M., Nash, S., Olbert, A.I., Uddin, M.G., 2025. Optimized intelligent learning for groundwater quality prediction in diverse aquifers of arid and semi-arid regions of India. *Clean. Eng. Technol.* 26, 100984. https://doi.org/10.1016/j.clet.2025.100984

Kim, S., Ando, T., Nakamura, Y., Hayashi, S., Kawaida, S., 2024. Clustering evaluation of water quality for various classes of in-flow rivers in connected brackish lakes. *Environ. Monit. Assess.* 196, 501. https://doi.org/10.1007/s10661-024-12670-7

Kini, K.R., Harrou, F., Madakyaru, M., Sun, Y., 2025. Enhancing fault detection in wastewater treatment plants: A multi-scale principal component analysis approach with the Kantorovich distance. *Environ. Sci. Water Res. Technol.* 11, 1215–1232. https://doi.org/10.1039/D4EW00831F

Korjus, K., Hebart, M.N., Vicente, R., 2016. An efficient data partitioning to improve classification performance while keeping parameters interpretable. *PLoS One* 11, e0161788. https://doi.org/10.1371/journal.pone.0161788

Kurniawan, T.A., Bandala, E.R., Othman, M.H.D., Goh, H.H., Anouzla, A., Chew, K.W., Aziz, F., Al-Hazmi, H.E., Nisa'ul Khoir, A., 2024. Implications of climate change on water quality and sanitation in climate hotspot locations: A case study in Indonesia. *Water Supply* 24, 517–542. https://doi.org/10.2166/ws.2024.008

Lap, B.Q., Phan, T.-T.-H., Nguyen, H. Du, Quang, L.X., Hang, P.T., Phi, N.Q., Hoang, V.T., Linh, P.G., Hang, B.T.T., 2023. Predicting Water Quality Index (WQI) by feature selection and machine learning: A case study of An Kim Hai irrigation system. *Ecol. Inform.* 74, 101991. https://doi.org/10.1016/j.ecoinf.2023.101991

Lee, J., Antonini, G., Al-Omari, A., Muller, C., Mathew, J., Bell, K., Pearce, J.M., Santoro, D., 2024. Electrochemical methods for nutrient removal in wastewater: *A review of advanced electrode materials, processes, and applications. Sustainability* 16, 9764. https://doi.org/10.3390/su16229764

Leong, W.C., Bahadori, A., Zhang, J., Ahmad, Z., 2021. Prediction of water quality index (WQI) using support vector machine (SVM) and least square-support vector machine (LS-SVM). *Int. J. River Basin Manag.* 19, 149–156. https://doi.org/10.1080/15715124.2019.1628030

Li, X., Ding, J., Ilyas, N., 2021. Machine learning method for quick identification of water quality index (WQI) based on Sentinel-2 MSI data: Ebinur Lake case study. *Water Supply* 21, 1291–1312. https://doi.org/10.2166/ws.2020.381

Lin, W., Hanyue, Y., Bin, L., 2022. Prediction of wastewater treatment system based on deep learning. *Front. Ecol.* 10, 1064555. https://doi.org/10.3389/fevo.2022.1064555

Liu, Y., Zheng, H., Zhao, J., 2025. Enhanced water quality prediction by LSTM and graph attention network (L-GAT): An analytical study of the Pearl River Basin. *Water Res.* X 28, 100383. https://doi.org/10.1016/j.wroa.2025.100383

Lopez-Betancur, D., Moreno, I., Guerrero-Mendez, C., Saucedo-Anaya, T., González, E., Bautista-Capetillo, C., González-Trinidad, J., 2022. Convolutional neural network for measurement of suspended solids and turbidity. *Appl. Sci.* 12, 6079. https://doi.org/10.3390/app12126079

Maddah, H.A., 2025. Design optimization and predictive modeling for TSS in mega surface WWTPs: A machine learning approach. *J. Environ. Manage.* 375, 124213. https://doi.org/10.1016/j.jenvman.2025.124213

Mahesh, N., Babu, J.J., Nithya, K., Arunmozhi, S.A., 2024. Water quality prediction using LSTM with combined normalizer for efficient water management. *Desalin. Water Treat.* 317, 100183. https://doi.org/10.1016/j.dwt.2024.100183

Mallick, J., Alqadhi, S., Alsubih, M., Ahmed, M.F.M., Abdo, H.G., 2025. Automated interpretation of deep learning-based water quality assessment system for enhanced environmental management decisions. *Appl. Water Sci.* 15, 110. https://doi.org/10.1007/s13201-025-02452-y

Manna, A., Biswas, D., 2023. Assessment of drinking water quality using water quality index: A review. *Water Conserv. Sci. Eng.* 8, 6. https://doi.org/10.1007/s41101-023-00185-0

Mohseni, U., Pande, C.B., Chandra Pal, S., Alshehri, F., 2024. Prediction of weighted arithmetic water quality index for urban water quality using ensemble machine learning model. *Chemosphere* 352, 141393. https://doi.org/10.1016/j.chemosphere.2024.141393

Mu, X., Jia, F., Qiu, S., Li, Y., Mei, N., Zhao, X., Han, B., Han, X., Zhang, J., Yao, H., 2025. Predicting and interpreting nitrogen removal performance and functional microbial abundance of single-stage partial nitrification and anammox system using machine learning methods. *Bioresour. Technol.* 437, 133119. https://doi.org/10.1016/j.biortech.2025.133119

Mukherjee, J., Lodh, B.K., Sharma, R., Mahata, N., Shah, M.P., Mandal, S., Ghanta, S., Bhunia, B., 2023. Advanced oxidation process for the treatment of industrial wastewater: A review on strategies, mechanisms, bottlenecks and prospects. *Chemosphere* 345, 140473. https://doi.org/10.1016/j.chemosphere.2023.140473

Nagpal, M., Siddique, M.A., Sharma, K., Sharma, N., Mittal, A., 2024. Optimizing wastewater treatment through artificial intelligence: recent advances and future prospects. *Water Sci. Technol.* 90, 731–757. https://doi.org/10.2166/wst.2024.259

Naik, M.R., Mahanty, B., Sahoo, S.K., Jha, V.N., Sahoo, N.K., 2022. Assessment of groundwater geochemistry using multivariate water quality index and potential health risk in industrial belt of central Odisha, India. *Environ. Pollut.* 303, 119161. https://doi.org/10.1016/j.envpol.2022.119161

Negi, D., Verma, S., Kuniyal, K., Daverey, A., 2025. Recent advances in biological nitrogen removal through anammox process - Bibliometric analysis and integrated technologies. *Environ. Eng. Res.* 31, 250466-0. https://doi.org/10.4491/eer.2025.466

Nishat, M.H., Khan, M.H.R.B., Ahmed, T., Hossain, S.N., Ahsan, A., El-Sergany, M.M., Shafiquzzaman, M., Imteaz, M.A., Alresheedi, M.T., 2025. Comparative analysis of machine learning models for predicting water quality index in Dhaka's rivers of Bangladesh. *Environ. Sci. Eur.* 37, 31. https://doi.org/10.1186/s12302-025-01078-w

Nure, J.F., Nkambule, T.T.I., 2023. The recent advances in adsorption and membrane separation and their hybrid technologies for micropollutants removal from wastewater. *J. Ind. Eng. Chem.* 126, 92–114. https://doi.org/10.1016/j.jiec.2023.06.034

Ortiz-Lopez, C., Bouchard, C., Rodriguez, M.J., 2024. Ensemble machine learning using hydrometeorological information to improve modeling of quality parameter of raw water supplying treatment plants. *J. Environ. Manage.* 362, 121378. https://doi.org/10.1016/j.jenvman.2024.121378

Pillonetto, G., Chen, T., Chiuso, A., De Nicolao, G., Ljung, L., 2022. Regularization of linear regression models, in: Regularized System Identification. Communications and Control Engineering. Springer, pp. 33–93. https://doi.org/10.1007/978-3-030-95860-2_3

Polesel, F., Andersen, H.R., Trapp, S., Plósz, B.G., 2016. Removal of antibiotics in biological wastewater treatment systems—A critical assessment using the Activated Sludge Modeling Framework for Xenobiotics (ASM-X). *Environ. Sci. Technol.* 50, 10316–10334. https://doi.org/10.1021/acs.est.6b01899

Prabu, P., Alluhaidan, A.S., Aziz, R., Basheer, S., 2025. Comparative analysis of machine learning models for detecting water quality anomalies in treatment plants. *Sci. Rep.* 15, 30453. https://doi.org/10.1038/s41598-025-15517-4

Qi, C., Yilmaz, E., Chen, Q., 2024. Background of machine learning, in: *Machine Learning Applications in Industrial Solid Ash.* Elsevier, pp. 93–130. https://doi.org/10.1016/B978-0-443-15524-6.00015-7

Rahman, A., Syeed, M.M.M., Karim, M.R., Fatema, K., Khan, R.H., Uddin, M.F., 2025. An optimized ensemble ML-WQI model for reliable water quality prediction by minimizing the eclipsing and ambiguity issues. *Appl. Water Sci.* 15, 113. https://doi.org/10.1007/s13201-025-02450-0

Rathi, B.S., Kumar, P.S., 2021. Application of adsorption process for effective removal of emerging contaminants from water and wastewater. *Environ. Pollut.* 280, 116995. https://doi.org/10.1016/j.envpol.2021.116995

Razzak, S.A., Sultana, N., Hossain, S.M.Z., Rahman, M.M., Yuan, Y., Hossain, M.M., Zhu, J., 2024. Hybrid machine learning modeling of nitrogen removal from wastewater using gas-liquid-solid circulating fluidized bed riser. *Chem. Eng. Res. Des.* 207, 295–307. https://doi.org/10.1016/j.cherd.2024.06.001

Rizkallah, L.W., 2025. Optimizing SVM hyperparameters for satellite imagery classification using metaheuristic and statistical techniques. *Int. J. Data Sci. Anal.* 20, 4945–4962. https://doi.org/10.1007/s41060-025-00762-7

Rossi, S., Capson-Tojo, G., Sànchez-Zurano, A., Carecci, D., Batstone, D.J., Acièn-Fernandez, G.F., Ficara, E., 2025. Recent advances and challenges in mechanistic modelling of photosynthetic processes for wastewater treatment. *Water Res.* 278, 123216. https://doi.org/10.1016/j.watres.2025.123216

Sak, G., Taşar, Ş., Dursun, G., 2025. Organic Pollutant Degradation Through Photocatalysis: Progress, Challenges, and Sustainable Solutions (Mini Review). *Appl. Sci.* 16, 204. https://doi.org/10.3390/app16010204

Santos, A.V., Amaral, M.C.S., Oliveira, S.M.A.C., 2024. Artificial neural networks and multivariate statistical process control to improve ammonia removal on membrane bioreactors treating refinery wastewater. *J. Water Process Eng.* 67, 106126. https://doi.org/10.1016/j.jwpe.2024.106126

Saravanan, A., Deivayanai, V.C., Kumar, P.S., Rangasamy, G., Hemavathy, R.V., Harshana, T., Gayathri, N., Alagumalai, K., 2022. A detailed review on advanced oxidation process in treatment of wastewater: Mechanism, challenges and future outlook. *Chemosphere* 308, 136524. https://doi.org/10.1016/j.chemosphere.2022.136524

Sarker, M.A.R., Chowdhury, M.A.H., Haque, M.T., Rahman, M.M., Meftaul, I.M., Jubayer, M.F., 2025. From data to decision: leveraging machine learning and water quality index for groundwater quality evaluation. *Sustain. Water Resour. Manag.* 11, 102. https://doi.org/10.1007/s40899-025-01276-7

Satyam, S., Patra, S., 2024. Innovations and challenges in adsorption-based wastewater remediation: A comprehensive review. *Heliyon* 10, e29573. https://doi.org/10.1016/j.heliyon.2024.e29573

Sharma, M., Yadav, A., Mandal, M.K., Pandey, S., Pal, S., Chaudhuri, H., Chakrabarti, S., Dubey, K.K., 2022. Wastewater treatment and sludge management strategies for environmental sustainability, in: Circular Economy and Sustainability. Elsevier, pp. 97–112. https://doi.org/10.1016/B978-0-12-821664-4.00027-3

Sheng, L., Lei, Z., Dzakpasu, M., Li, Y.-Y., Li, Q., Chen, R., 2020. Application of the anammox-based process for nitrogen removal from anaerobic digestion effluent: A review of treatment performance, biochemical reactions, and impact factors. *J. Water Process Eng.* 38, 101595. https://doi.org/10.1016/j.jwpe.2020.101595

Singh, Dhananjay, Singh, Deepak, Mishra, V., Kushwaha, J., Sengar, M., Sinha, S., Singh, S., Giri, B.S., 2024. Strategies for biological treatment of waste water: A critical review. *J. Clean. Prod.* 454, 142266. https://doi.org/10.1016/j.jclepro.2024.142266

Srivastava, R.R., Singh, P.K., 2023. Selection of factors affecting integrated municipal wastewater treatment and reuse network: an interpretive structural modelling (ISM) approach. *Environ. Dev. Sustain.* 25, 9137–9161. https://doi.org/10.1007/s10668-022-02428-x

Sukmana, H., Bellahsen, N., Pantoja, F., Hodur, C., 2021. Adsorption and coagulation in wastewater treatment – Review. *Prog. Agric. Eng. Sci.* 17, 49–68. https://doi.org/10.1556/446.2021.00029

Sun, J., Xu, Y., Yang, H., Liu, J., He, Z., 2024. Machine learning facilitated the conceptual design of an alum dosing system for phosphorus removal in a wastewater treatment plant. *Chemosphere* 351, 141154. https://doi.org/10.1016/j.chemosphere.2024.141154

Sun, Y., Li, J., Xu, Y., Zhang, T., Wang, X., 2023. Deep learning versus conventional methods for missing data imputation: A review and comparative study. *Expert Syst. Appl.* 227, 120201. https://doi.org/10.1016/j.eswa.2023.120201

Suphawan, K., Chaisee, K., 2021. Gaussian process regression for predicting water quality index: A case study on Ping River basin, Thailand. *AIMS Environ. Sci.* 8, 268–282. https://doi.org/10.3934/environsci.2021018

Tang, W., Wang, J., Yin, J., Han, Y., Zhang, C., Wu, W., Zhang, J., 2025. Emerging technologies and recent advances in the advanced treatment of petrochemical wastewater: A comprehensive review. *Carbon Hydrog.* 27, 444–462. https://doi.org/10.1002/cbh2.70039

Toivonen, E., Räsänen, E., 2024. Time-series analysis approach to the characteristics and correlations of wastewater variables measured in paper industry. *J. Water Process Eng.* 61, 105231. https://doi.org/10.1016/j.jwpe.2024.105231

Tran, H.N., 2023. Adsorption technology for water and wastewater treatments. *Water* 15, 2857. https://doi.org/10.3390/w15152857

Trianni, A., Negri, M., Cagno, E., 2021. What factors affect the selection of industrial wastewater treatment configuration? *J. Environ. Manage.* 285, 112099. https://doi.org/10.1016/j.jenvman.2021.112099

Uddin, M.G., Nash, S., Olbert, A.I., 2021. A review of water quality index models and their use for assessing surface water quality. *Ecol. Indic.* 122, 107218. https://doi.org/10.1016/j.ecolind.2020.107218

UN- Habitat, WHO, 2021. Progress on wastewater treatment: Global status and acceleration needs for SDG indicator 6.3.1.

Vanrolleghem, P.A., Khalil, M., Serrao, M., Sparks, J., Therrien, J.-D., 2025. Machine learning in wastewater: opportunities and challenges – "not everything is a nail!" *Curr. Opin. Biotechnol.* 93, 103271. https://doi.org/10.1016/j.copbio.2025.103271

Verma, S., Daverey, A., 2026. Biocarrier selection for anammox enrichment under realistic conditions: Plastic brush vs. polyethylene sponge. *Ind. Biotechnol.* 00, 1–12. https://doi.org/10.1177/15509087251397126

Verma, S., Daverey, A., 2024. Nitrogen removal from urea effluent in an anammox sequencing batch biofilm reactor with valorised dishwashing scrubber as biocarrier. *J. Taiwan Inst. Chem. Eng.* 166, 105502. https://doi.org/10.1016/j.jtice.2024.105502

Verma, S., Daverey, A., 2023. Valorization of dishwashing scrubber as biocarrier for the enrichment of anammox bacteria under realistic conditions. *Geomicrobiol. J.* 0, 1–12. https://doi.org/10.1080/01490451.2023.2218363

Verma, S., Daverey, A., 2021. Anammox process: role of reactor systems for its application and implementation in wastewater treatment plants, in: *Integrated and Hybrid Process Technology for Water and Wastewater Treatment*. Elsevier, pp. 273–292. https://doi.org/10.1016/B978-0-12-823031-2.00005-7

Wan, X., Li, X., Wang, X., Yi, X., Zhao, Y., He, X., Wu, R., Huang, M., 2022. Water quality prediction model using Gaussian process regression based on deep learning for carbon neutrality in papermaking wastewater treatment system. *Environ. Res.* 211, 112942. https://doi.org/10.1016/j.envres.2022.112942

Wang, D., Thunéll, S., Lindberg, U., Jiang, L., Trygg, J., Tysklind, M., Souihi, N., 2021. A machine learning framework to improve effluent quality control in wastewater treatment plants. *Sci. Total Environ.* 784, 147138. https://doi.org/10.1016/j.scitotenv.2021.147138

Wang, F., Wang, Y., Zhang, K., Hu, M., Weng, Q., Zhang, H., 2021. Spatial heterogeneity modeling of water quality based on random forest regression and model interpretation. *Environ. Res.* 202, 111660. https://doi.org/10.1016/j.envres.2021.111660

Waqas, S., Harun, N.Y., Sambudi, N.S., Arshad, U., Nordin, N.A.H.M., Bilad, M.R., Saeed, A.A.H., Malik, A.A., 2022. SVM and ANN Modelling Approach for the Optimization of Membrane Permeability of a Membrane Rotating Biological Contactor for Wastewater Treatment. *Membranes (Basel)*. 12, 821. https://doi.org/10.3390/membranes12090821

Wei, Q., Chen, Y., Zhang, H., Yin, H., Xu, Z., 2026. Interpretable modeling for effluent organic load control and optimization in wastewater treatment. *Bioresour. Technol.* 439, 133327. https://doi.org/10.1016/j.biortech.2025.133327

Wei, Q., Xu, Z., Yin, H., 2025. Enhanced nitrogen prediction and mechanistic process analysis in high-salinity wastewater treatment using interpretable machine learning approach. *Bioresour. Technol.* 426, 132393. https://doi.org/10.1016/j.biortech.2025.132393

WHO, 2017. Guidelines for drinking-water quality. World Health Organization, Geneva.

Xu, S., Lu, B., Baldea, M., Edgar, T.F., Wojsznis, W., Blevins, T., Nixon, M., 2015. Data cleaning in the process industries. *Rev. Chem. Eng.* 31. https://doi.org/10.1515/revce-2015-0022

Yakamercan, E., Bhatt, P., Aygun, A., Adesope, A.W., Simsek, H., 2023. Comprehensive understanding of electrochemical treatment systems combined with biological processes for wastewater remediation. *Environ. Pollut.* 330, 121680. https://doi.org/10.1016/j.envpol.2023.121680

Yang, H., Kong, J., Hu, H., Du, Y., Gao, M., Chen, F., 2022. A Review of Remote Sensing for Water Quality Retrieval: Progress and Challenges. *Remote Sens.* 14, 1770. https://doi.org/10.3390/rs14081770

Yang, L., Shami, A., 2020. On hyperparameter optimization of machine learning algorithms: Theory and practice. *Neurocomputing* 415, 295–316. https://doi.org/10.1016/j.neucom.2020.07.061

Yuan, Q., Wang, X., Xu, D., Liu, H., Zhang, H., Yu, Q., Bi, Y., Li, L., 2025. Machine Learning-Assisted Catalysts for Advanced Oxidation Processes: Progress, Challenges, and Prospects. *Catalysts* 15, 282. https://doi.org/10.3390/catal15030282

Yun, J., Yu, Y., Tao, C., Zhai, M., Zhang, H., Chen, Y., Li, H., Zhang, B., Ma, J., 2025. Machine learning-based optimization of enhanced nitrogen removal in a full-scale urban wastewater treatment plant with ecological combination ponds. *Water Res.* 285, 123976. https://doi.org/10.1016/j.watres.2025.123976

Zamfir, F.-S., Carbureanu, M., Mihalache, S.F., 2025. Application of Machine Learning Models in Optimizing Wastewater Treatment Processes: *A Review. Appl. Sci.* 15, 8360. https://doi.org/10.3390/app15158360

Zhang, C., Yu, Q., He, Y., Wu, G., Fan, F., Xu, K., Ren, H., Geng, J., 2025. Decoding pharmaceutical removability in full-scale wastewater treatment plants via machine learning model integrating modified electrophilicity index. *Water Res.* 286, 124218. https://doi.org/10.1016/j.watres.2025.124218

Zhao, Y., Fan, D., Li, Y., Yang, F., 2022. Application of machine learning in predicting the adsorption capacity of organic compounds onto biochar and resin. *Environ. Res.* 208, 112694. https://doi.org/10.1016/j.envres.2022.112694

Zheng, H., Hou, S., Liu, J., Xiong, Y., Wang, Y., 2024. Advanced Machine Learning and Water Quality Index (WQI) Assessment: Evaluating Groundwater Quality at the Yopurga Landfill. *Water* 16, 1666. https://doi.org/10.3390/w16121666

Zheng, T., Zhang, Z., Liu, Y., Zou, L., 2025. Recent Progress in Catalytically Driven Advanced Oxidation Processes for Wastewater Treatment. *Catalysts* 15, 761. https://doi.org/10.3390/catal15080761

Zhou, M., Chen, J., Yu, S., Chen, B., Chen, C., Shen, L., Li, B., Lin, H., 2023. The coupling of persulfate activation and membrane separation for the effective pollutant degradation and membrane fouling alleviation. *Chem. Eng. J.* 451, 139009. https://doi.org/10.1016/j.cej.2022.139009

Zhu, J., Jiang, Z., Feng, L., 2022. Improved neural network with least square support vector machine for wastewater treatment process. *Chemosphere* 308, 136116. https://doi.org/10.1016/j.chemosphere.2022.136116

Zhu, M., Wang, J., Yang, X., Zhang, Y., Zhang, L., Ren, H., Wu, B., Ye, L., 2022. A review of the application of machine learning in water quality evaluation. *Eco-Environment Heal.* 1, 107–116. https://doi.org/10.1016/j.eehl.2022.06.001

Zou, R., Tang, K., Hambly, A.C., Chhetri, R.K., Andersen, H.R., Zhang, Y., 2022. Elimination of recalcitrant micropollutants by medium pressure UV-catalyzed bioelectrochemical advanced oxidation process: Influencing factors, transformation pathway and toxicity assessment. *Sci. Total Environ.* 828, 154543. https://doi.org/10.1016/j.scitotenv.2022.154543

Index

For Product Safety Concerns and Information please contact our
EU representative GPSR@taylorandfrancis.com Taylor & Francis
Verlag GmbH, Kaufingerstraße 24, 80331 München, Germany